Molecular Biology and Biotechnology of Extremophiles

Molecular Biology and Biotechnology of Extremophiles

Edited by

R.A. HERBERT
Reader in Microbiology
Department of Biological Sciences
University of Dundee
Dundee

and

R.J. SHARP
Assistant Head
Division of Biotechnology
Centre for Applied Microbiology and Research
Porton Down
Salisbury

Blackie
Glasgow and London

Published in the USA by
Chapman and Hall
New York

Blackie & Son Limited.
Bishopbriggs, Glasgow G64 2NZ
and
7 Leicester Place, London WC2H 7BP

Published in the USA by
Chapman and Hall
a division of Routledge, Chapman and Hall, Inc.
29 West 35th Street, New York, NY 10001-2291

British Library Cataloguing in Publication Data

Molecular biology and biotechnology of
extremophiles.
I. Herbert, R.A.   II. Sharp, R.J.
574.8

ISBN 0-216-93153-3

Library of Congress Cataloging-in-Publication Data

Molecular biology and biotechnology of extremophiles / edited by R.A
Herbert and R.J. Sharp.
     p.   cm.
Includes bibliographical references and index.
ISBN 0-412-03241-4
1. Molecular microbiology.   2. Microbial biotechnology.
3. Extreme environments.   I. Herbert, R.A.   II. Sharp, R.J.
QR74.M63 1992
576—dc20
                                                    91-25198
                                                    CIP

Typeset by Thomson Press (India) Ltd., New Delhi, India
Printed in Great Britain by St Edmundsbury Press, Bury St Edmunds, Suffolk

Preface

It is now well recognised that many environments considered by man to be extreme are colonised by micro-organisms which are specifically adapted to these ecological niches. These organisms not only survive but actively grow under such conditions. A diverse range of bacteria, cyanobacteria, algae and yeasts has now been isolated from these habitats which are extreme in terms of temperature, pH, salinity and pressure as well as species which are resistant to radiation and toxic chemicals. Whilst originally considered to be mere 'scientific curiosities', it is now generally accepted that many have considerable biotechnological and commercial significance. Recently the term 'extremophile' has been used to describe these organisms.

Over the past twenty years extensive studies of the ecology, physiology, taxonomy and molecular biology of these micro-organisms have been undertaken. These have resulted in a complete reassessment of our concept of microbial evolution. The identification of the Archaeobacteria as the third kingdom of living organisms has given considerable impetus to extremophile research and is presenting many new challenges.

Extremophiles provide a valuable resource for exploitation in novel biotechnological processes and in developing our understanding of how biomolecules are stabilised when subjected to extreme conditions. They have provided thermostable enzymes for application in industrial processes, are used as diagnostic enzymes and have applications in molecular biology. They also provide models for protein engineers attempting to determine the basis of protein stability. In the field of bioremediation they have considerable application in the removal of heavy metals and radionuclides from waste waters, the biodegradation of toxic pollutants, and the treatment of domestic, agricultural and industrial effluents as well as the microbial desulphurisation of coal to reduce sulphur emissions. In the search for alternative energy resources to replace fossil fuels, they offer potential for the large scale production of ethanol, organic solvents, methane and hydrogen. Extremophiles have also been exploited for many years in the leaching of metals from low grade ores and more recently in precious metal recovery.

The overall aim of this book is to provide a comprehensive review of our current understanding of the mechanisms that these micro-organisms have evolved to survive and grow in particular extreme environments and how these may be, or are being, exploited for biotechnological processes. Whilst this volume is aimed primarily at established research workers in universities, research institutes and industry we hope that it will also provide a stimulating introduction to the subject for senior undergraduates and postgraduates who are interested in organisms that inhabit unusual and extreme environments. In future years these organisms may form the basis of new biotechnologically based industries. The whole subject is currently topical, scientifically stimulating and of immense practical significance.

R.A. Herbert
R.J. Sharp

Contributors

Professor P.L. Bergquist Department of Cellular and Molecular Biology, University of Auckland, Private Bag, Auckland, New Zealand

Dr D.A. Cowan Department of Biochemistry, University College London, Gower Street, London WC1 6BT, UK

Dr G.M. Gadd Department of Biological Sciences, University of Dundee, Dundee DD1 4HN, UK

Dr E.A. Galinski Institüt für Mikrobiologie, Rheinische-Friedrich-Wilhelms Universität, 5300 Bonn 1, Germany

Dr W.D. Grant Department of Microbiology, University of Leicester, Leicester LE1 7RH, UK

Professor K. Horikoshi Department of Applied Microbiology, The Riken Institute, Wako-Shi, Saitama 351, Japan

Dr W.J. Ingledew Department of Biochemistry and Microbiology, University of St Andrews, St Andrews KY16 9AL, UK

Dr P. Luton Microbiology and Environment Group, Division of Biotechnology, Centre for Applied Microbiology and Research, Porton Down, Salisbury, UK

Dr C.I. Masters Department of Pathology, Uniformed Services University of the Health Sciences, Maryland 20814-4799, USA

Dr A. Maule Microbiology and Environment Group, Division of Biotechnology, Centre for Applied Microbiology and Research, Porton Down, Salisbury, UK

Dr N.P. Minton Molecular Genetics Group, Division of Biotechnology, Centre for Applied Microbiology and Research, Porton Down, Salisbury, UK

Professor H.W. Morgan School of Science and Technology, University of Waikato, Private Bag, Hamilton, New Zealand

Professor B.E.B. Moseley Institute of Food Research, Agricultural and Food Research Council, Reading Laboratory, Shinfield, Berkshire RG2 9AT, UK

Dr P.R. Norris Department of Biological Sciences, University of Warwick, Coventry CV4 7AL, UK

Dr J.D. Oultram Molecular Genetics Group, Division of Biotechnology, Centre for Applied Microbiology and Research, Porton Down, Salisbury, UK

Dr D. Prieur Observatoire Océanologique de Roscoff, Université Pierre et Marie Curie, Place Georges Teissier, 2680 Roscoff, France

Dr N.J. Russell Department of Biochemistry, University of Wales College of Cardiff, PO Box 903, Cardiff CF1 1ST, UK

Dr M.D. Smith Life Technologies Inc., Research Products Division, Gaithersburg, Maryland, USA

Dr B.J. Tindall Deutsche Sammlung für Mikroorganismen und Zellculturen GmbH, Mascheroder Weg 1b, D3300 Braunschweig/Stockheim, Germany

Contents

1 Biochemistry and molecular biology of the extremely thermophilic archaeobacteria   1
D.A. COWAN

  1.1 Introduction   1
  1.2 Archaeobacterial phylogeny   2
  1.3 Ecology of the thermophilic Archaea   9
      1.3.1 Morphology of the Archaea   13
      1.3.2 Physiology and biochemistry of the Archea   13
      1.3.3 Enzymes   23
      1.3.4 Structural macromolecules   27
      1.3.5 Lipids and lipid biosynthesis   29
      1.3.6 Molecular genetics of the extremely thermophilic Archaea   34
  References   38

2 The molecular genetics and biotechnological application of enzymes from extremely thermophilic eubacteria   44
P.L. BERGQUIST and H.W. MORGAN

  2.1 Introduction   44
  2.2 Aerobic eubacteria   45
      2.2.1 The molecular biology and genetics of *Thermus*   45
      2.2.2 *Thermus* aquaticus DNA polymerase   45
      2.2.3 *Thermus aquaticus* restriction–modification system   47
      2.2.4 Expression of other genes from *Thermus* genomic libraries   48
      2.2.5 Proteinases from *Thermus* spp.   49
      2.2.6 Genetic transfer and plasmids in *Thermus*   52
      2.2.7 A repetitive sequence in *Thermus thermophilus*   54
      2.2.8 Promoter regions and other control sequences in *Thermus*   54
      2.2.9 Genes and proteins from thermophilic strains of *Bacillus*   55
      2.2.10 α-Amylases from *Bacillus* spp.   56
      2.2.11 Pullulanases and related enzymes from thermophilic bacilli   57
      2.2.12 Other genes and genetic systems in *Bacillus stearothermophilus* 58
  2.3 Anaerobic eubacteria   59
      2.3.1 Cloning of genes involved in cellulose hydrolysis   60
      2.3.2 Hemicellulose hydrolysis   62
      2.3.3 Starch hydrolysis   65
      2.3.4 Pullulanases   67
  2.4 Other enzymes from anaerobic thermophiles   71
      2.4.1 *Thermotoga*   71
  References   72

3 Biotechnological prospects for halophiles and halotolerant micro-organisms   76
E.A. GALINSKI and B.J. TINDALL

  3.1 Introduction   76
  3.2 Micro-organisms in the food industry   77

3.2.1   Food spoilage   77
3.2.2   Fermentation products   78
3.2.3   Single-cell protein (SCP)   80
3.2.4   Food colouring/flavouring   82
3.3   Production of commercially useful compounds   82
3.3.1   Biological fermentation processes at high salinities   82
3.3.2   Pharmaceutical compounds   87
3.3.3   Polymers   89
3.3.4   Enzymes   93
3.3.5   Compatible solutes   94
3.4   Future aspects   103
3.4.1   Environmental biotechnology   103
3.4.2   Agricultural aspects   104
3.4.3   Fuel from renewable sources   106
References   108

4 Acidophilic bacteria: adaptations and applications   115
P.R. NORRIS and W.J. INGLEDEW

4.1   Introduction   115
4.2   Constraints on growth at acid pH   116
4.2.1   Chemiosmotic considerations   116
4.2.2   Considerations of the conditions in the periplasm and the implications
        for its processes   125
4.3   The diversity of the extreme acidophiles   127
4.3.1   Iron- and sulphur-oxidising acidophiles   127
4.3.2   Phylum- and group-specific traits?   129
4.4   The bacterial extraction of metals from mineral sulphides   132
4.4.1   Factors influencing the selection of bacteria for mineral-leaching processes 133
4.5   Molecular genetic studies of acidophiles   135
4.5.1   The development of genetic systems for acidophiles   136
4.5.2   Gene transfer   137
4.6   Concluding comments: diversity, identification and applied molecular biology 138
References   139

5 Alkaliphiles: ecology and biotechnological applications   143
W.D. GRANT and K. HORIKOSHI

5.1   Introduction   143
5.1.1   Ecology and environments   143
5.1.2   Alkaliphile diversity   146
5.1.3   Alkaliphile physiology   151
5.2   Alkaliphiles and industry   153
5.2.1   Enzymes   153
5.2.2   Spirulina   158
5.2.3   Secretion vectors   159
5.2.4   Future trends   159
References   160

6 Physiology and biotechnological potential of deep-sea bacteria   163
D. PRIEUR

6.1   Introduction   163
6.2   Deep-sea bacteria   164
6.3   Hydrothermal vents   172
6.3.1   Distribution of vent fields and their main features   172

6.3.2 Chemical features of hydrothermal fluids and expected metabolisms 174
6.3.3 Abundance and activity of bacteria in sea water 175
6.3.4 Bacterial communities on inert surfaces and bacterial mats 176
6.4.5 Main features of mesophilic bacteria isolated from sea water and surfaces 177
6.3.6 Invertebrate-associated bacteria 178
6.4 Biotechnology of deep-sea bacteria 194
References 197

7 Physiology and molecular biology of psychrophilic micro-organisms 203
N.J. RUSSELL

7.1 What are psychrophiles and psychrotrophs? 203
7.2 Microbial types of psychrophiles 205
7.3 Ecology of psychrophiles and psychrotrophs 206
7.3.1 Food 206
7.3.2 Terrestrial and aquatic ecosystems 208
7.4 Molecular mechanisms of adaptation to low temperature 211
7.4.1 Lipids, membrances and nutrient uptake 211
7.4.2 Proteins and protein synthesis 215
7.5 Biotechnological uses and potential of psychrophiles 219
References 221

8 Molecular biology and biotechnology of microbial interactions with organic and inorganic heavy metal compounds 225
G.M. GADD

8.1 Introduction 225
8.2 Physiology of metal-microbe interactions 226
8.3 Molecular biology of heavy metal tolerance 230
8.3.1 Plasmid-mediated bacterial heavy metal resistance 230
8.3.2 Metal-binding proteins of fungi 237
8.4 Biotechnological aspects of metal-microbe interactions 240
8.4.1 Microbial removal and recovery of heavy metals and radionuclides 240
References 252

9 Molecular biology of radiation-resistant bacteria 258
M.D SMITH, C.I. MASTERS and B.E.B. MOSELEY

9.1 Introduction 258
9.2 Types of radiation 258
9.3 Radiation resistance of bacterial species 260
9.4 Repair of radiation damage: mutation rates and mutagens 260
9.5 The biology of the Deinobacteriaceae 261
9.5.1 Radiobiology 262
9.5.2 Molecular description of *D. radiodurans* 264
9.5.3 Genome of the deinococci 265
9.5.4 Transformation of the deinococci 267
9.6 Shuttle plasmids between *D. radiodurans* and *E. coli* 270
9.7 Gene expression in the deinococci 271
9.7.1 The DNA repair genes *mtcA, mtcB, uvsC, uvsD* and *uvsC* 271
9.7.2 The heterologous DNA repair gene *denv* of bacteriophage T4 272
9.7.3 The HPI gene 272
9.7.4 The *leuB* gene 272
9.7.5 The *trp* and *asp* genes 273
9.8 Prospects in molecular biology 273

9.9   Biotechnology of the deinococci   273
        9.9.1   Radiation exposure measure   274
        9.9.2   Restriction endonucleases   274
        9.9.3   Z-DNA-binding protein   274
        9.9.4   DNA repair enzymes   275
        9.9.5   Membrane-bound exoenzymes   275
        9.9.6   DNA polymerase   275
        9.9.7   Manganese   276
9.10 Prospects for biotechnology   276
References   277

10   Obligate anaerobes and their biotechnological potential   281
N.P. MINTON, A. MAULE, P. LUTON and J.D. OULTRAM

10.1   Introduction   281
10.2   The industrial exploitation of anaerobic fermentations   282
        10.2.1   The acetone/butanol/ethanol fermentation   282
        10.2.2   Methane generation   286
        10.2.3   Other fermentations   287
        10.2.4   Anaerobic mixed culture fermentations   288
10.3   Anaerobic disposal of organic wastes   289
        10.3.1   Landfilling of wastes   290
        10.3.2   Anaerobic digesters   292
        10.3.3   Xenobiotic breakdown   294
10.4   Anaerobic biotransformations   296
        10.4.1   Fermentation manipulation   297
        10.4.2   Stereospecific, reductive biotransformations   298
10.5   Anaerobic enzymes as industrial products   299
10.6   Molecular genetics of anaerobic bacteria   303
        10.6.1   Host–vector system development   303
        10.6.2   Gene cloning   309
        10.6.3   Recombinant manipulation of anaerobes   312
        10.6.4   Gene probes for the detection of anaerobes   314
References   315

Subject index   321

Species index   327

1 Biochemistry and molecular biology of the extremely thermophilic archaeobacteria

D.A. COWAN

1.1 Introduction

'Thermophilic' implies a sphere of ecological novelty and, to the biochemist, a diversity of subtle and ingenious biochemical and physiological adaptations which permits survival in an environment which we, from our anthropomorphic viewpoint, consider to be impossibly hostile. 'Archaeobacteria' engenders the impression of ancient life forms, offering the possibility of discovering new biochemical processes and macromolecules together with a mechanism for probing the very origins of life. The realities of the study of this group of organisms are not so far from the expectations. As this chapter will demonstrate, much of the structure and function of these organisms is indeed highly novel, while characteristics common to the other two primary kingdoms, the prokaryotes and the eukaryotes, do give insights into the origins and evolution of life.

This review cannot possibly cover the topic of the thermophilic archaeo-bacteria in great depth. For more detail, the reader is directed to a number of excellent and comprehensive reviews on this rapidly expanding field [1–9].

The first of the thermophilic archaeobacteria to be revealed to the microbiological community was isolated by Brock's group in 1970 [10]. First thought to be a thermophilic example of the mycoplasma (and named *Thermoplasma* accordingly), this microbiological oddity was assigned to a more appropriate taxonomic position only after the proposal of the third primary kingdom (the Archaeobacteria) by Carl Woese and George Fox [11]. Since that time the collection of new genera and species of thermophilic archaeobacteria has burgeoned, largely through the efforts of Wolfram Zillig of the Max Planck Institute for Biochemistry in Munich and Karl Stetter of the University of Regensberg. Even with some 20 genera and 35 species, the rate of discovery of new thermophilic archaeobacteria is not likely to diminish for some time to come.

A strong thread of incentive has dominated the search for new thermo-philic archaeobacteria, the incentive to define experimentally the maximum temperature capable of supporting life. A steady rise in the maximum bacterial growth temperature (Table 1) reflects the increased interest of the scientific community together with improvements in laboratory techniques (media and equipment) and sampling technology (terrestrial, shallow marine and

Table 1.1   Historical development of thermophilic microbiology.

Period	Growth temperature range (°C)	Representative organisms
Pre-1960s	37–65	*Bacillus stearothermophilus*
1960s–70s	60–85	*Bacillus caldolyticus* *Thermus* spp. *Thermoplasma acidophilum* *Sulfolobus* spp.
1980s	85–105	*Desulfurococcus* spp. *Thermoproteus tenax* *Pyrodictium* spp.

deep submarine). Following the realisation that life could indeed exist at temperatures approaching the boiling point of water, predictions of the upper temperature limit of life have clustered at around 150°C. Despite relatively easy access to liquid environments of up to 400°C, the limit of bacterial survival is currently stalled at around 120°C. It must be assumed that either this represents a natural limit at which some essential (and irreplaceable) biochemical component(s) becomes limiting, or that some advance in technology which will access the next quantum step is yet to be achieved.

A report that the abyssal 'black smoker' environment has yielded bacteria capable of growing at 250°C [12] has been largely discounted. Repeated failures to reproduce the original findings, coupled with reports of the production of cell-like artefacts under similar conditions [13] and questions of the stability of critical macromolecules [14], have caused the validity of the original data to be queried. Suffice to say that, if by some chance the original report is indeed valid, such organisms must rely on a biochemistry very different from that with which we are familiar.

1.2   Archaeobacterial phylogeny

The original reclassification by Woese and Fox [11] of the two primary kingdoms (the Prokaryotes and Eukaryotes) into three primary kingdoms (Eubacteria, Archaeobacteria and Eukaryotes) was based on comparisons of partial nucleic acid sequences derived from 16S and 18S rRNAs. The 16S rRNAs (and 18S rRNAs in eukaryotes) are universal in distribution, readily isolated and purified, easily sequenced and contain highly conserved sequences. This makes the rRNAs well suited for use as molecular chronometers and appropriate for measurement over large phylogenetic distances.

The methodology for obtaining partial rRNA sequences is relatively simple [11]. Purified rRNAs are digested with T1 RNase and the digests separated by two-dimensional electrophoresis, giving a 'fingerprint' of six- to 12-base oligonucleotides. After sequencing individual oligonucleotides, a catalogue characteristic of the particular organism is obtained. By comparing the catalogue from one organism with that from another, it is possible to generate an *association coefficient* (S_{AB}), where:

$$S_{AB} = 2N_{AB}/(N_A + N_B)$$

N_A and N_B are the total number of nucleotides in sequences of hexamers or larger in catalogues from organisms A and B, and N_{AB} is the total nucleotide complement in sequences common to both organisms. The association coefficient is thus a measure of the number of mutations in each organism, assuming that the sequence is derived from a common ancestor. The evolutionary distance between the organisms is reflected in a timescale only if the mutation rate is similar in both organisms.

Further developments in technology now permit complete rRNA sequences to be used for phylogeny. The greater accuracy generated by the more extensive data and the more advanced computational methods allows the more detailed assignment of branch orders and lengths. However, while modern computational methods have simplified the synthesis of phylogenetic relationships, they have served to introduce both added complexity and contention. For example, the computational methods of parsimony analysis [15, 16] and distance-matrix treeing [17] generate trees of similar structure but with some variations in the detailed positioning of certain genera [18]. Each responds differently to one of the most fundamental problems of treeing; the presence of varying evolutionary rates (different mutation rates), both among different lineages and within the different positions in the sequence. Parsimony analysis makes no correction for variable evolutionary rates, thus tending to introduce artefactual similarities between distantly related sequences [19]. On the other hand, distance-matrix analysis incorporates a correction factor for variable evolutionary rates, but assumes equal rates of change for all positions in the sequence [19]. Modifications to this method [20] have improved the reliability of the distance-matrix analysis method.

Using these methods in the analysis of complete 16S/18S rRNA sequences, detailed phylogenetic trees indicating the relationships between the archaeobacteria, the eukaryotes and the eubacteria have been derived (Figure 1.1). This is an example of a typical 'unrooted' tree, in which no indication of the specific nature of the 'progenote' (or ancestral cell type) is implicit in the construction. The implication of the branch lengths is that the archaeobacteria are phylogenetically related more closely to the eubacteria than to the eukaryotes. Phylogenetic trees can be 'rooted' by the inclusion of data from an appropriate 'outgroup' [7, 21]. The position of the group is determined by comparing the sequences of pairs of paralogous genes that diverged from each other before the three primary lineages. One approach [22] is to use a set of aboriginally duplicated genes, i.e. genes in which duplication occurred before separation of the primary kingdoms. The rooted evolutionary tree

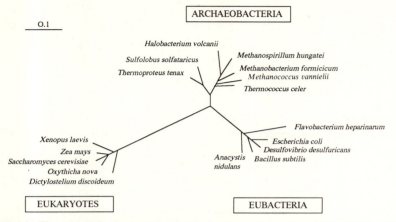

Figure 1.1 The unrooted phylogenetic tree of the three primary kingdoms. Redrawn from C.R. Woese and G.J. Olsen, *System. Applied Microbiol.* **7** (1986) 161, with the kind permission of the authors. The distance measure bar corresponds to 0.1 mutational events per sequence position.

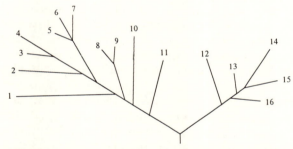

Figure 1.2 A rooted archaeobacterial phylogenetic tree. Modified from C.R. Woese, *Microbiol. Rev.* **51** (1987) 221, with the kind permission of the author. Phylogenetic data are based on 16S rRNA sequence comparisons. The root was imposed by the use of outgroup consensus sequences. Numbered groups refer to: 1, *Thermoplasma*; 2, *Methanospirillum*; 3, *Methanosarcina*; 4, FS-1; 5, *Halobacterium volcani*; 6, *Halococcus morrhuae*; 7, *Halobacterium cutirubrum*; 8, *Methanobacterium thermautotrophicum*; 9, *Methanobacterium formicicum*; 10, *Methanococcus vannielii*; 11, *Thermococcus celer*; 12, *Thermoproteus tenax*; 13, *Pyrodictium occultum*; 14, *Sulfolobus solfataricus*; 15, *Desulfurococcus mobilis*; and 16, *Hyperthermus butylicum*.

(Figure 1.2) of Woese and co-workers suggest that the progenote may have been thermophilic, anaerobic and have other characteristics typical of the current S-dependent archaeobacteria [7].

This topology is supported by the analysis of a number of other molecular chronometers. These include the 5S rRNAs [23], comparisons of amino acid sequences of DNA-dependent RNA polymerase subunits [18] and glyceraldehyde-3-phosphate dehydrogenases [24], ribosomal antibiotic sensitivity patterns [25] and duplicated genes (elongation factors and ATPase subunits) [22]. Each of these phylogenetic analyses shows a tripartite evolutionary tree

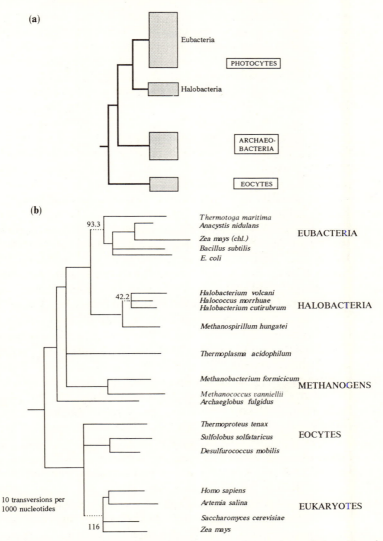

Figure 1.3 The 'eocyte' phylogenetic topology, based on ribosomal morphology (a) and 16S rRNA sequence data (b). (a) and (b) are redrawn from J.A. Lake, M.W. Clark, E. Henderson, S.P. Fay, M. Oakes, A. Scheinmann, J.P. Thornber and R.A. Mah, *Proc. J. Natl. Acad. Sci. USA* **82** (1985) 3716, and J.A. Lake, *Can. J. Microbiol.* **35** (1989) 109, respectively, with the kind permission of the authors.

topology, although there are substantial variations in the branch lengths and therefore the relationships between the three superkingdoms.

The phylogeny of the primary kingdoms is not, however, without contention. An alternative topology presented by James Lake in 1984 [26, 27] was derived from comparative morphologies of ribosomal subunits. This evolutionary tree (Figure 1.3a) differs primarily in the composition of the three monophyletic groupings; the *photocytes* (eubacteria and halobacteria), *archaeobacteria* (methanogens) and *eocytes* (eukaryotes and S-dependent archaeobacteria). This version has been strongly criticised on several grounds [7], not the least significant being that the analysis was based on structural features which may reflect changes in both ribosomal RNA and protein components and which may not be directly related to fundamental evolutionary changes.

More recently, Lake [28] has reassessed rRNA sequence data using an alternative computation (evolutionary parsimony) and generated an evolutionary tree which is similar to his original eocyte topology (Figure 1.3b). All extant organisms are structured into two superkingdoms, the *Parkaryotes* (eubacteria, halobacteria and methanogens) and the *Karyotes* (eocytes and eukaryotes).

These two divergent views on fundamental phylogeny are currently very much a matter of dispute. Claims that maximum parsimony methods are invalidated by their sensitivity to unequal evolution rate artefacts [29] have been countered by assertions that evolutionary parsimony is subject to alignment variations and that the 'eocyte' topology is an artefact resulting from the selection of the database used in the analyses [30]. The ability of even the informed scientist to reconcile these alternative hypotheses is limited. The complexity of the arguments is such that a detailed knowledge of the computational methods and the nature of the data utilised is a prerequisite to any critical analysis of the value of the alternatives.

In the most recent publication on urkingdom structure, Woese has proposed a completely new system of nomenclature for the tripartite phylogenetic tree (ref. 31 and Figure 1.4). The three urkingdoms (which are to be termed *domains*) become the Bacteria, the Eucarya and the Archaea. The last is comprised of two separate lineages (or kingdoms); the Euryarchaeota (in common use, the euryarchaeotes or euryotes) which includes the methanogens, the extreme halophiles and certain related extreme thermophiles, and the Crenarchaeota (in common usage, crenarchaeotes or chrenotes) which is entirely comprised of thermophilic species, referred to previously and variously as the 'thermoacidophiles', the 'S-dependent archaeobacteria', the 'eocytes' or just the 'extremely thermophilic archaeobacteria'. Since thermophily occurs on both branches of the Archaea, it is presumed that this is indicative of the ancestral phenotype. Time alone will determine the acceptability of this new classification.

As the chapter title suggests, this discussion will encompass all the Archaea capable of existing at thermophilic temperatures. Representatives of this group are not only found in the crenotes but also exist within the methanogenic group of the euryotes. No extremely thermophilic representatives of the extreme halophiles have yet been identified. Since the old terminology

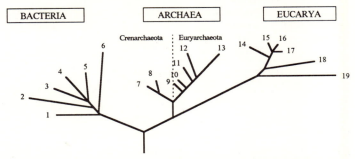

Figure 1.4 A new nomenclature for the three primary kingdoms presented on the rooted phylogenetic tree. Redrawn from C.R. Woese, O. Kandler and M.L. Wheelis, *Proc. Natl. Acad. Sci. USA* **87** (1990) 4576, with the kind permission of the authors. Numbered groups refer to: 1, the Thermotogales; 2, the flavobacteria and relatives; 3, the cyanobacteria; 4, the purple bacteria; 5, the Gram-positive bacteria; 6, the green non-sulphur bacteria; 7, *Pyrodictum* species; 8, *Thermoproteus* species; 9, the Thermococcales; 10, the Methanococcales; 11, the Methanobacteriales; 12, the Methanomicrobiales; 13, the extreme halophiles; 14, the animals; 15, the ciliates; 16, the green plants; 17, the fungi; 18, the flagellates; and 19, the microsporidia.

of S-dependent archaeobacteria is phenotypically derived and has no basis in phylogenetic taxonomy, it will not be used henceforth.

The thermophilic representatives of the Archaea comprise some eight families and a total of 19 genera (Table 1.2). This is a diverse collection of organisms with a wide variation of morphological, physiological and biochemical characteristics. These are brought together purely for their capacity to survive at very high temperatures, and such similarities as exist result from inevitable adaptive pressures induced by the requirements for survival at such temperatures, i.e. *thermophily*.

Some mention must be made of the particular terminology which has developed in the field of high-temperature microbiology. Strictly, the terms *thermophile* and *extreme thermophile* apply to organisms having growth minima, optima and maxima within defined ranges (Table 1.3). Under this arbitrary classification, all the organisms named in Table 1.2 (other than the thermophiles *Sulfolobus acidocaldarius* and *Thermoplasma acidophilum*) are extreme thermophiles. With the increasing number of organisms capable of growing optimally at temperature much in excess of 70–75°C, the need for an additional category became evident. The term *caldoactive* [69] attained limited popularity but has been effectively replaced by *ultrathermophilic* [55] and *hyperthermophilic* [39], particularly with reference to those organisms capable of optimal growth at or above 90°C. For the sake of simplicity, the term thermophilic will be used throughout this chapter to imply the ability to grow at high temperature, without presuming any particular temperature preference.

Table 1.2   Species diversity of the thermophilic Archaea (including the thermophilic representatives of both crenotes and euryotes).

Order	Genus	Species	Reference
Sulfolobales	*Sulfolobus*	*acidocaldarius*	32
		*solfataricus**	33
		shibatae	34
	Metallosphaera	*sedula*	35
	Acidianus	*brierleyi†*	36, 37
		infernus	37
	Desulfurolobus	*ambivalens*	38
	Hyperthermus	*butylicus*	39
Thermoproteales	*Thermoproteus*	*tenax*	40
		neutrophilus	41
		uzoniensis	42
	Pyrobaculum	*icelandicus*	43
		organophicum	43
	Thermofilum	*pendens*	44
		librum	5
	Desulfurococcus	*mobilis*	45
		mucosus	45
		amylolyticus	46
		strain $Tok_{12}S_1$	47, 48
		strains S and SY	49
	Staphylothermus	*marinus*	50
Thermococcales	*Thermococcus*	*celer*	51
		stetteri	52
		littoralis	53
	Pyrococcus	*furiosus*	54
		woesei	55
	Caldococcus	*littoralis*	56
Pyrodictales	*Pyrodictium*	*occultum*	57
		brockii	57
		abyssum	K. Stetter, personal communication
	Thermodiscus	*maritimus*	58
Archaeglobales	*Archaeglobus*	*fulgidus*	59
		profundus	60
Thermoplasmales	*Thermoplasma*	*acidophilum‡*	10
		volcanium	61
Methanobacteriales	*Methanobacterium*	*thermoautotrophicum*	62
	Methanothermus	*fervidus*	63
		sociabilis	64
Methanococcales	*Methanococcus*	*jannaschii*	65
		thermolithotrophicus	66
		igneus	67
	Methanopyrus	Not yet named	68

*Originally assigned the genus name of *Caldariella solfataricus* [33].
†Recently reclassified from *Sulfolobus* [37].
‡Originally classified as a thermophilic mycoplasma.

Table 1.3 Nomenclature of high-temperature bacteria (modified from ref. 70).

Term	Growth minimum (°C)	Growth optimum (°C)	Growth maximum (°C)
Thermophile	> 30	> 50	> 60
Extreme thermophile	> 40	> 65	> 70
Hyperthermophile	Not defined	> 100	Currently < 115

1.3 Ecology of thermophilic Archaea

Thermophilic habitats (or biotopes) in the temperature range 40–70°C are extremely common; solar-heated soil, decaying vegetation, industrial waste heat, self-heating coal tips and ore tailing and so forth. However, biotopes with temperatures of greater than 70°C, where most thermophilic Archaea are found, are relatively uncommon. They differ markedly in the origins and nature of the heat source, both of which influence the characteristics of the biotope and therefore the diversity of the biota.

The most common 'high-temperature' thermal habitats are of geothermal origin, usually associated with tectonically active zones. These are widely if sparsely distributed over the earth's surface. Groundwater, penetrating to depths up to 3000 m [71], is heated by magma. Thermal expansion forces the superheated water to the surface, where it issues (often boiling because of the reduction in pressure) in the form of *thermal springs, hot pools* or *geysers.* Where the water supply to the heat source is low or the heat source is very near the surface, the water often emerges as steam, in what are known as *fumaroles.*

In general, the pH of thermal biotopes falls into a bimodal distribution: pH 1.0–2.5 and pH 6.0–8.5. These two ranges reflect the chemical compositions of the two major buffering components: sulphuric acid, with a pK_a of 1.8 and sodium carbonate and bicarbonate, with pK_as of 6.3 and 10.2. The different origins of the two buffering systems and their consequences in the mineralisation of thermal waters generate very different environments (ref. 72 and Figure 1.5).

Acid thermal waters are often sulphur- and iron-rich but otherwise poorly mineralised. The acid (primarily sulphuric acid) is derived from hydrogen sulphide, carried to the surface from geological sulphide deposits by geothermal steam. The hydrogen sulphide is oxidised to sulphur on contact with oxygen in the upper soil profile. Further oxidation of the sulphur by (primarily) mesophilic sulphur-oxidising bacteria generates sulphuric acid. This is leached to pooling areas by surface or shallow groundwater, where it may be heated by geothermal steam. Such acid thermal sites often have a low water flow and are subject to substantial temperature shifts in response to changes in both groundwater flow and steam supply. The lack of mineralisation reflects

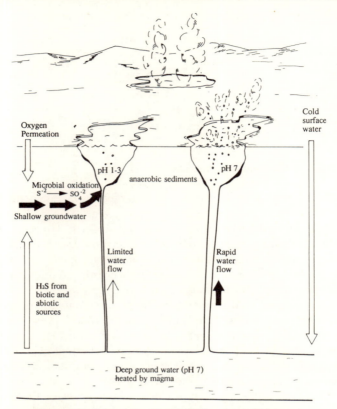

Figure 1.5   Origins of acid and alkaline thermal biotypes.

the fact that the hydrological cycle is brief and water has not penetrated the depths of the crustal zone.

Alkaline thermal sites are usually highly mineralised, and the basins in which they are sited are often heavily encrusted with silicates. This reflects the deep penetration of the source groundwater and the reactivity of superheated water. The effect of input from shallow acid groundwater is minimised by the continual and substantial flow of neutral or alkaline waters from the depths. For the same reason, alkaline thermal pools are also much more stable with respect to temperature, pH and volume and may have lifetimes of hundreds of years or more [73].

The marine equivalent of the alkaline thermal spring differs only in that the water erupting is saline and that temperatures of above 100°C can be reached because of the added hydrostatic pressure. The biotope is much more localised because of the absence of a pooling area. A specialised version of this biotope is found at great depths in ocean rifts such as the Galapagos

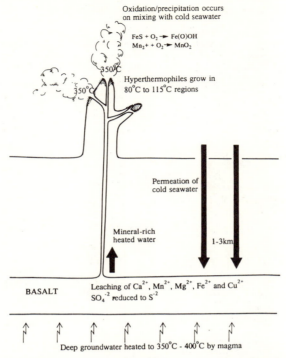

Figure 1.6 Origins and geochemistry of 'black smoker' thermal environments.

Rift and the East Pacific Rise (2500 m–2600 m), the Juan de Fuca Ridge (1570 m), the Guaymas Basin (2000 m) and the base of the Florida escarpment (3266 m) [74]. Because of the reactivity of superheated saline water, the waters issuing from these deep-sea hydrothermal vents are very highly mineralised, and the precipitation of insoluble metal salts from the issuing vents has given rise to the term *black smokers* (Figure 1.6). Temperatures in the region 350–400°C have been recorded in the vent waters, although very rapid mixing with the ambient waters (around 4°C) generates very steep temperature gradients.

The chemistry of the outflow waters of black smokers is complex, being based on the solubilisation of deep basalts and subsequent reductive chemical changes [75]. Oxidative changes occur only on contact between the superheated vent flow and the cold oxygenated abyssal waters.

Each of the thermal biotopes described above has a representative fauna (Table 1.4). Some genera appear to predominate in particular environments, for example *Methanococcus* in deep-sea hydrothermal vents [76] and *Desulfurococcus/Thermoproteus* in neutral/alkaline surface thermal springs [49].

Table 1.4   Representative genera from the various thermal biotopes.

Biotope	Temperature range (°C)	pH range	Typical genera
Self-heating coal/ore pile	50–60	Around 2	*Thermoplasma*
Neutral silicaceous pool	Up to 100	5.0–9.5	*Thermoproteus* *Pyrobaculum* *Thermofilum* *Desulfurococcus* *Methanothermus*
Acid thermal pool	Up to 100	1.0–3.5	*Sulfolobus* *Acidianus* *Metallosphaera*
Shallow marine vent	Up to 103	5.0–7.0	*Thermococcus* *Pyrodictium* *Pyrococcus* *Thermodiscus* *Staphylothermus* *Archaeglobus*
Deep-sea thermal vent	4–350	Around 7	*Methanococcus* *Methanopyrus* *Desulfurococcus*

Archaeal biomass levels in thermal waters can be surprisingly high. Cell densities in free solution are relatively low but large numbers of cells are found adhering to silicate surfaces, particulate matter, detritus, etc. Estimates of *Thermoproteus*-like cells in a 97.5°C thermal pool in Rotorua, New Zealand, were 1.1×10^3/ml and 4.3×10^4/ml for pool water and sediment respectively [49]. Sampling from surface sites is generally straightforward (given the inherent dangers in working in an environment potentially comprised of boiling water or mud, fragile silicate terraces and toxic fumes). The primary requirements for successful sample collection are the acquisition of a volume of sediment, the retention of anaerobic status in the sample (for the strictly anaerobic Archaea), and the need to buffer very low-pH samples to nearer neutrality. Despite some efforts, there is no evidence of organisms which are highly sensitive to temperature drop and thus no incentive to keep thermal water samples heated between the sampling site and the laboratory.

Sampling from shallow marine sites requires only the addition of scuba equipment and some syringe-like device for collecting samples below the sediment surface or within the heated water stream. For deep-sea hydro-thermal systems, a higher level of technology is essential. Using the submersible *Alvin*, remote-controlled manipulators and various ingenious sampling and coring devices, sampling from black smokers and similar thermal sources has been very successful (see, for example, refs. 65, 68, 76 and chapter 6).

1.3.1 Morphology of the Archaea

The Archaea demonstrate a diversity of structural morphologies typical of any other diverse group of micro-organisms (Table 1.5). These range from the typically rod-shaped and coccoid structures of *Thermoproteus* and *Desulfurococcus* to the highly variable *Thermoplasma*, in which the absence of a cell wall produces a wide variety of coccoid, filamentous and disc- and club-shaped cells. A number of extremely thermophilic spirilla have been cultured under laboratory conditions (H.W. Morgan and R.M. Daniel, personal communication), but whether these are representatives of the bacterial or archaeal kingdoms is not yet known.

1.3.2 Physiology and biochemistry of the Archaea

Amongst the Archaea are represented many of the various physical characteristics typical of eubacteria (Table 1.6). Typically, the anaerobic representatives grow either chemolithoautotrophically, utilising carbon dioxide as a sole carbon source and deriving energy by sulphur oxidation of molecular hydrogen, or heterotrophically by sulphur respiration of various carbon and energy sources. The former is viewed as a primaeval mode of metabolism [58], especially since high-temperature, anaerobic, aquatic environments rich in carbon dioxide and hydrogen may have predominated in the earth's primaeval biotype. Some examples of the Archaea, including isolates of *Desulfurococcus, Pyrococcus, Thermococcus* and *Pyrodictium*, are capable of growing anaerobically in the absence of sulphur and/or hydrogen, suggesting the presence of fermentative metabolism. The only true fermentative hyperthermophilic archaeobacterium currently known is *Hyperthermus butylicus* [39], which ferments peptides with the production of carbon dioxide, 1-butanol, propanol and acetic and phenylacetic acids. Hydrogen sulphide is formed as a growth-supplementary energy source but not by sulphur respiration. The Sulfolobales show some interesting metabolic disparities. Isolates of the genus *Sulfolobus* have been shown to grow either aerobically by the oxidation of molecular sulphur or anaerobically using sulphur reduction [77]. *Desulfurolobus ambivalans* [38] is also capable of contrary modes of growth, either aerobic oxidation of sulphur or anaerobic reduction of sulphur with molecular hydrogen using carbon dioxide as a sole carbon source in both cases. The distinctive metabolism of sulphate reduction is present in the fermentative anaerobe *Archaeolobus fulgidus* [78]. Thiosulphate and sulphite (but not inorganic sulphur) can also be utilised as terminal electron acceptors, and a variety of electron donors, including hydrogen, carbon dioxide, short-chain alcohols and simple organic acids, are suitable as substrates [78]. This organism generates small quantities of methane and, while lacking some of the co-factors typical of the true methanogens [79], is considered to hold a phylogenetic position between the Crenotes and the Euroytes [79].

Table 1.5 Morphological characteristics of the thermophilic Archaea.

Genus	Cell shape	Cell numbers	Cell size	Gram reaction	Motility	Flagella	Pili
Acidianus	Irregular cocci	Single; attach to particulates	0.5–2 µm wide	Negative	No	No	Yes
Archaeglobus	Irregular cocci	Single	0.4–1 µm wide	Negative	Weak	Monopolar, polytrichous	No
Desulfurococcus	Coccoid, may form giant cells	Single, sometimes paired	1 µm wide		Yes (*D. mobilis*) No (*D. mucosus*)	Monopolar, polytrichous (*D. mobilis*)	No
Desulfurolobus	Coccoid, lobed	Single	?	?	No	No	No
Hyperthermus	Irregular spheres	Single or clumped	1.5 µm diameter	?	No	No	Numerous
Metallosphaera	Irregular cocci	Single; attach to particulates	0.8–1.2 µm wide	Negative	No	No	Yes
Pyrobaculum	Rods; 'golf clubs'*	Aggregates; 'rafts'	1.5–8 µm long 0.5 µm wide	Negative	Yes	Bipolar, polytrichous	No
Pyrococcus	Spherical, some elongated or constricted	Single, pairs	0.5–2 µm diameter	Negative	Yes	Monopolar, polytrichous	No
Pyrodictium	Disc- or dish-shaped, irregular	Network connected by a network of fibres	0.3–2.5 µm diameter 0.2 µm thick	?	No	No	Thin filaments
Staphylothermus	Coccoid, irregular	Single, short chains	0.5–1 µm diameter (also 15-µm diameter giant cell)	Negative	No	No	No

Sulfolobus	Coccoid, irregular, lobed	Single	0.8–2 μm diameter	?	Weak (*S. shibatae*)	No	Yes
Thermococcus	Irregular spheres and diploforms	Various	0.3–3 μm diameter	?	Various	No (*T. littoralis*) Monopolar polytrichous (*T. celer*) Both (*T. stetteri*)	?
Thermodiscus	Disc-shaped	Single or clumped	0.3–3 μm diameter	?	No	No	?
Thermofilum	Thin stiff rods with some branches and terminal spheroids	Single	0.2–0.35 μm diameter 1–100 μm long	Negative	No	No	Terminal
Thermoplasma	Irregular and variable including cocci, discs, clubs, filaments and buds	Clumps	0.1–5 μm diameter	?	?	Monopolar, monotrichous	Yes
Thermoproteus	Aseptate rods, sometimes branched; 'golf clubs'	Various	1–80 μm long 0.4 μm wide	Negative	No	No	Yes

*'Golf clubs' are spheres protruding laterally from ends and sharp bends of rod-shaped cells.

Table 1.6 Physiological characteristics of the thermophilic Archaea.

Organism	Temperature range (°C)	Temperature maximum (°C)	pH optimum	Oxygen status	Metabolism/energy yield	Nutritional requirements
Acidianus infernus	65–95	90	2.0	Aerobic or anaerobic	Obligate autotroph, either aerobic ($S^\circ \rightarrow SO_4^{2-}$) or anaerobic ($S^\circ \rightarrow S^{2-}$)	Hydrogen
Archaeoglobus fulgidus	60–95	83	6.5	Obligate anaerobe	Chemolithoautotrophic on H_2, CO_2 and $S_2O_3^-$; chemo-organotrophic on complex media; SO_4^{2-} reduction (S° not reduced)	
Desulfurococcus spp.	85–90	85–88	5.5–6.5	Obligate anaerobes	Obligate heterotrophs; anaerobic S° respiration; possible fermentation	Growth stimulated by S°
Desulfurolobus ambivalens	Up to 87	81	2.5	Aerobic or anaerobic	Obligate chemolithoautotroph; with CO_2 as sole C source will oxidise S° with O_2 or reduce S° with H_2.	
Hyperthermus butylicus	Up to 108	95–106	7.0	Obligate anaerobe	Fermentative; energy also obtained from $H_2 + S^\circ \rightarrow H_2S$ (no S° respiration)	Requires sodium chloride at 17 g/l
Metallosphaera sedula	50–80	75	1–4.5	Aerobe	Facultative autotroph ($S^\circ/S^{2-} \rightarrow SO_4^{2-}$) or heterotrophic growth on complex media	Efficient extraction of heavy metals from ores
Pyrobaculum icelandicum	74–102	100	6.0	Obligate anaerobe	Facultative autotroph; H_2/S autotrophy or heterotrophy on complex media	
Pyrobaculum organotrophicum	78–102	102	6.0	Obligate anaerobe	Obligate heterotroph	
Pyrococcus spp.	70–103	100–103	6.5–7.5	Obligate anaerobes	Obligate heterotroph; S° respiration; fermentation? (*P. furiosus*)	S° stimulates *P. furiosus* growth

Organism			pH		Metabolism	Notes
Pyrodictium spp.	82–110	115	5.5	Anaerobic	Obligate $H_2/S°$ autotrophy	Growth of P. brockii but not P. occultum stimulated by yeast extract
Staphylothermus marinus	65–98	92	6.5	Obligate anaerobe	Obligate heterotroph with S° respiration	
Sulfolobus spp.	60–90	75, S. acidocaldarius; 81, S. shibatae; 87, S. solfataricus	3.0	Aerobic or microaerophilic	Facultative autotrophs; $S° \rightarrow SO_4^{2-}$, $Fe[III]$, $S^{2-} \rightarrow S°$; $Fe[III] \rightarrow Fe[II]$ under microaerophilic conditions	Grow on complex media
Thermococcus celer	75–97	88	5.8	Obligate anaerobe	Obligate heterotroph; S° respiration of complex media	Growth stimulated by S°
Thermodiscus maritimus	75–98	88	5.5	Obligate anaerobe	Facultative autotroph; S° respiration; fermentation?	
Thermofilum spp.	55–100	80, T. librum; 85–90, T. pendens	5.0	Obligate	Obligate heterotroph; S° respiration	T. pendens requires a polar lipid fraction from T. tenax for growth
Thermoplasma spp.	33–67	59	1–2	Facultative aerobes	Obligate heterotroph; S° respiration during anaerobic growth	
Thermoproteus neutrophilus	80–97	85	5.5–7.5	Obligate anaerobe	Obligate $H_2/S°$ autotroph	
Thermoproteus tenax	80–92	88	5.5	Obligate anaerobe	Facultative autotroph; either $H_2/S°$ autotrophy or respiration of complex media	
Thermoproteus uzoniensis	<75–95	90	5.6	Obligate anaerobe	Obligate heterotroph; ferments peptides with S° reduction to H_2S	Growth stimulated by S°

Of the Archaea, only the Sulfolobales demonstrate aerobic metabolism, a physiological characteristic which may reflect their relatively recent evolutionary divergence. The prevalence of strict anaerobiosis in the majority of the Archaea probably indicates both the nature of the primaeval habitat [58] and the fact that the oxygen solubility in boiling water is so low that there will not have been a strong evolutionary pressure for the acquisition of aerobic metabolic systems in the very high-temperature aquatic and marine habitats.

1.3.2.1 *Central metabolism.* It is tempting to speculate that the central metabolic pathways operating in the more phylogenetically 'primitive' representatives of the Archaea may represent early stages in the evolutionary development of these pathways. Certain features of the glucose catabolism pathway of *Sulfolobus* would appear to support this contention. The glycolytic pathway appears to be absent in the Archaea and, at least in *Sulfolobus* species and in *Thermoplasma acidophilum*, glucose is oxidised via a non-phosphorylating modification of the Entner–Doudoroff pathway [9] (Figure 1.7). The simplicity of this pathway, with the absence of substrate-level phosphorylation, the low net yield of ATP and the optional use of either NAD^+ or $NADP^+$ as electron carriers, argues strongly for this pathway as a primitive energy conservation mechanism. The fate of glyceraldehyde is uncertain: in *Thermoplasma acidophilum* but not in *Sulfolobus*, glyceraldehyde is converted to a second molecule of pyruvate by the sequential action of an $NADP^-$-dependent dehydrogenase, glycerate kinase, enolase and pyruvate kinase [9]. Acetyl CoA appears to be the immediate fate of pyruvate in all the Archaea studied. The conversion is catalysed by pyruvate oxidoreductase with a [4SFe–4S] ferredoxin as an electron acceptor.

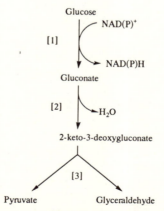

Figure 1.7   Glucose catabolism in the Sulfolobales. The enzymes involved in the sequences are (1) glucose dehydrogenase, (2) gluconate dehydratase and (3) aldolase.

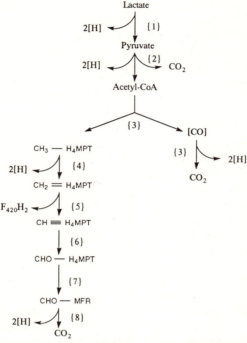

Figure 1.8 Proposed pathway for the oxidation of lactate to carbon dioxide by *A. fulgidus* (from ref. 80). Abbreviations are: CH_3–H_4MPT, methyltetrahydromethopterin; CH_2=H_4MPT, methylenetetrahydromethopterin; $CH \equiv H_4MPT$, methyltetrahydromethopterin; CHO–H_4MPT, formyltetrahydromethopterin: CHO–MFR, formyl-methanofuran; F_{420}–H_2, reduced coenzyme F_{420}. Enzymes are (1) lactate dehydrogenase, (2) pyruvate dehydrogenase, (3) carbon monoxide dehydrogenase, (4) methylenetetrahydromethopterin reductase, (5) methylenetetrahydromethopterin dehydrogenase, (6) methylenyltetrahydromethopterin cyclohydrolase, (7) formyltetrahydromethopterin-methanofuran formyltransferase, and (8) formylmethanofuran dehydrogenase.

The oxidative pathways in the anaerobic Crenotes are likely to be highly variable. For example, *Archaeglobus fulgidus*, an anaerobic sulphate-reducing archaeobacterium which grows on lactate and sulphate, appears to contain a 'carbon monoxide dehydrogenase' pathway [80]. In this proposed sequence of reactions, lactate is oxidised to three molecules of carbon dioxide (Figure 1.8) using a C_1 pathway as an alternative to the citric acid cycle. Demonstration of the ability of cell extracts to catalyse the exchange of carbon dioxide with the carbonyl group of acetyl CoA is strong evidence of the presence of the key enzyme of the pathway, the carbon monoxide dehydrogenase.

Gluconeogenesis in the aerobic Archaea may operate by a reversal of the non-phosphorylated Entner–Doudoroff pathway (although this has yet to be substantiated). In those organisms in which studies have been carried out, the enzymes required for a classical gluconeogenic pathway are apparently lacking [9].

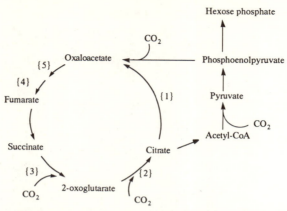

Figure 1.9   The reductive citric acid cycle of *Sulfolobus* species and *T. neutrophilus*. The enzyme activities identified in these organisms currently include citrate synthase (1), isocitrate dehydrogenase (2), 2-oxoglutarate : ferredoxin oxidoreductase (3), fumarase [81] and (4) malate dehydogenase (5).

The presence and role of citric acid cycle enzymes in the Archaea appear to depend on the phenotypic characteristics of the species, whether aerobic or anaerobic, autotrophic or heterotrophic. For example, in the aerobic heterotroph *Thermoplasma acidophilum* the citric acid cycle appears to operate in an oxidative mode. Alternatively, in autotrophically grown *Thermoproteus neutrophilus* or in *Sulfolobus acidocaldarius* grown either autotrophically or heterotrophically, the citric acid cycle operates in a reductive mode for carbon dioxide fixation (Figure 1.9). In *Thermoproteus neutrophilus* grown heterotrophically on acetate, the fumarate reductase is 'repressed' and an incomplete cycle operates [9].

The mechanisms of energy generation in the majority of the Crenotes are, at present, poorly understood. The only exception is the aerobic *Sulfolobus*. F_0/F_1-type ATPases have been isolated from *S. solfataricus* and *S. acidocaldarius* and have been studied in some detail [81,82]. It is now accepted that, at least in these organisms, ATP is generated by oxidative phosphorylation coupled with proton transport [83,84]. Various components of a putative electron transport system have been identified, including NADH dehydrogenase [85], a novel quinone *caldariellaquinone* [86], an a-type cytochrome [87], cytochrome aa_3 [88] and an a-type terminal oxidase [89]. In contrast, only substrate-level phosphorylation is operative in *Thermoplasma acidophilum* as a source of ATP, utilising the reaction [9]:

$$\text{Acetyl CoA} + \text{ADP} + \text{P}_i \Rightarrow \text{acetate} + \text{CoA} + \text{ATP}$$

The pathways responsible for electron transport and ATP formation in the sulphur-respiring anaerobic Archaea have yet to be elucidated.

Few studies have been carried out on nitrogen metabolism in the Archaea. An aspartate aminotransferase has been characterised [90], but there is currently no evidence to suggest the critical role of this or any other amino acid as an intermediary in either amino biosynthesis or nitrogen excretion. Suggestions that the roles of the amide amino acids in nitrogen metabolism might well be reduced in extreme thermophiles as a consequence of their relatively high thermolability have yet to receive support (or otherwise) from experimental data.

Few data are available on the other pathways of nitrogen metabolism. One exception is the detailed study on the arginine biosynthesis pathways in the aerobe *S. solfataricus* and the anaerobe *Pyrococcus furiosus*. Van de Castelle *et al.* [91] demonstrated that the former uses a linear version (Figure 1.10) of the mesophilic bacterial pathway but that the latter possesses only the enzymes required for the last three steps. The arginine biosynthesis pathways in both organisms demonstrate characteristics which might be thought of as 'primitive': the truncation of the pathway in *P. furiosus* and, in

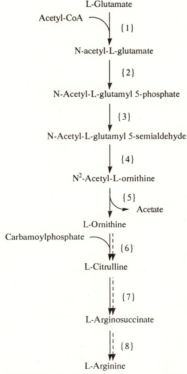

Figure 1.10 Arginine biosynthesis pathways in *S. solfataricus* (———) and *P. furiosus* (————). The enzymes identified as involved are *N*-acetylglutamate synthetase (1), *N*-acetylglutamate 5-phosphotransferase (2), *N*-acetylglutamate 5-semialdehyde dehydrogenase (3), *N*-acetylornithine 5-aminotransferase (4), acetylornithase (5), ornithine carbamoyl transferase (6), arginosuccinate synthetase (7), and arginosuccinase (8).

Table 1.7   Enzymes from the extremely thermophilic Archaea which have been isolated and/or investigated in some detail.

Enzyme	Organism	Reference
s-Adenosylmethionine synthetase	*Sulfolobus solfataricus*	92
Alcohol dehydrogenase	*Sulfolobus solfataricus*	93
Aminopeptidase	*Sulfolobus solfataricus*	94
α-Amylase	*Pyrococcus furiosus* *Sulfolobus solfataricus*	95,96
ATPase		97,81
ATP sulphurylase	*Sulpholobus solfataricus* *Archaeglobus fulgidus*	98
Aspartate aminotransferase	*Sulfolobus solfataricus*	99
Citrate synthase	*Sulfolobus solfataricus* *Thermoplasma acidophilum* *Sulfolobus acidocaldarius*	100–102
Carboxylesterase	*Sulfolobus acidocaldarius*	103
Dihydrolipoamide dehydrogenase	*Thermoplasma acidophilum*	104
DNA polymerase	*Thermoplasma acidophilum* *Sulfolobus acidocaldarius* *Sulfolobus solfataricus*	105–107
Fumarase	*Sulfolobus solfataricus*	108
α-Glucosidase	*Pyrococcus furiosus* *Thermococcus celer*	96,109
β-Galactosidase	*Sulfolobus solfataricus*	110,111
β-Glucosidase	*Thermococcus celer*	96
Glucose dehydrogenase	*Thermoplasma acidophilum* *Sulfolobus solfataricus*	112,113
Glyceraldehyde-3-phosphate dehydrogenase	*Sulfolobus acidocaldarius* *Thermoproteus tenax* *Pyrococcus woesei*	114
Hydrogenase	*Pyrococcus furiosus*	115
Isocitrate dehydrogenase	*Sulfolobus acidocaldarius*	116
Malate dehydrogenase	*Thermoplasma acidophilum* *Sulfolobus acidocaldarius*	117
Malic enzyme	*Sulfolobus solfataricus*	118
NADH dehydrogenase	*Sulfolobus acidocaldarius*	85
2-Oxoacid ferredoxin oxidoreductase	*Thermoplasma acidophilum* *Sulfolobus acidocaldarius* *Thermodiscus maritimus*	119
Polyphosphate kinase	*Sulfolobus acidocaldarius*	120
Protease	*Thermococcus celer* *Pyrococcus furiosus* *Desulfurococcus mucosus* *Sulfolobus acidocaldarius*	48,96,121–125

Table 1.7 (*Continued*)

Enzyme	Organism	Reference
Pullulanase	*Pyrococcus furiosus*	126
RNA polymerase	*Thermoplasma acidophilum*	127
	Sulfolobus acidocaldarius	
	Thermoproteus tenax	
	Desulfurococcus mucosus	
Succinate thiokinase	*Thermoplasma acidophilum*	102
Sulphur oxygenase	*Acidianus brierleyi*	128
Topoisomerase I	*Sulfolobus acidocaldarius*	129
Transglucosylase	*Desulfurococcus mucosus*	96

S. solfataricus, the lack of the ornithine acetyltransferase-catalysed energy-conserving cyclic reaction typical of many eukaryotes, eubacteria and methanogens. The involvement of carbamoyl phosphate in the *P. furiosus* pathway raises interesting questions on how this extremely unstable metabolite is protected: the half-life in aqueous solution at the *Pyrococcus* growth temperature is less than 1 second [91]. The very high levels of ornithine carbamoyltransferase activity reported in *P. furiosus* extracts [91] might well have evolved as a mechanism to minimise carbamoyl phosphate pool sizes.

1.3.3 *Enzymes*

A wide range of archaeobacterial enzymes, both intracellular and extracellular, has been investigated by the classical procedures of isolation, purification and structural and functional characterisation. Table 1.7 presents a reasonably comprehensive survey of publications on archaeobacterial enzymes in which some structural as well as functional data are available. Reasons for targeting these enzymes include their accessibility, their suitability as models for investigating protein thermostability and their potential as biocatalysts in modern biotechnology.

A consistent characteristic of all these enzymes is their high level of thermostability. The general positive correlation between the *thermophilicity* of the source organism and *thermostability* of both intracellular and extra-cellular proteins has been demonstrated frequently, both for protein populations [130] and for many different individual purified proteins. The trend of increasing thermostability shown for extracellular proteases isolated from a range of mesophiles, thermophiles, extreme thermophiles and ultrathermophiles (Table 1.8) is typical of many other enzymes studied. It can be assumed with almost total confidence that a protein from a thermophilic source will be more thermostable than the 'homologous' enzyme from a mesophilic

Table 1.8 The thermostability of proteases isolated from micro-organisms from different thermal biotopes.

Organism	Growth temperature optimum (°C)	Protease	T_M (°C)	Half-life at (min)	°C	Reference
Bacillus subtilis	37	Subtilisin		10	60	131
Bacillus stearothermophilus	55			15	87	131
Bacillus thermoproteolyticus	60	Thermolysin	86	60	80	131
Thermus aquaticus	72	Caldolysin	92	>30h	80	132
				15	95	132
Thermococcus celer	88	Protease		40	95	96
Desulfurococcus mucosus	85	Archaelysin	98	70–90	95	124
Sulfolobus solfataricus	85	'Neutral protease'		40	95	96
Pyrococcus furiosus	94	66 kDa protease		>33h	98	122
		'Pyrolysin' complex	>115*	17	105	123

*Connaris, H., Cowan, D.A. and Sharp R.J., unpublished results.

source. There is, however, no justification for assuming that all proteins from thermophilies will be more stable than all proteins from mesophiles!

The molecular mechanisms responsible for enhanced protein thermostability and the structural and functional consequences of protein thermostability have been the subject of considerable interest. A growing body of evidence, reviewed recently by Jaenicke and Závodszky [133], supports the view that the marginal conformational stability of proteins (expressed as a free energy difference between folded and unfolded states of only 30–60 kJ/mol) can be perturbed by relatively minor changes in primary structure. The evolutionary consequence is a multitude of strategies in the thermostabilisation of proteins in thermophilic organisms, in which the optimisation of intramolecular interactions, packing densities, interiorisation of hydrophobic residues and surface exposure of hydrophilic residues will all play a part in the overall thermodynamic stability of the protein. This conclusion has been supported by a growing body of evidence from protein engineering studies where diverse point mutations can enhance (or reduce) protein thermostability [134–136].

There are consequences, both structural and functional, in artificially enhancing protein thermostability. For example, it is likely that increasing the thermal stability of a protein may result in reduced conformational flexibility. Depending on locality and extent of changes in conformational flexibility, this may result in significant (and sometimes detrimental) consequences with respect to biological function. There is now evidence from various sources including proteolysis studies [137], ^{1}H–^{2}H exchange studies [138], X-ray diffraction [139] and calculations of flexibility profiles based on normalised B-values [140] that at room temperature, for example, a thermophilic protein will be less flexible than its mesophilic equivalent. It is also now believed that, at their respective growth temperatures, similar proteins from both mesophilic and thermophilic sources will possess similar levels of molecular flexibility, a consequence of the fact that molecular flexibility is critical for function [141, 142]. This perhaps explains the general observation that functionally similar proteins from micro-organisms of different thermal biotopes tend to possess similar catalytic rates *at their respective growth temperatures*. The only notable exceptions to this particular rule are examples where the substrate is rendered more accessible/susceptible to catalytic attack at the higher reaction temperature. Thus the observation that the specific activities of thermophilic proteinases are considerably higher than those of mesophilic proteinases [143] is attributable to the temperature-enhanced susceptibility of the polymeric substrate and not to any particular intrinsic property of the enzymes. There is a common misconception that thermophilicity will be accompanied by the benefits of corresponding high catalytic rates. While this would be perfectly reasonable if the temperature–kinetic relationship of Arrhenius was the only factor governing the action of enzymes, the functional constraints imposed by the structural requirements of a stable three-dimensional macromolecule obviously play a major role.

It has been noted on many occasions that thermostable proteins tend to demonstrate enhanced resistance to denaturation by such agents as organic solvents, detergents and chaotropic compounds such as urea and guanidine hydrochloride [93, 94, 108, 118, 122, 131, 144]. Thermostable proteins are also more resistant to degradation by proteolysis [137]. While the mechanisms of protein unfolding by organic solvents, detergents, etc. are poorly understood, the obvious fact that thermostable proteins show enhanced resistance to all these agents implies that some aspect of the unfolding pathway is common to all (see Figure 1.11). It seems obvious that the first step of protein unfolding must be the rapid reversible conformational transition which is a reflection of the normal flexibility of the protein. Loss of tertiary structure via denaturation (by temperature, detergent, solvent, etc.) or proteolysis will only progress from the unfolded conformer. Restriction of the initial *reversible* conformational transitions, a consequence of thermo-stabilisation, will proportionally reduce the tendency of the protein to partake in further *irreversible* unfolding steps. The resistance of thermostable proteins to proteolysis supports this scheme (Figure 1.11) since it is well known that only unfolded proteins are susceptible to proteolytic attack.

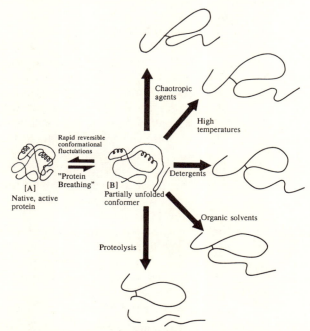

Figure 1.11   A schematic view of reversible protein unfolding and irreversible protein denaturation. Thermostabilisation increases molecular rigidity and reduces the occurrence of the reversibly unfolded form (B) which is the 'substrate' for further unfolding/degradation processes.

There has been much speculation about the consequences of protein thermostability in respect of substrate specificity. On one hand, it has been suggested that the enhanced rigidity of thermostable protein structures might enhance substrate selectivity. On the other, it has been proposed that the kinetic consequences of high temperature might make active sites more fluid, thus reducing substrate specificity. Both these views ignore the now accepted view that variations in the flexibility of the polypeptide in the region of the active site are probably highly constrained by the requirements of catalysis. Nevertheless, some enzymes from the hyperthermophilic Archaea do appear to demonstrate broader specificity than their mesophilic counterparts. The examples of dual co-factor specificity displayed by *Pyrococcus woesei* glyceraldehyde-3-phosphate dehydrogenase [114], glucose dehydrogenases from *Thermoplasma acidophilum* and *Sulfolobus solfataricus* [112, 113] and *T. acidophilum* malate dehydrogenase [117] may indicate some relaxation of specificity, but it must be recorded that this has also been observed in a limited number of mesophilic eubacteria and archaeobacteria. Other archaeo-bacterial enzymes seem to possess broader substrate specificity than is typical for mesophilic equivalents (e.g. *S. solfataricus* alcohol dehydrogenase [93] and *S. acidocaldarius* glycosyltransferase [45]), but there are insufficient data available at present to draw any serious conclusions.

1.3.4   *Structural macromolecules*

1.3.4.1   *S-layers.*   The cell walls of the extremely thermophilic archaeobacteria are much simpler than those of most eubacteria and other archaeobacteria. They are composed solely of an 'S-layer' structure—a regular two-dimensional array of glycoprotein monomers which covers the entire surface of the cell in the formation of a three-dimensional protein network (Figure 1.12). The association between the S-layer and the plasma is sufficiently intimate that there is effectively no periplasmic space. This cell wall structure is so simple that it is difficult not to think of it as a primitive precursor of the much more complex eukaryotic cell wall structures. However, it is worth noting that the non-thermophilic archaeobacteria possess quite a heterogeneous range of alternative cell wall structures, including pseudomurein, multiple S-layers and protein fibrils [145].

The S-layers of the crenotes are reasonably heterogeneous in composition and construction. The Thermoproteales all possess monolayers of glycoprotein monomers, generally arranged in hexagonal array [146]. The hexagonal array of subunits is responsible for the helical topology observed in electron micrographs of many archaeobacterial isolates. *Desulfurococcus* species appear to possess two tiers of S-layer monomers, one of which is organised in a tetragonal rather than hexagonal array [45]. The S-layers of both *Thermodiscus maritimus* and *Pyrodictium occultum* are sufficiently thick to contain more than one layer, but the latter yields only a single protein species

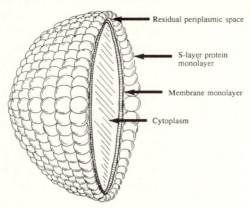

Residual periplasmic space

S-layer protein monolayer

Membrane monolayer

Cytoplasm

Figure 1.12 Schematic representation of a single S-layer structure comprising a crenote cell wall.

after sodium dodecylsulphate (SDS) dissociation. S-layer protein subunits from different archaeobacteria vary from 20 to 17 kDa and have degrees of glycosylation ranging from none (*Acidianus brierleyi*) to 20% (*Thermoproteus tenax*) [145]. There appears to be no consistency in amino acid composition, although it has been reported that the *Sulfolobus* S-layers are enriched in hydroxyl amino acids [145].

A striking characteristic of some crenote S-layers is the extraordinarily high chemical and thermal resistance of the S-layer matrices. Standard techniques for the isolation of S-layer complexes from *T. tenax* include such severe protocols as 'sonication followed by boiling for 30 minutes in 2% SDS' [145]. In most instances these procedures result in dissociation of S-layer cell envelope into monomeric units although, in the case of *T. tenax*, all attempts to dissociate the matrix into monomers have been unsuccessful [148]. The general observation that more severe chemical treatments are necessary to dissociate the S-layer matrix monomers than are required to dissociate the matrix from the underlying plasma membrane has led to the conclusion that the interactions between subunits are considerably stronger than those between S-layer matrix and the membrane.

Another characteristic of S-layer monomers, presumably related to their resistance to dissociation, is their tendency to reassemble spontaneously into large two-dimensional arrays [45]. The strength of the subunit interaction and the tendency for spontaneous reassembly implies particularly strong intermolecular non-covalent bonds between subunits. In the case of the non-dissociatable *T. tenax* S-layer matrix, the possibility of post-translational covalent cross-links should not be discounted. However, elucidation of molecular detail of these interactions must await detailed structural data.

S-layer protein is the single most abundant polypeptide species in the thermophilic archaeobacteria. Furthermore, biosynthesis will be continual throughout most of the growth cycle in response to cell division and growth.

The S-layer protein gene must therefore be very highly expressed and, as such, is likely to be an excellent target for studies of the control of gene expression in the Archaea.

The interaction between the S-layer and the cell membrane is intimate. In the case of *T. tenax*, the S-layer network appears to be supported on protein 'pillars' [148] which contact the surface of the plasma membrane. In consequence, there is little or no periplasmic space. In the absence of any other cell wall component the S-layer is presumed to be solely responsible for the maintenance of cell shape [148]. Although there is considerable variation in the cell morphology of many of the thermophilic archaeobacteria, with the formation of 'golf clubs' [40] and other protuberances, they nevertheless adhere to general shape categories (see above). It is notable that *Thermoplasma acidophilum*, which lacks a cell wall altogether, has a highly variable cell morphology typical of the cell wall-less mycoplasmas [10].

It has been suggested that S-layers act as a molecular sieve, limiting access of molecules to the cell membrane [149]. However, in *S. acidocaldarius* the pore sizes have been estimated at 2.5–5 nm in diameter [150], dimensions which will only restrict the passage of large macromolecules such as globular proteins of 20 kDa or greater [151].

Other possible functions of the S-layer might include cell–cell interactions, such as would be necessary in intercell communication or cell adhesion, and cell–particle interactions, such as adsorption to inorganic surfaces and interactions with phage particles. There is little direct evidence to support or refute these suggestions, although a specific alignment of S-layers in associated cells of the eubacterium *Deinococcus* has been reported [151]. Such a non-random interaction might indicate the formation of 'connexons', continuous channels between two cells which could facilitate chemical communication and possibly aid the transfer of genetic material [151].

1.3.5   *Lipids and lipid biosynthesis*

The lipid structures of the Archaea are unique and are thus a diagnostic taxonomic determinant of this superkingdom. However, within the different phyla of the Archaea, there is a wide variation of structural detail and complexity [152, 153]. The most notable characteristic of archaeal glycerolipids is the absence of the 'classical' ester linkages between the glycerol backbone and fatty acyl moiety. These are replaced with ether linkages between the backbone structure (glycerol, tetritol and nonitol—for details, see below) and an isoprenoid alcohol.

The stability of the ether bond is clearly much greater than that of the relatively labile ester bond. This comparison invites the inevitable conclusion that deletion of the ester linkage is an evolutionary adaptation designed to protect membrane structures from chemical degeneration induced by high temperature and low pH. However, ether-linked lipids are common to all

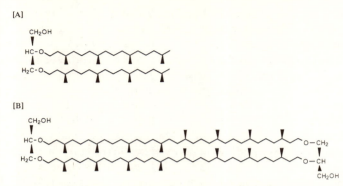

Figure 1.13 Basic isoprenoid lipid structures of the extremely thermophilic Archaea. (A) 2,3-di-*O*-phytanyl-*sn*-glycerol; (B) glyerol-dialkyl-glycerol tetraether.

three phenotypic groups of the Archaea, including both thermophilic and mesophilic representatives. This might argue for the role of the extremely thermophilic archaeobacteria as the evolutionary progenitors of the Archaea (although not necessarily for the Eucarya).

In the extremely thermophilic archaeobacteria, the other remarkable structural feature of the glycerolipids is the presence of isoprenoid diethers and tetraethers (Figure 1.13). The basic structure consists of a 72-membered ring with 18 chiral centres formed by two 2,3-*sn*-glycerol moieties linked by ether bonds to two isoprenoid C_{40} diols. The C_{40} isoprenoid chains are derived from the head-to-head ($\omega-\omega'$) linkage of two C_{20} O-phytanyl residues. Both diether-based and tetraether-based structures occur in the extremely thermophilic Archaea, although tetraethers predominate to a large extent in several of the species studied [152].

Two novel variations on the glycerol backbone have been reported. These include tetritol and nonitol (also named calditol [154]), a branched nine-carbon polyol (Figure 1.14). Tetritol is present as a minor component, while nonitol is frequently present in place of one glycerol unit. The glycerol–nonitol tetraethers make up a high proportion of the total tetraether component in *Sulfolobus* cells [152].

A variety of other constituents is found in the polar lipid fraction of the thermophilic archaeobacteria. The free hydroxyl groups of the tetraether lipids are known to be both phosphorylated and glycosylated [152]. Sulphated lipids have also been detected. Two well-characterised glycolipid components of *Sulfolobus solfataricus*, which together make up about 75% of the total polar lipid content, are diglycosyl–diglycerol tetraether and glycosylglycerol–nonitol (Figure 1.15).

Various non-polar lipids have been identified in the extremely thermophilic archaeobacteria, comprising between 7 and 20% of the total lipid [152]. These include squalenes (C_{30} hexaisoprenoids, C_{25} pentaisoprenoids and C_{20}

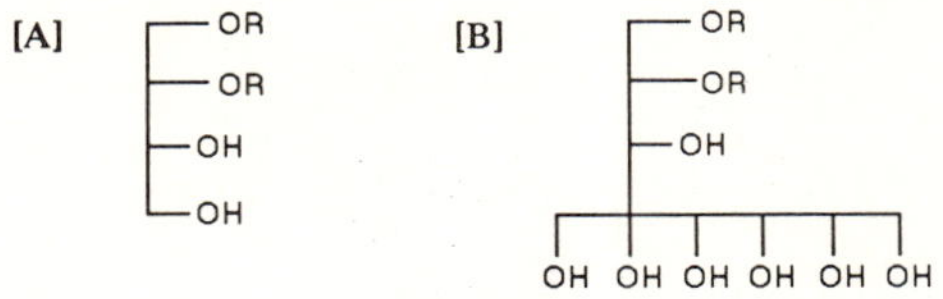

[A]
OR
OR
OH
OH
[B]
OR
OR
OH
OH OH OH OH OH OH

R represents a phytanyl, biphytanyl or modified biphytanyl chain.

Figure 1.14 Tetritol (A) and nonitol (B) backbones of archaeobacterial ether-linked lipids.

[A]
CH_2OH
HC-O
H_2C-O
O—CH_2
O—CH
CH_2
O
Gal
Glc
[B]
CH_2OH
HC-O
H_2C-O
O
O
OH
OH OH OH OH OH OH
Glc

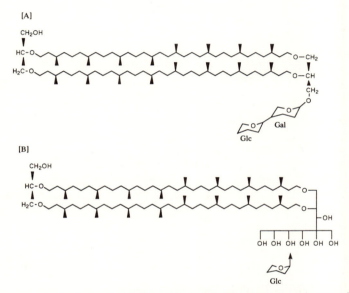

Figure 1.15 *S. solfataricus* glycolipids: (A) *β*Glc*p*-*β*Gal*p*-*O*-(diglyceryl tetraether) and (B) *β*Glc*p*-(glycerylnonityl tetraether)-OH.

tetraisoprenoids) showing varying degrees of unsaturation [152] and various alkylbenzenes, the latter being present only in small quantities. Both *Thermoplasma acidophilum* and *S. solfataricus* contain their own novel lipoquinones, respectively thermaplasmaquinone [155] and caldariellaquinone [156].

Membranes containing a high percentage of tetraether lipids will be both chemically and physically stable. Chemical stability is undoubtedly imparted by the presence of the resistant ether linkages. Physical stability will be derived from the absence of a cleavage plane such as found at the junction of the bilayer of eukaryotic membranes. The mobility of lipid components will be reduced by the limitation to the degrees of freedom imposed by a bipolar structure [157]. In the thermophilic archaeobacteria, some control over the degree of membrane fluidity is apparently maintained by modification to

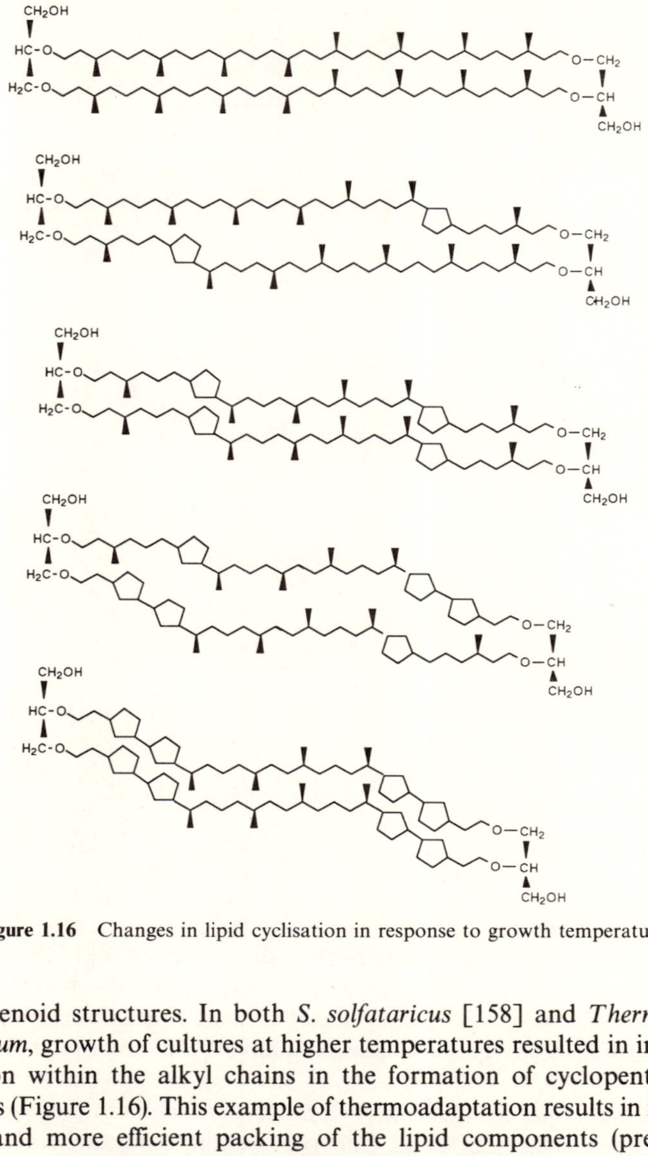

Figure 1.16   Changes in lipid cyclisation in response to growth temperature.

the isoprenoid structures. In both *S. solfataricus* [158] and *Thermoplasma acidophilum*, growth of cultures at higher temperatures resulted in increasing cyclisation within the alkyl chains in the formation of cyclopentane ring structures (Figure 1.16). This example of thermoadaptation results in increased rigidity and more efficient packing of the lipid components (presumably resulting in enhanced hydrophobic interactions) with the increased frequency of the cyclopentane rings [159].

The biosynthesis of archaeobacterial lipids has attracted considerable attention because of the unusual covalent linkages present: the ether and the ω–ω' biphytanyl bonds. As the result of a considerable body of work, De Rosa and Gambacorta have proposed a generalised framework for the synthesis

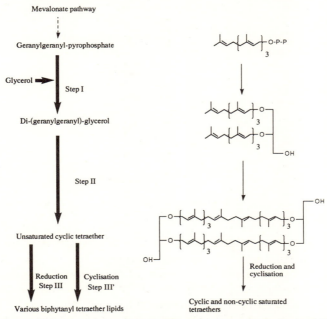

Mevalonate pathway

Geranylgeranyl-pyrophosphate

Glycerol ➤ Step I

Di-(geranylgeranyl)-glycerol

Step II

Unsaturated cyclic tetraether

Reduction   Cyclisation
Step III   Step III'

Various biphytanyl tetraether lipids

Reduction and cyclisation

Cyclic and non-cyclic saturated tetraethers

Figure 1.17 Biosynthesis of biphytanyl tetraether lipids in extremely thermophilic Archaea.

of diphytanyl tetraether lipids [157]. This synthesis pathway (Figure 1.17) utilises geranylgeranyl pyrophosphate, synthesised by the classical mevalonate pathway [152], as a precursor in the alkylation of glycerol (step I, Figure 1.17). Only in *Methanobacterium thermoautotrophicum* have the enzymes responsible for ether synthesis been identified. In this organism two prenyl transferases, one cytosolic and one membrane-associated, have been shown to catalyse the formation of the ether linkages [160]. The first catalyses the condensation between geranylgeranyl pyrophosphate and 1-glycerophosphate, while the second alkylates the remaining 2-hydroxyl of the 3-*O*-alkyl glyceryl-1-phosphate.

Head-to-head linkage of the geranylgeranyl ether structures produced an unsaturated macrocyclic tetraether structure (step II, Figure 1.17). This is a highly unusual reaction, having no parallel in terpene chemistry. Little is known of the mechanism and it is not currently clear whether the formation of the C_{40} biphytanyl chain occurs before or after formation of the ether linkages between the glycerol backbone and the C_{20} phytanyl structure. There is some indirect evidence from studies on a mesophilic methanogen that the precursor is a C_{20} chain linked to glycerol [161]. Nothing is currently known of the enzymology of this clearly most interesting reaction.

A combination of reductive and cyclisation reactions (steps III and III', Figure 1.17) is presumed subsequently to yield the various linear and

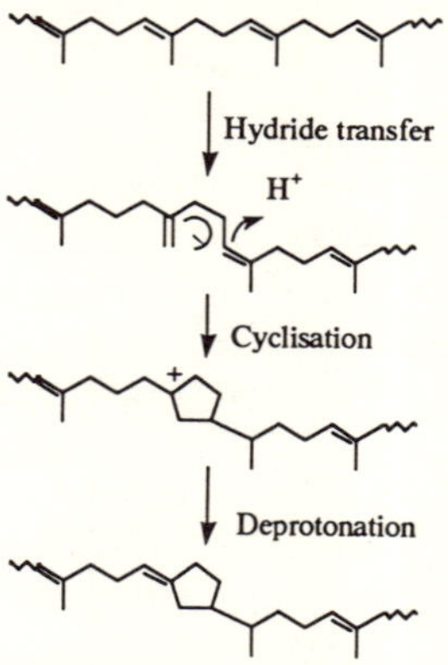

Figure 1.18   Proposed mechanism for cyclopentane ring formation in *S. solfataricus* tetraether lipids.

cyclopentane-containing isoprenoid structures which form the basis of polar lipids in the membranes of the extremely thermophilic Archaea. The regular disposition and symmetrical arrangement of the cyclopentane rings implies a cyclisation mechanism operating in a defined and reproducible manner on the tetraether lipid structure. A mechanism has been proposed by Trincone *et al.* [162] based on the ^{13}C labelling patterns of *S. solfataricus* tetraether lipids after growth in the presence of ^{13}C/^{2}H-labelled mevalonic lactone precursors. The mechanism (Figure 1.18) requires unsaturated precursors and implies a hydride transfer step. The possibility of a completely different cyclisation mechanism such as via hydroxylation of the unsaturated chain has not yet been ruled out [162].

1.3.6   *Molecular genetics of the extremely thermophilic Archaea*

1.3.6.1   *Genome structure and organisation.* The gene structure and organisation of the Archaea have been extensively and recently reviewed [3, 163–165]. In general, the molecular genetic characteristics of the euryotes (the more extensively studied group) are paralleled in the crenotes.

Genome size is typical of the 'prokaryotes' [163, 166, 167], ranging from 0.8 to 3×10^9 base pairs. In some archaeal species at least, the nuclear material is present as a large single chromosome. While this is typical of the eubacterial kingdom, the presence in some crenotes of a nucleosome structure is reminiscent of the eukaryotes. Small basic 'histone-like' proteins have been isolated from a number of the Archaea [168–171]. The protein *Hta* from *Thermoplasma acidophilum* [169] has been shown to interact with DNA to form small defined structures [172], but many of the other so-called histone-like proteins may have *in-vivo* functions unrelated to the formation of basic nuclear structure [173].

Extrachromosomal DNA elements are not widely distributed in the crenotes. Both *Thermoproteus tenax* and *Sulfolobus* have been shown to possess DNA viruses [174, 175] TTV1-TTV-4 and SSV1 respectively. SSV1 was originally thought to be a plasmid but was subsequently found to be a UV-inducible prophage. Only one 'true' plasmid has been reported [38]. The genetic element pSL10 from *Desulfurococcus ambivalens*, while only present as a plasmid and never in the integrated state, appears to be related to SSV1. For a more detailed review of this topic, the reader is referred to two excellent recent reviews [176, 177].

1.3.6.2 *tRNA and RNA*. A number of studies have contributed to a preliminary understanding of the organisation of genes in the crenotes. Early work on genes encoding rRNAs indicated that they occur as a single rRNA 'operon' containing the 16S and 23S rRNA genes, linked and co-transcribed in that order [165]. The 5S rRNA genes (either one or two copies) are unlinked to the 16S/23S rRNA operon but may be situated locally. In *Thermoplasma acidophilum*, the three rRNA genes are unlinked [178], a unique characteristic which possibly reflects the uncertain taxonomic status of this organism. All Archaea also contain 7S RNA, analogous to the eucaryal 7S and *Escherichia coli* 4.5S RNAs [179]. The organisation of a limited number of rRNA genes has now been investigated in considerable detail (for example ref. 180).

Several tRNA genes from crenotes have been cloned and sequenced. While some are located in the proximity of rRNA genes, others occur independently, either as single transcriptional units or as gene clusters [165]. Many tRNA genes and at least one rRNA gene in the Archaea contain introns. The location of these introns is variable either 3' to the anticodon (as in Eucarya) or in unique positions such as 5' to the anticodon or in the variable loop [165]. A large intron in the 23S rRNA gene of *Desulfurococcus mobilis* [181] was successfully spliced by cell extracts of other crenotes which lacked this intron [187], suggesting that other introns probably exist in other genes. A most exciting result has recently come to light with a report (D. Comb, personal communication) that two introns have been found in the *Thermococcus littoralis* DNA polymerase gene. This is the first example of introns in non-Eucarya protein genes.

The first crenote protein gene to be cloned and sequenced was reported in 1988 [165], since when similar publications have appeared with increasing frequency (Table 1.9). The initial conclusion (but one which was not guaranteed *a priori*) was that the Archaea use the classical genetic code.

The emphasis with most studies of archaeal structural gene sequences has been on a comparison with equivalent features in bacteria and eukaryotes. In terms of overall organisation of genes and their expression signals, the Archaea resemble bacteria more than the eukaryotes, although there are a number of features reminiscent of the latter. Translation start sites are usually initiated by AUG, although the sequences GUG and UUG have been

Table 1.9 Cloning of crenote protein structural genes*.

Enzyme/protein	Organism	Introns in open reading frame	Putative transcription initiation sites	Putative transcription termination signals	Putative ribosome-binding site	Expression	Homology	Reference
Aspartate aminotransferase	*Sulfolobus solfataricus*	No	Box A($_{-25}$TTTATA$_{-20}$) Box B($_{-2}$TTGT$_{+2}$)	Pyrimidine-rich sequence	TGTGGT	No	Eukaryotic	183
ATPase subunits	*Sulfolobus acidocaldarius*	No	NA	TTTTTAT	TGGTGA; GTGGTAA; GAGAC GAGGGGA GGTG	No	NA†	184
Citrate synthase	*Thermoplasma acidophilum*	No	Box A(TTTAT/AATA) Box B(A/TTGA)	TTTT?	NA	*Escherichia coli*	Eukaryotic	183
DNA polymerase	*Thermococcus littoralis*	Yes	NA	NA	NA	Yes	NA	‡
DNA polymerase	*Pyrococcus furiosus*	Yes?	NA	NA	NA	Yes?	NA	§
EF-2 protein	*Sulfolobus acidocaldarius*	No	Box A($_{-49}$TTTAAA$_{-44}$) Box B($_{-42}$CTTAAA$_{-37}$)	TTTTTTTT	GGTTAG	No	Eukaryotic	186
β-Galactosidase	*Sulfolobus solfataricus*	No	NA	NA	NA	*Escherichia coli*	None	187

Glutamine synthetase	Sulfolobus solfataricus	No	Box A($_{-28}$TTTAAA$_{-23}$)	Oligo(T)	NA	No	Eubacterial (eukaryotic promoter)	188
Glyceraldehyde-3-phosphate dehydrogenase	Pyrococcus woesei	No	Box A($_{-26}$TTATAA$_{-20}$)	T_6	GAGGT	Escherichia coli	NA	114
Protease (thermopsin)	Sulfolobus acidocaldarius	No	AAAGCTTATATA or AAATTATTTAAA	T_6ACT_5	GTGAT	No	None	125
RNA polymerase	Sulfolobus acidocaldarius	No	Box A($_{-26}$TTATTA$_{-20}$) Box B(TG)	TTTTT	AGAG	No	Eukaryotic	189

*Several proteins from the *Thermoproteus tenax* virus TTV1 have also been cloned and sequenced [190–192].
†NA data not available.
‡Cloned by New England Biolabs, MA, USA (Comb, personal communication).
§Cloned by Stratagene Ltd, La Jolla, USA (Mathur, personal communication).

observed. Complementary sequences at the 3′ end of the 16S rRNA are usually observed. Sequences resembling the bacterial 'Shine–Dalgarno' sequence are detected in the proximity of the initiation codon, but the positioning is variable, sometimes overlapping the initiation site or even positioned downstream. However, the similarity between the archaeal and bacterial translation mechanisms is emphasised by the successful expression of archaeal protein genes in *E. coli* [110].

Promoter sequences located upstream of open reading frames in archaeal genes appear to consist of two AT-rich regions. Box A is a conserved AT-rich region of 7–8 nucleotides, positioned 18–25 nucleotides upstream of the transcription initiation site, with a consensus sequence of $TTTA^A/_TA$ [193]. This is equivalent to the bacterial Pridnow box and the eucaryal TATA box and has been shown to be critical for the specific initiation of transcription [194]. The similarity between box A and the eucaryal RNA polII promoters is consistent with the structural homology between the archaeal RNA polymerases and the eucaryal RNA polymerases II and III [195, 196]. However, a sequence showing homology to the −35 element of the bacterial promoter has been observed upstream of tRNA promoters in *Thermococcus celer* [197].

The second archaeal promoter sequence (box B) is a less strongly conserved three- to four-nucleotide sequence positioned in the vicinity of the initiation codon. The putative consensus sequence is $AT^T/_GAC$ [194].

Transcription termination in archaeal genes may occur at oligo(dT) sequences, but other signals including oligo(dA) sequences and stem and loop structures may also be implicated [165, 198–200].

In general, the size and complexity of the archaeal genome is less than in either the bacterial or eucaryal kingdoms. This is exemplified by the absence of repetitive sequences, the limited length of transcript leader sequences and, in some instances, the duplication of both transcription and translation initiation signals. These features, and the exhibition of both eucaryal and bacterial characteristics, are consistent with the primordial and progenitorial nature of the kingdom.

References

1. C.R. Woese, *Scientific American* **244** (1981) 94.
2. O. Kandler, *Archaebacteria*, Gustav Fischer, Stuttgart, (1982) 366 pp.
3. C.R. Woese and R.S. Wolfe, (eds.) *Archaebacteria*, Vol. VIII of *The Bacteria*, Academic Press, London (1985) 581 pp.
4. O. Kandler and W. Zillig, *Archaebacteria' 85*, Gustav Fischer, Stuttgart, New York (1986) 434 pp.
5. K.O. Stetter, in *Thermophiles: General, Molecular and Applied Microbiology*, ed. Brock T.D., John Wiley & Sons, New York (1986) ch. 3.
6. C.A. Fewson, *Biochem. Education* **14** (1986) 103.
7. C.R. Woese and G.J. Olsen, *Syst. Appl. Microbiol.* **7** (1986) 161.
8. C.R. Woese, *Microbiol. Rev.* **51** (1987) 221.

9. M.J. Danson, *Adv. Microb. Physiol.* **29** (1988) 165.
10. G. Darland, T.D. Brock, W. Samsonoff, and F. Conti, *Science* **170** (1970) 1416.
11. C.R. Woese and G.E. Fox, *Proc. Natl. Acad. Sci. USA* **74** (1977) 5088.
12. J.A. Barross and J.W. Deming, *Nature* **303** (1983) 423.
13. J.D. Trent, R.A. Chastain and A.A. Yayanos, *Nature* **307** (1984) 737.
14. R.H. White, *Nature* **310** (1984) 430.
15. J. Felsenstein, in *Cladistics: Perspectives in the Reconstruction of Evolutionary History*, eds. T. Ducan and T.F. Stuessy, Columbia University Press, New York (1984) 169.
16. W. Hennig, *Phylogenetic Systematics*. University of Illinois Press, Urbana (1966).
17. W.M. Fitch and E. Margoliash, *Science* **155** (1967) 279.
18. W. Zillig, H-P. Klenk, P. Palm, G. Pühler, F. Gropp, R. Garrett and H. Leffers, *Can. J. Microbiol.* **35** (1989) 73.
19. L. Achenbach-Richter, R. Gupta, K.O. Stetter and C.R. Woese, *System. Appl. Microbiol.* **9** (1987) 34.
20. G.J. Olsen, *Cold Spring Harbor Symp. Quant. Biol.* **52** (1987) 825.
21. L. Achenbach-Richter, R. Gupta, W. Zillig and C.R. Woese, *System. Appl. Microbiol.* **10** (1988) 231.
22. N. Iwabe, K-I. Kuma, M. Hasegawa, S. Osawa and T. Miyata, *Proc. Natl. Acad. Sci. USA* **86** (1989) 9355.
23. J. Wolters and E.A. Erdmann, *J. Mol. Evolution* **24** (1986) 152.
24. R. Hensel, P. Zwickl, S. Fabry, J. Land and P. Palm, *Can. J. Microbiol.* **35** (1989) 81.
25. R. Amils, L. Ramirez, J.L. Sanz, I. Marin, A.G. Pisabarro and D. Urena, *Can. J. Microbiol.* **35** (1989) 141.
26. J.A. Lake, E. Henderson, M.W. Clark and M. Oakes, *Proc. Natl. Acad. Sci. USA* **81** (1984) 3786.
27. J.A. Lake, M.W. Clark, E. Henderson, S.P. Fay, M. Oakes, A. Scheinman, J.P. Thornber and R.A. Mah, *Proc. Natl. Acad. Sci. USA* **82** (1985) 3716.
28. J.A. Lake, *Nature* **331** (1988) 184.
29. J.A. Lake, *Can. J. Microbiol.* **35** (1989) 109.
30. G.J. Olsen and C.R. Woese, *Can. J. Microbiol.* **35** (1989) 119.
31. C.R. Woese, O. Kandler and M.L. Wheelis, *Proc. Natl. Acad. Sci. USA* **87** (1990) 4576.
32. T.D. Brock, K.M. Brock, R.T. Belly and R.L. Weiss, *Arch. Mikrobiol.* **84** (1972) 54.
33. M. de Rosa, A. Gambacorta and J.D. Bu'Lock, *J. Gen. Microbiol.* **86** (1975) 156.
34. D. Grogan, P. Palm and W. Zillig, *Arch. Microbiol.* **154** (1990) 594.
35. G. Huber, C. Spinnler, A. Gambacorta and K.O. Stetter, *System. Appl. Microbiol.* **12** (1989) 38.
36. C.L. Brierley and J.A. Brierley, *Can. J. Microbiol.* **19** (1973) 183.
37. A. Segerer, A. Neuner, J.K. Kristjansson and K.O. Stetter, *Int. J. System. Bacteriol.* **36** (1986) 559.
38. W. Zillig, S. Yeats, I. Holz, A. Bock, M. Rettenberger, F. Gropp and G. Simon, *System. Appl. Microbiol.* **8** (1986) 197.
39. W. Zillig, I. Holz, D. Janekovic, H.-P. Klenk, E. Imsel, J. Trent, S. Wunderl, V.H. Forjaz, R. Coutinho and T. Ferreira, *J. Bacteriol.* **172** (1990) 3959.
40. W. Zillig, K.O. Stetter, W. Schafer, D. Janekovic, S. Wunderl, I. Holz and P. Palm, *Zbl. Bakt. Hyg., I Abt. Orig.* **C2** (1981) 205.
41. K.O. Stetter and W. Zillig, in *Archaebacteria*, Vol. VIII of *The Bacteria*, eds. C.R. Woese and R.S. Wolfe, Academic Press, London (1985) ch. 2.
42. E.A. Bonch-Osmolovskaya, M.L. Miroshnichenko, N.A. Kostrikina, N.A. Chernych and G.A. Zavarzin, *Arch. Microbiol.* **154** (1990) 556.
43. R. Huber, J.K. Kristjansson and K.O. Stetter, *Arch. Microbiol.* **149** (1987) 95.
44. W. Zillig, A. Gierl, G. Schreiber, S. Wunderl, D. Janekovic, K.O. Stetter and H.P. Klenk, *System. Appl. Microbiol.* **4** (1983) 79.
45. W. Zillig, K.O. Stetter, D. Prangishvilli, W. Schafer, S. Wunderl, D. Janekovic, I. Holz and P. Palm, *Zbl. Bakt. Hyg., I Abt. Orig.*, **C3** (1982) 304.
46. E.A. Bonch-Osmolovskaya, A.I. Slesarev, M. L. Miroshnichenko, T.P. Svetlichnaya, and V.A. Alekseev, *Microbiology* **57** (1988) 78.
47. B.K.C. Patel, P.M. Jasperse-Herst, H.W. Morgan and R.M. Daniel, *N.Z. J. Marine Freshwater Res.* **20** (1986) 439.

48. D.A. Cowan, K.A. Smolenski, R.M. Daniel and H.W. Morgan, *Biochem. J.* **247** (1987) 121.
49. H.W. Jannasch, C.O. Wirsen, S.J. Molyneaux and T.A. Langworthy, *Appl. Environ. Microbiol.* **54** (1988) 1203.
50. G. Fiala, K.O. Stetter, H.W. Jannasch, T.A. Langworthy and J. Madon, *System. Appl. Microbiol.* **8** (1986) 106.
51. W. Zillig, I. Holz, D. Janekovic, W. Schafer and W.D. Reiter, *System. Appl. Microbiol.* **4** (1983) 88.
52. M.L. Miroshnichenko, E.A. Bonch-Osmolovskaya, A. Neuner, N.A. Kostrikina, N.A. Chernych and V.A. Alekseev, *System. Appl. Microbiol.* **12** (1989) 257.
53. A Neuner, H.W. Jannasch, S. Belkin and K.O. Stetter, *Arch. Microbiol.* **153** (1990) 205.
54. G. Fiala and K.O. Stetter, *Arch. Microbiol.* **145** (1986) 56.
55. W. Zillig, I. Holz, H.P. Klenk, J. Trent, S. Wunderl, D. Janekovic, E. Insel and B. Haas, *System. Appl. Microbiol.* **9** (1987) 62.
56. V.A. Svetlichny, A.I. Slesarev, T.P. Svetlichnaya and G.A. Zavarzin, *Mikrobiologia* **56** (1987) 831 (in Russian).
57. K.O. Setter, H. Konig and E. Stackenbrandt, *System. Appl. Microbiol.* **4** (1983) 535.
58. F. Fischer, W. Zillig, K.O. Stetter and G. Schreiber, *Nature* **301** (1983) 511.
59. K.O. Stetter, *System. Appl. Microbiol.* **10** (1988) 172.
60. S. Burggraf, H. W. Jannasch, B. Nicolaus and K.O. Stetter, *Syst. Appl. Microbiol.* **13** (1990) 24.
61. A. Segerer, T.A. Langworthy and K.O. Stetter, *System. Appl. Microbiol.* **10** (1988) 161.
62. J.G. Zeikus and R.S. Wolfe, *J. Bacteriol.* **109** (1972) 707.
63. K.O. Stetter, M. Thomm, J. Winter, G. Wildgruber, H. Huber, W. Zillig, D. Janekovic, H. Konig, P. Palm and S. Wunderl, *Zbl. Bakt. Hyg., I Abt. Orig.* **C2** (1981) 166.
64. G. Lauerer, J.K. Kristjansson, T.A. Langworthy, H. Konig and K.O. Stetter, *System. Appl. Microbiol.* **8** (1986) 100.
65. W.J. Jones, M.J. Leigh, F. Mayer, C.R. Woese and R.S. Wolfe, *Arch. Microbiol.* **136** (1983) 254.
66. H. Huber, M. Thomm, H. Konig, G. Thies and K.O. Stetter, *Arch. Microbiol.* **132** (1982) 47.
67. S. Burggraf, H. Fricke, A. Neuner, J. Kristjansson, P. Rouvier, L. Mandelco, C.R. Woese and K.O. Stetter, *System. Appl. Microbiol.* **13** (1990) 263.
68. H. Huber, M. Kurr, W.H. Jannasch and K.O. Stetter, *Nature* **342** (1989) 833.
69. U.J. Heinen and W. Heinen, *Arch. Mikrobiol.* **82** (1972) 1.
70. R.A.D. Williams, *Sci. Prog., Oxf.* **62** (1975) 373.
71. W.R. Keefer, *The Geologic Story of Yellowstone National Park*, US Geological Survey Bull. (1972) 1347.
72. T.D. Brock, *System. Appl. Microbiol.* **7** (1986) 213.
73. T.D. Brock, in *Thermophiles: General, Molecular and Applied Microbiology*, ed. T.D. Brock, John Wiley & Sons, New York (1986) ch. 1.
74. J.F. Grassle, *Science* **229** (1985) 713.
75. H.W. Jannasch and M.J. Mottl, *Science* **229** (1985) 717.
76. W.J. Jones, C.E. Stugard and H.W. Jannasch, *Arch. Microbiol.* **151** (1989) 312.
77. A. Segerer, K.O. Stetter and F. Klink, *Nature* **313** (1985) 787.
78. G. Zellner, E. Stackenbrandt, H. Kneifel, P. Messner, U.B. Sleytr, E. Conway de Macario, H-P. Zabel, K.O. Stetter and J. Winter, *System. Appl. Microbiol.* **11** (1989) 151.
79. L. Achenbach-Richter, K.O. Stetter and C.R. Woese, *Nature* **327** (1987) 348.
80. D. Möller and R.K. Thauer, *Arch. Microbiol.* **153** (1990) 251.
81. M. Lubben, H. Lundsorf and G. Schafer, *Biol. Chem. Hoppe-Seyler* **369** (1988) 1259.
82. J. Konishi, K. Denda, T. Oshima, T. Wakagi, E. Uchida, Y. Ohsumi, Y. Anraku, T. Matsumoto, T. Wakabayashi, Y. Mukohata, K. Ihara, K-I. Inatomi, K. Kato, T. Ohta, W. Allison and M. Yoshida, *J. Biochem.* **108** (1990) 554.
83. S. Anemüller, M. Lübben and G. Schäfer, *FEBS Lett.* **193** (1985) 83.
84. R. Moll and G. Schäfer, *FEBS Lett.* **232** (1988) 359.
85. T. Wakao, T. Wakagi and T. Oshima, *J. Biochem.* **102** (1987) 255.
86. M. De Rosa, S. De Rosa, A. Gambacorta, L. Minale, R.H. Thomson and R.D. Worthington, *J. Chem. Soc. Perkin I* (1977) 653.
87. T. Wakagi and T. Oshima, *Syst. Appl. Microbiol.* **7** (1986) 342.
88. S. Anemüller and G. Schäfer, *FEBS Lett.* **244** (1989) 451.

89. T. Wakagi, T. Yamauchi, T. Oshima, M. Müller, A. Azzi and N. Sone, *Biochem. Biophys. Res. Comm.* **165** (1989) 1110.
90. G. Marino, G. Nitti, M.I. Arnone, G. Sannia, A. Gambacorta and M. De Risam *J. Biol. Chem.* **263** (1988) 12305.
91. M. Van de Castelle, M. Demarez, C. Legrain, N. Glansdorff and A. Pierard, *J. Gen. Microbiol.* **136** (1990) 1177.
92. M. Porecelli, G. Cacciapuoti, M. Carteni-Farina and A. Gambacorta, *Eur. J. Biochem.* **177** (1988) 273.
93. R. Rella, C.A. Raia, M. Pensa, F.M. Pisani, A. Gambacorta, M. De Rosa and M. Rossi. *Eur. J. Biochem.* **167** (1987) 475.
94. M. Hanner, B. Redl and G. Stöffler, *Biochim. Biophys. Acta* **1033** (1990) 148.
95. R. Koch, P. Zablowski, A. Spreinat and G. Antranikian, *FEMS Microbiol. Lett.* **71** (1990) 21.
96. J.M. Bragger, R.M. Daniel, T. Coolbear and H. W. Morgan, *Appl. Microbiol. Biotech.* **31** (1988) 556.
97. J. Konishi, T. Wakagi, T. Oshima and M. Yoshida, *J. Biochem.* **102** (1987) 533.
98. C. Dahl, H-G. Koch, O. Keuken and H.G. Truper, *FEMS Microbiol. Lett.* **67** (1990) 27.
99. M.I. Arnone, M.V. Cubellis, G. Nitti, G. Sannia, and G. Marino, *Italian J. Biochem.* **263** (1988) 12305.
100. G. Löhlein-Werhahn, P. Goepfert and H. Eggerer, *Biol. Chem. Hoppe-Seyler* **369** (1988) 109.
101. W. Grossbütter and H. Görisch, *System. Appl. Microbiol..* **6** (1985) 119.
102. M.J. Danson, S.C. Black, D.L. Woodland and P.A. Wood, *FEBS Lett.* **179** (1985) 120.
103. H. Sobek and H. Gorisch, *Biochem. J.* **250** (1988) 453.
104. L.D. Smith, S. Bungard, M.J. Danson and D.W. Hough, *Biochem. Soc. Trans.* **15** (1987) 1097.
105. D.Z. Chinchaladze, D.A. Prangishvili, L.A. Kachabaeva and M.M. Zaalishvili, *Mol. Biol.* **19** (1986) 1193.
106. C. Elie, A.M. De Recondo and P. Forterre, *Eur. J. Biochem.* **178** (1989) 619.
107. M. Rossi, R. Rella, M. Pensa, S. Bartolucci, M. De Rosa, A. Gambacorta, C.A. Raia and N. Dell'Aversano, *System. Appl. Microbiol.* **7** (1986) 336.
108. S. Puchegger, B. Redl and G. Stoffler, *J. Gen. Microbiol.* **136** (1990) 1537.
109. H.R. Costantino, S.H. Brown and R.M. Kelly, *J. Bacteriol.* **172** (1990) 3654.
110. F.M. Pisani, R. Rella, C. A. Raia, C. Rozzo, R. Nucci, A. Gambacorta, M. De Rosa and M. Rossi, *Eur. J. Biochem.* **187** (1990) 321.
111. L.D. Smith, N. Budgen, S.J. Bungard and M.J. Danson, *Biochem. J.* **261** (1989) 973.
112. P. Giardina, M-G. De Baisi, M. De Rosa, A. Gambacorta and V. Buonocore, *Biochem. J.* **239** (1986) 517.
113. M.V. Cubellis, C. Rozzo, P. Montecucchi and M. Rossi, *Gene* **94** (1990) 89.
114. P. Zwickl, S. Fabry, C. Bogedain, A. Haas and R. Hensel, *J. Bacteriol.* **172** (1990) 4329.
115. F.O. Bryant and M.W. Adams, *J. Biol. Chem.* **264** (1989) 5070.
116. M.J. Danson and P.A. Wood, *FEBS Lett.* **172** (1984) 289.
117. H. Görisch, T. Hartl, W. Grobebüter and J.J. Stezowski, *Biochem. J.* **226** (1985) 885.
118. A. Guagliardi, M. Moracci, G. Manco, M. Rossi and S. Bartolucci, *Biochim. Biophys. Acta* **957** (1988) 301.
119. L. Kerscher, S. Nowitzki and D. Oesterhelt, *Eur. J. Biochem.* **128** (1982) 223.
120. R. Skórko, J. Osipiuk and K.O. Stetter, *J. Bacteriol.* **171** (1989) 5162.
121. M. Fusek, X-L. Lin and J. Tang, *J. Biol. Chem.* **265** (1990) 1496.
122. I.I. Blumentals, A.S. Robinson and R.M. Kelly, *Appl. Environ. Microbiol.* **56** (1990) 1992.
123. H. Connaris, D.A. Cowan and R.J. Sharp, *J. Gen. Microbiol.* **137** (1991) 1193.
124. R. Eggen, A. Geerling, J. Watts and W.M. De Vos, *FEMS Microbiol. Lett.* **71** (1990) 17.
125. X-L. Lin and J. Tang, *J. Biol. Chem.* **265** (1990) 1490.
126. S.H. Brown, H.R. Costantino and R.M. Kelly, *Appl. Environ. Microbiol.* **56** (1990) 1985.
127. D. Prangishivilli, W. Zillig, A. Gierl, L. Biesert and I. Holz, *Eur. J. Biochem.* **122** (1982) 471.
128. T. Emmel, W. Sand, W.A. Konig and E. Bock, *J. Gen. Microbiol.* **132** (1986) 3415.
129. A. Kikuchi and K. Asai, *Nature* **309** (1984) 677.
130. R.K. Owusu and D.A. Cowan, *Enzyme Microb. Technol.* **11** (1989) 568.
131. D.A. Cowan, R.M. Daniel and H.W. Morgan, *Trends Biotechnol.* **3** (1985) 68.
132. D.A. Cowan and R.M. Daniel, *Biochem. J.* **24** (1982) 2053.
133. R. Jaenicke and P. Závodszky, *FEBS Lett.* **268** (1990) 344.
134. J. Biro, S. Fabry, W. Dietmaier, C. Bogedain and R. Hensel, *FEBS Lett.* **275** (1990) 130.

135. Y. Ncsoh and T. Sekiguchi, *Trends Biotechnol.* **8** (1990) 16.
136. C. Ganter and A. Plückthun, *Biochemistry* **29** (1990) 9395.
137. R.M. Daniel, D.A. Cowan, H.W. Morgan and M.P. Curran, *Biochem. J.* **207** (1982) 641.
138. G. Wagner and K. Wütrich, *J. Mol. Biol.* **130** (1979) 31.
139. H. Frauenfelder, G.A. Petsko and D. Tsernoglou, *Nature* **280** (1979) 558.
140. M. Vihinen, *Prot. Engineer.* **1** (1987) 477.
141. R. Jaenicke, *Prog. Biophys. Mol. Biol.* **47** (1987) 237.
142. W.S. Bennet and R. Huber, *Crit. Rev. Biochem.* **15** (1983) 291.
143. D.A. Cowan, R.M. Daniel and H.W. Morgan, *Int. J. Biochem.* **19** (1987) 741.
144. V. Buonocore, O. Sgambati, M. De Rosa, E. Esposito and A. Gambacorta, *J. Appl. Biochem.* **2** (1980) 390.
145. H. König, R. Skorko, W. Zillig and W-D. Reiter, *Arch. Microbiol.* **132** (1982) 297.
146. U.B. Sleytr and P. Messner, *Ann. Rev. Microbiol.* **37** (1983) 311.
147. O. Kandler and H. König, in *Archaebacteria*, Vol. VIII of *The Bacteria*, eds. C.R. Woese and E.S. Wolfe, Academic Press, London (1985) ch. 9.
148. I. Wildhaber and W. Baumeister, *EMBO J.* **6** (1987) 1475.
149. U.B. Sletyr, P. Messner, M. Sara and D. Pum, *System. Appl. Microbiol.* **7** (1986) 310.
150. K.A. Taylor, J.F. Deatherage and L.A. Amos, *Nature* **299** (1982) 840.
151. W. Baumeister and R. Hegerl, *FEMS Microbiol. Lett.* **36** (1986) 119.
152. T.A. Langworthy, in *Archaebacteria*, Vol. VIII of *The Bacteria*, eds. C.R. Woese and R.S. Wolfe, Academic Press, London (1985) ch. 10.
153. M. De Rosa, A. Gambacorta and A. Gliozzi, *Microbiol. Rev.* **50** (1986) 70.
154. M. De Rosa, S. De Rosa, A. Gambacorta and J.D. Bu'Lock, *Phytochemistry* **19** (1980) 249.
155. M.D. Collins and T.A. Langworthy, *System. Appl. Microbiol.* **4** (1983) 295.
156. M. De Rosa, S. De Rosa, A. Gambacorta, L. Minale, R.H. Thomson and R.D. Worthington, *J. Chem. Soc. Perkin Trans.* **1** (1977) 653.
157. M. De Rosa and A. Gambacorta, *Chimicaoggi* **May** (1989) 37.
158. M. De Rosa, E. Esposito, A. Gambacorta, B. Nicholaus and J.D. Bu'Lock, *Phytochemistry* **19** (1980) 827.
159. A. Gliozzi, G. Paoli, M. De Rosa and A. Gambacorta, *Biochim. Biophys. Acta* **735** (1983) 234.
160. D.L. Zhang, L. Daniels and C.D. Poulter, *J. Am. Chem. Soc.* **112** (1990) 1264.
161. C.D. Poulter, T. Aoki and L. Daniels, *J. Am. Chem. Soc.* **110** (1988) 2620.
162. A. Trincone, A. Gambacorta and M. De Rosa, in *Microbiology of Extreme Environments and its Potential for Biotechnology*, eds. M.S. De Costa, J.C. Duarte and R.A.D. Williams, Elsevier, Amsterdam (1989) 180.
163. W.F. Doolittle, in *Archaebacteria*, Vol. VIII of *The Bacteria*, eds. C.R. Woese and R.S. Wolfe, Academic Press, London (1985) ch. 13.
164. W. Zillig, W. Reiter, P. Palm, F. Gropp, G. Pühler and H.P. Klenk, *Eur. J. Biochem.* **173** (1988) 473.
165. J.W. Brown, C.J. Daniels and J.N. Reeve, *CRC Crit. Rev. Microbiol.* **16** (1989) 287.
166. K.M. Noll, *J. Bacteriol.* **171** (1989) 6270.
167. A Yamaguchi and T. Oshima, *Nucleic Acids Res.* **18** (1990) 1133.
168. D.G. Searcy, *Biochim. Biophys. Acta* **395** (1975) 535.
169. M. Thomm, K.O. Stetter and W. Zillig, in *Archaebacteria*, ed. O. Kandler, Gustav Fischer, Stuttgart, (1982) 128.
170. D.A. Cowan, *Biochem. Soc. Trans.* **15** (1987) 640.
171. G.R. Green, D.G. Searcy and R.J. Delange, *Biochim. Biophys. Acta* **741** (1983) 251.
172. D.G. Searcy and D.B. Stein, *Biochim. Biophys. Acta* **609** (1980) 180.
173. M.S. Schmid, *Cell* **63** (1990) 451.
174. D. Janekovic, S. Wunderl, I. Holz, W. Zillig, A. Gierl and H. Neumann, *Mol. Gen. Genet.* **192** (1983) 39.
175. A. Martin, S. Yeats, D. Janekovic, W-D. Reiter, W. Aicher and W. Zillig, *EMBO J.* **3** (1984) 2165
176. W. Zillig, W-D. Reiter, P. Palm, F. Gropp, H. Neumann and M. Rettenberger, in *Viruses of Archaebacteria*, ed. R. Calendar, Plenum Press, New York (1988) 517.
177. W-D. Reiter, W. Zillig and P. Palm, *Adv. Virus Res.* **34** (1988) 143.
178. J. Tu and W. Zillig, *Nucleic Acids Res.* **10** (1982) 7231.
179. B.P. Kaine, *Mol. Gen. Genet* **22** (1990) 315.

180. J. Kjems, H. Leffers, T. Olesen, I. Holtz and R.A. Garret, *System Appl. Microbiol.* **13** (1990) 117.
181. J. Kjems and R.A. Garret, *Nature* **318** (1985) 675.
182. J. Kjems and R.A. Garret, *Cell* **54** (1988) 693.
183. M.V. Cubellis, C. Rozzo, G. Nitti, M.I. Arnone, G. Marino and G. Sannia, *Eur. J. Biochem.* **186** (1990) 375.
184. K. Denda, J. Konishi, T. Oshima, T. Date and M. Yoshida, *J. Biol. Chem.* **264** (1989) 7119.
185. K.J. Sutherland, C.M. Henneke, P. Towner, D.W. Hough and M.J. Danson, *Eur. J. Biochem.* **194** (1990) 839.
186. J. Schroder and F. Klink, *Eur. J. Biochem.* **195** (1991) 321.
187. S. Little, P. Cartwright, C. Campbell, A. Prenneta, J. McChesney, A. Mountain and M. Robinson, *Nucleic Acids Res.* **19** (1989) 7980.
188. A.M. Sanangelantoni, D. Barbarini, G. Dipasquale, P. Cammarano and O. Tiboni, *Mol. Gen. Genet.* **221** (1990) 187.
189. G. Pühler, F. Lottspeich and W. Zillig, *Nucleic Acids Res.* **17** (1989) 4517.
190. H. Neumann, V. Schwass, C. Eckershorn and W. Zillig, *Mol. Gen. Genet.* **217** (1989) 105.
191. H. Neumann and W. Zillig, *Nucleic Acids Res.* **17** (1989) 9475.
192. H. Neumann and W. Zillig, *Nucleic Acids Res.* **18** (1990) 195.
193. W-D. Reiter, P. Palm and W. Zillig, *Nucleic Acids Res.* **16** (1988) 1.
194. U. Hüdeophl, W-D. Reiter and W. Zillig, *Proc. Natl. Acad. Sci. USA* **87** (1990) 5851.
195. J. Huet, R. Schnabel, A. Santenac and W. Zillig, *EMBO J.* **2** (1983) 1291.
196. G. Pühler, H. Lefers, F. Gropp, P. Palm, H.P. Klenk, F. Lottspeich, R.A. Garret and W. Zillig, *Proc. Natl. Acad. Sci. USA* **86** (1989) 4569.
197. D.E. Culham and R.N. Nazar, *Mol. Gen. Genet.* **212** (1988) 382.
198. W-D. Reiter, P. Palm and W. Zillig, *Nucleic Acids Res.* **16** (1988) 2445.
199. D.E. Culham and R.N. Nazar, *Mol. Gen. Genet.* **216** (1989) 412.
200. R.A. Zimmerman and H.K. Ree, *Nucleic Acids Res.* **18** (1990) 4471.

2 The molecular genetics and biotechnological application of enzymes from extremely thermophilic eubacteria

P.L BERGQUIST and H.W. MORGAN

2.1 Introduction

Interest in thermophilic bacteria has grown steadily since the pioneering studies of thermal environments by Brock and co-workers. The potential use of thermophilic bacteria, or enzymes derived from them, in biotechnology was commented on by Brock [1]. This potential will be enhanced as an increasingly diverse range of bacterial species are reported in thermal environments. As yet however, thermophiles have not superseded their mesophilic counterparts in any major existing biotechnological application. The use of DNA polymerase derived from *Thermus aquaticus* underlies the development of a new technology [2], and may indicate that the true potential of these organisms will lie in novel applications rather than substitution in existing biotechnology.

The immediate attraction in the use of thermophiles has been the inherent stability of their proteins, particularly enzymes. The property of thermostability is often associated with increased stability to a range of other denaturing agents, and so the enzymes can generally be regarded as more robust. The ability to function at high temperatures ($> 70°C$) may confer additional advantages: for example, for proteinases, the substrate itself may denature and be more susceptible to attack; for lipases, most substrates would exist in liquid rather than solid phase; and for fermentations, volatile end-products, such as ethanol, may more easily be recovered.

Thermophilic proteins produced in large quantities in mesophiles retain their temperature stability since the major component of thermostability is genetically encoded. An increasing number of thermophilic enzymes have been cloned and expressed in alternative, usually mesophilic, hosts. This trend may in part stem from a belief that yields from the native organism are low. Certainly, apparent production rates of extracellular enzymes seem to be much lower than those from mesophiles, and there is still some confusion over the widely held belief that thermophilic bacteria have higher cell maintenance requirements and lower cell yields [3,4]. Cloning genes for thermophilic enzymes into mesophiles does produce one immediate benefit in that the thermostable gene product can be readily purified and contaminating activities in the host removed by simple heat denaturation [5,6]. This observation should not preclude more detailed research on the growth and physiology of thermophilic cultures; quite significant increases in enzyme yield or growth rate can be obtained with minor modifications to culture conditions [3,7,8].

It is not within the scope of this chapter to review comprehensively all suggested applications for thermophilic bacteria; reviews by Da Costa *et al.* [9] and Wiegel and Ljungdahl [10] provide further reading on this aspect.

2.2   Aerobic eubacteria

Two genera of aerobic eubacteria have been most widely studied—*Thermus* (mainly *T. aquaticus*) and *Bacillus* (mainly *B. stearothermophilus*). The taxonomy of both these groups is uncertain. Whereas Brock [11] originally suggested that *T. aquaticus* is the only recognised *Thermus* species, more recent work has pointed to the existence of at least two and possibly more species [12]. The taxonomy of thermophilic bacilli has recently been reviewed by Sharp *et al.* [13].

2.2.1   *The molecular biology and genetics of* Thermus

The biotechnological applications and molecular genetics of extremely thermophilic bacteria were reviewed several years ago [14]. Since that time, there has been an increasing interest in thermophilic bacteria and, in particular, in the enzymes they produce. In large part, this interest has been engendered by the discovery of the utility of *Thermus aquaticus* DNA polymerase in the polymerase chain reaction (PCR—reviewed in ref. 2). Although many papers have been published which use the PCR, there is still only minor interest in the genetics of extreme thermophiles.

2.2.2   Thermus aquaticus *DNA polymerase*

Although several genes from *Thermus* had been cloned in *Escherichia coli*, it was the polymerase chain reaction (PCR) and the production of cloned *Taq* DNA polymerase that attracted most scientific attention. There is a significant history of attempts to clone genes encoding thermostable proteins in mesophilic strains, but none of these had the long-term significance of the successful production of the thermostable DNA polymerase.

The PCR reaction was invented by Mullis and Faloona [15] and is an *in-vitro* method for primer-directed amplification of specific sequences, defined at their 5′ and 3′ ends by specific synthetic oligonucleotide primers. The basic procedure incorporates initial denaturation of the target DNA sample. Renaturation with the specific primers in excess favours the formation of the primer–template complex when the temperature is lowered in the renaturation step and a copy is synthesised from each initial target strand during the extension step. The cycle of melting of the DNA, reannealing with template and extension is repeated, resulting in an exponential increase in the amount of target defined by the primers. After 30–40 repetitions of this cycling procedure, microgram amounts of the specific target DNA are generated.

The limitation of the procedure is that some sequence information must be available at sites flanking the particular target site. There are various ways of overcoming partial sequence information (e.g. anchored PCR [16]). The particular advantage enjoyed by *Taq* polymerase in the PCR reaction is its half-life at the denaturing temperature ($T_{1/2}$ at 95°C, 40 minutes), which means that the enzyme does not have to be added repeatedly, as in earlier versions of the procedure using the Klenow polymerase. The PCR has found a variety of uses in diagnostic medicine, anthropology, taxonomy and molecular biology. In the last area, it has been used to introduce deletions and insertions into DNA, at the same time as the amplification of the initial sequence [17]. A general review of results and applications is provided in ref. 2, which should be consulted for detailed applications. Other thermophilic polymerases have been used for this purpose and as a replacement for Klenow polymerase and Sequenase for sequence analysis of DNA [18].

Several prior publications to the Lawyer *et al.* [19] paper described the difficulties of cloning and direct expression of *Thermus* genes (reviewed in ref. 14). These early reports suggested that expression of *Thermus* genes was the result of the fortuitous creation of a promoter sequence that was recognised in *Escherichia coli*. Sequence analysis of the *Thermus aquaticus* restriction endonuclease–modification methylase region showed no sequences recognisable as *E. coli* promoters [20]. Croft *et al.* [21] showed that the *leuB* region of *Thermus* was preceded by sequences recognised by *E. coli* RNA polymerase but not by a ribosomal binding site. These sequences were common to several *Thermus* strains and allowed the expression of low but sufficient levels of β-isopropyl malate dehydrogenase in complementation experiments. Whether or not these sequences are used as promoters in *Thermus* is unknown, primarily because of the difficulty of reintroducing DNA manipulated in *E. coli* back into *Thermus* (see later).

The cloning and expression of *Taq* polymerase is an object lesson in the difficulties of manipulating thermophile genes and their products, which may show minimal enzymatic activity at the usual growth temperature of the cloning host (37°C). The expression of the DNA polymerase in the host organism was low and polymerase activity could not be demonstrated with recombinant bacteriophage produced from a λ library of the *T. aquaticus* genomic DNA. It was necessary to produce antibody to the 94 kDa enzyme for screening of an expression library in λgt11 to identify bacteriophage-carrying epitopes of the enzyme [19]. The *Thermus* fragment was then used to identify bacteriophage in the λCh35:*Taq* library that showed homology to the probe. Two *Hind*III fragments were found to be necessary for production of *Taq* polymerase, and the gene was assembled in two parts and sequenced. In a novel application, the PCR itself was used in the confirmation of the structure of polymerase gene [19]. Further manipulations were required to remove a stop site upstream of the gene which prevented translational read-through from the *lacZ* promoter of the cloning vector.

As expected from previous work with *Thermus thermophilus*, *T. flavus* and *T. caldophilus*, the DNA has a high G:C content and there is an extreme bias towards G or C in the third position of codons [14]. Significant sequence similarity can be seen with other DNA polymerases such as *E. coli* DNA polymerase I and T7 DNA polymerase at both the N-terminal and C-terminal regions of *Taq* polymerase. One of the major features of *Taq* polymerase is its lack of a $3' \rightarrow 5'$ exonuclease activity (which in the case of the *E. coli* Klenow fragment brings about degradation of single- and double-stranded DNA from its 3'OH end). However, this lack of proof-reading ability has only a minor effect on misincorporation of incorrect deoxyribonucleotide triphosphates [22].

The biotechnological significance of the polymerase chain reaction and the role of thermostable polymerases in amplification and sequencing reactions cannot be overestimated. This technology has in many cases superseded cloning, and has provided facile methods for the generation of recombinant genes and plasmids that would be complicated and tedious to produce by other methods. For example, the PCR has been used to generate deletions in front of a thermophilic xylanase gene to allow better control and overall expression from a heat-inducible promoter sequence [23]. The appropriate deletions could not be obtained using conventional Bal31 deletion procedures. Others have constructed mutant and chimeric genes using the polymerase chain reaction [17] and have used the enzyme to sequence the PCR products. Several ingenious procedures have been developed to introduce mutations at internal sites in genes and for the construction of entire plasmids carrying the desired mutations or structural alterations [24]. There are several procedures available for medical and microbiological diagnosis, including the detection of human immunodeficiency virus [25], detection and typing of human genital papillomia viruses [26], diagnosis of retinoblastoma [27] and the detection of human cytomegalovirus [28].

2.2.3 Thermus aquaticus *restriction–modification system*

Thermus aquaticus possesses a type II restriction–modification system that recognises the palindromic sequence TCGA. The restriction endonuclease makes a double-strand cleavage between the T and C residues and the *Taq*I methylase modifies the A residue on each strand at the N^6 position [20]. An earlier review summarised preliminary work on the fortuitous overexpression of the *Taq* methylase in *Escherichia coli* [14].

Recombinant plasmids carrying the *Taq* methylase genes were selected for their methylation phenotype by digesting genomic libraries of *Thermus aquaticus* in *E. coli* with *Taq*I restriction endonuclease and transforming into recipient cells. This procedure destroys molecules that are unmethylated, and survivors of a *Pst*I-generated library carried a 3.5 kb *Pst*I fragment in common and produced *Taq* methylase but not *Taq* endonuclease [20]. Survivors from a *Bam*HI library expressed both enzymes and carried a common 5.5 kb *Bam*HI fragment. Sequence analysis allowed the identification of two adjacent open reading frames that could be correlated with endonuclease and methylase activity. From the orientation of the inserted DNA fragments it was concluded that expression was occurring from vector promoters. Slatko *et al.* [20] concluded that the sequence that promotes the transcription of the endonuclease gene in *Escherichia coli* is fortuitous and possibly without function in *Thermus aquaticus*.

The *Taq*I restriction endonuclease was cloned downstream of an inducible promoter as a gene fusion with the signal sequence of alkaline phosphatase [29]. A series of molecular genetic modifications of the cloned DNA allowed a significant increase in the level of expression. It was found that the amino-terminal portion of the endonuclease does not seem to be critical for its function as it tolerates seven additional amino acids from the fusion plus at least one different amino acid at the beginning of the endonuclease sequence. Once sufficient enzyme had been produced to allow recognition and selection of plasmids carrying the genes, the amount of enzyme produced in *E. coli* could be increased by appropriate recombinant DNA manipulations using *E. coli* expression sequences. In both this example and in the case of *Taq* polymerase, once an assay for the gene product has been established there seems to be no barrier to the expression of the thermophilic protein in *E. coli*. Furthermore, in all cases examined, the thermophilic enzyme maintains its thermal stability after synthesis in *E. coli*.

2.2.4 *Expression of other genes from* Thermus *genomic libraries*

The expression of the *leuB* from *Thermus thermophilus* HB8 and sequence analysis of the *leuB* gene have been reviewed previously [14]. Nicholls *et al.* [30] have cloned and sequenced a genomic *Hind*III fragment carrying the malate dehydrogenase gene and the succinyl CoA synthetase alpha subunit from *Thermus aquaticus* B. Only the latter sequence has been published: it shows the familiar *Thermus* third-position codon G–C bias and has substantial homology at the protein level with rat and *E. coli* succinyl CoA synthetase. These two enzymes are also assumed to be translated from a polycistronic mRNA.

The *tufA* structural genes of *Thermus thermophilus* were identified by using the *E. coli* elongation factor EF–Tu gene as a probe. The sequence of the gene was translated into amino acids and the putative protein showed 70%

homology with the *E. coli* elongation factor [31]. Although there was a ribosomal binding site upstream of the initiation codon, there were no recognisable *E. coli*-like promoter sequences and the gene was not expressed in *E. coli*, although two related ribosomal proteins were expressed in minicells. The gene for *Thermus thermophilus* HB8 elongation factor G has also been cloned and showed the familiar high G:C content in the third position of codons and had significant homology with the *E. coli* elongation factor G gene.

Yakhnin *et al.* [32] have sequenced the equivalent of *E. coli* ribosomal proteins S12 and S7 from *Thermus thermophilus*. These genes show 69% (S12) and 52% (S7) homology with the *E. coli* B sequences and appear to be part of a ribosomal proteins operon which includes the genes for elongation factors EF-G and EF-Tu. No 16S RNA gene was found in the intercistronic region between the two ribosomal protein genes, as is found in *E. coli*.

The L-lactate dehydrogenase gene from *Thermus caldophilus* GK24 was cloned in *E. coli* [33]. The nucleotide sequence was determined and similar upstream 5' sequences were found to the *T. thermophilus leuB* gene, which may represent a *Thermus* promoter sequence. The *Thermus caldophilus* gene was used as a probe to identify the L-lactate dehydrogenase of *Thermus aquaticus* YT-1 which was isolated as a 2.4 kb *Sac*I fragment in a pUC vector. No enzyme was expressed by this plasmid, but recloning into a high-level expression vector under the *tac* promoter allowed production of the L-lactate dehydrogenase at about 10% of the *E. coli* cellular protein [34]. The determination of the complete nucleotide sequence allowed comparison of the putative protein sequences of the *T. caldophilus* and *T. aquaticus* L-lactate dehydrogenases, which showed 87% identity and differed at 23 sites in charge and polarity, and these changes are presumed to be related to kinetic differences between the two enzymes.

The gene coding for a thermostable pullulanase from *Thermus* strain AMD-33 was cloned in an expression vector in *E. coli*. Pullulanase activity was found to be independent of the presence of IPTG, the inducer of the *tac* promoter carried by the vector [35]. The temperature and pH profiles of the recombinant enzyme and the pullulanase isolated from *Thermus* AMD-33 were virtually identical.

2.2.5   *Proteinases from* Thermus *spp.*

Many isolates of *Thermus* produce extracellular serine proteinases [36]. Although the enzymes are thermostable and have high specific activities, the culture yield is too low to consider industrial application [37]. Increases in yield of at least three orders of magnitude would be required to match proteinase production by currently used mesophiles. Such increases seemingly can be met only by cloning and overproduction in a suitable host.

However, our understanding of the growth and physiology of these relatively new isolates is still insubstantial when compared with many mesophiles. In a detailed examination of proteinase production by *Thermus* strain Rt41A, Janssen *et al.* [8] demonstrated that careful control of medium composition allowed increases of yield of nearly 100-fold over conventional culture conditions. In particular, the chelation of calcium ions by medium constituents, especially phosphate, had to be avoided. Calcium ions are necessary for proteinase thermostability at the growth temperature. When a medium composition was devised that contained only anabolic amounts of phosphate, such that the residual phosphate concentration at maximal cell yield was less than 100 μM, the proteinase was essentially completely stable in the fermenter broth. Under these conditions, enzyme production by *Thermus* Rt41A was proportional to cell yield and the upper limit to proteinase production was the result of foaming in the fermenter at high cell densities. It is now apparent that previous studies which have reported on low yields of proteinase have in practice been recording the balance of high production rates against high denaturation rates [38, 39]. The potential for further increases in yield under different fermenter configurations remains to be established.

This enzyme is extremely stable to heat and detergents and is a valuable adjunct to the polymerase chain reaction. It can be used to prepare DNA suitable for PCR amplification by incubating small amounts with blood or tissue samples for 30 minutes at 94°C. The enzyme breaks down cellular structures and is itself inactivated during this period [40]. It has been used for the specific amplification of diagnostic fragments for cystic fibrosis and papillomavirus type 16, but appears to be quite general in its action on bacteria and lower metozoa to tissues and fluids from mammals. The enzyme lends itself to automation of the PCR reaction for diagnostic purposes.

The gene for the proteinase from *Thermus* Rt41A has been isolated by using a forward primer synthesised from the determined amino-terminal sequence of the mature enzyme and a highly conserved sequence for serine proteases around *his*-72 (Figure 2.1). A reverse primer was synthesised which matched a conserved sequence for alkaline serine proteases adjacent to the active site serine (Figure 2.1, position 221). The product of the polymerase chain reaction was used to identify the appropriate restriction enzyme fragments in the genome. Sequence determination shows considerable homology at the DNA and protein level to aqualysin I and thermitase (Figure 2.1).

The serine proteinase from *Thermus aquaticus* has been the subject of several attempts for cloning into *E. coli*. Kwon *et al.* [41] used oligonucleotide probes based on the amino-terminal sequence of purified aqualysin I from *Thermus aquaticus* YT-1 to isolate and clone a 1.1 kb *Pst*I fragment that contained the sequence of the mature enzyme. The nucleotide sequence showed that the fragment contained a single open reading frame without a stop codon, and by inference from other Gram-negative bacteria they

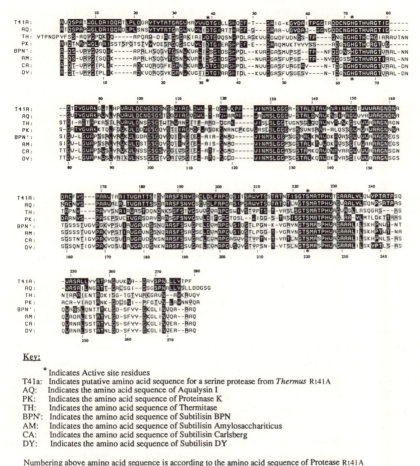

Key:

* Indicates Active site residues

T41a: Indicates putative amino acid sequence for a serine protease from *Thermus* Rt41A
AQ: Indicates the amino acid sequence of Aqualysin I
PK: Indicates the amino acid sequence of Proteinase K
TH: Indicates the amino acid sequence of Thermitase
BPN':   Indicates the amino acid sequence of Subtilisin BPN
AM: Indicates the amino acid sequence of Subtilisin Amylosacchariticus
CA: Indicates the amino acid sequence of Subtilisin Carlsberg
DY: Indicates the amino acid sequence of Subtilisin DY

Numbering above amino acid sequence is according to the amino acid sequence of Protease Rt41A
Numbering below amino acid sequence is according to the amino acid sequence of Subtilisin.

Figure 2.1 Comparison of amino acid sequence similarities between subtilisin-like serine proteases and an alkaline serine protease from *Thermus* Rt41A (G. Munro and R. McHale, unpublished). Identical amino acids are indicated by reverse font, and similar amino acids are indicated by the stippled font. The University of Wisconsin Computer Group definitions of similar amino acids have been followed.

concluded that the enzyme was synthesised with three domains, an amino-terminal leader involved in inner membrane transport, the protease and a carboxy-terminal domain that was required for extracellular secretion. The mature enzyme was not expressed in *E. coli*.

Aqualysin I is produced as a large precursor, and the mature enzyme lacking the amino-terminal preprosequence accumulated in the membrane fraction of *Escherichia coli* [42]. Treatment of the membrane fraction at 65°C released the active enzyme into the soluble fraction. The entire gene was cloned under a controlled promoter since it was anticipated that direct expression of the intact gene product would be lethal to the cell. The gene was cloned in two parts and then assembled under the control of an inducible promoter. The sequence of the complete gene showed that the amino terminus of the deduced protein sequence had the characteristics of a signal peptide sequence, and at the 5' end in the non-coding area an *E. coli*-like promoter sequence could be identified. This promoter may have allowed constitutive expression of aqualysin and effectively made the intact gene unclonable. Aqualysin appears to be unusual in that there seem to be four domains in the precursor structure, and both amino- and carboxy-terminal processing takes place to give the mature enzyme of $M_r = 28\,000$.

In an interesting approach derived from this work, Takagi *et al.* [43] have used site-directed mutagenesis to introduce cysteine substitutions to allow the formation of a disulphide bond in subtilisin E, using the information obtained on the location of the two cysteine residues in aqualysin I. The mutant subtilisin carrying cys residues at positions 61 and 98 appeared to form a disulphide bond spontaneously in the *E. coli* expression system, and the half-life and optimal temperature of enzyme action was increased significantly over wild type. These results suggested that it was possible to enhance the thermostability of subtilisin without changing the catalytic efficiency of the enzyme.

Aqualysin shows significant homology at the protein level with other subtilisins such as proteinase K, thermitase, subtilisin BPN', subtilisin *Amylosacchariticus*, subtilisin Carlsberg and subtilisin DY. These enzymes are produced by several species of *Bacillus*, *Thermoactinomyces* (thermitase) and the fungus *Tritirachium albus* (proteinase K). All have the conserved Asp-32, His-64 and Ser-221 of the active cleft of subtilisin BPN' and maintain the identity of several subsites (see Figure 2.1).

The complete amino acid sequences of thermitase and proteinase K are known, along with aqualysin they are related to the subtilisins but contain cysteine residues. Almost every property of subtilisin has been altered by site-directed, random and cassette mutagenesis [44], and two high-resolution X-ray crystallographic structures are known. Subtilisin variants produced in the laboratory have their equivalents in nature, for example Asn-218 → Ser is present in proteinase K, and both mutant and natural variants are known with increased temperature stability. Thus the isolation of even more stable related subtilisin-like enzymes is of considerable practical and theoretical interest.

2.2.6   *Genetic transfer and plasmids in* Thermus

A previous review summarised the occurrence of plasmids and the demonstration of chromosomal transformation in *Thermus*. Since that time, five small plasmids from 13.5 to 8.3 kb have been identified in *Thermus aquaticus* which are different from the plasmid complement of *Thermus thermophilis* HB8 [45]. No correlations could be established between the presence of the plasmids and phenotypic or physiological behaviour. Koyama and Furukawa [46] cloned and sequenced the tryptophan synthetase genes of *Thermus thermophilus* HB27, but these genes proved to be unsuitable as a selective marker for genetic experiments because of their high frequency of recombination into the chromosome of a *trpB* mutant to give marker replacement and the loss of the plasmid portion of the construct. *Thermus* strain T2 is only distantly related to *Thermus thermophilus* HB8, so they cloned the tryptophan synthetase genes from T2 into the *E. coli* plasmid pUC13 as a *Sac*I–*Bgl*III fragment [46]. This plasmid, pKA216, was then cloned into the single *Bgl*III site of the *Thermus thermophilus* HB8 plasmid pTT8 to give a shuttle vector capable of replicating in *E. coli* or *Thermus*. This shuttle plasmid transformed a *T. thermophilus trpB* mutant to Trp$^+$ at the low but detectable frequency of 10^4 transformants per microgram of plasmid DNA.

Koyama *et al.* [47] have also reported the cloning of an α- and a β-galactosidase gene from *Thermus* T2 in *Escherichia coli* and their reintroduction into *Thermus thermophilus* HB27 using a shuttle vector composed of pUC18 from *E. coli* and pTT8 from *Thermus thermophilus*. Expression of the β-galactosidase in *Thermus* was observed directly using a chromogenic substrate.

A major problem for plasmid transformation is that no antibiotic resistance genes are known to be carried by *Thermus* plasmids. We endeavoured to construct a similar shuttle vector to that described above using the pTT8 plasmid, the *Thermus leu* promoter region, joined to either a *Bacillus stearothermophilus* kanamycin or tetracycline resistance gene, and pUC19 for replication in *E. coli*. We were unable to obtain reproducible transformation of *Thermus thermophilus* HB8 with our shuttle plasmids (M.K. Ashby and P.L. Bergquist, unpublished). The low frequency of transformation observed by Koyama and Furukawa [46, 48] and previous negative results with *Thermus leu* genes (reviewed in ref. 14) suggests that a restriction/modification phenomenon is involved with thermophilic DNA that has been replicated in *E. coli* and only a strong selective system allows plasmid transformation to be revealed. The additional problem of the high frequency of homologous recombination as observed by Koyama and co-workers [46, 48] further complicates experimental protocols that rely on co-transformation of chromosomal markers.

The conjugative transposon Tn916 has been transferred from *Bacillus subtilis* to *Thermus aquaticus* by broth matings at 48°C [49]. The transposon could be transformed back into *Bacillus subtilis*, although at a lower frequency. Transfer appears to be by way of conjugation rather than transformation since it was unaffected by DNase. This procedure offers promise of a more extensive genetic analysis in at least the lower-temperature-range *Thermus* strains, since Tn916 can be modified by recombinant DNA techniques to accommodate inserted DNA without loss of transfer ability.

2.2.7   *A repetitive sequence in* Thermus thermophilus

We observed that some phenotypic changes from the protease$^+$ to protease$^-$ condition observed in *Thermus* strain T351 could be best explained by the occurrence of insertion sequences (D.R. Love, M.B. Streiff and P.L. Bergquist, unpublished). Since a minimal medium had been established for *Thermus thermophilus*, we examined this strain for the occurrence of repeated sequences in the genome that might represent insertion sequences or the terminal portions of transposons [50]. One plasmid from a genomic library of *T. thermophilus* HB8 hybridised to multiple genomic bands in a Southern blot experiment, suggesting that it contained a repetitive DNA element. Four plasmids were identified which carried copies of the element. These plasmids were found to contain overlapping portions of the same part of the genome found in the original plasmid. This insertion sequence (IS1000) was 1196 bp long, and another plasmid carried two copies in tandem separated by 256 bp. The sequence is present on the genome of *Thermus thermophilus* HB8 and HB27 and some New Zealand *Thermus* isolates but was not present in other *Thermus* strains or on the *Thermus thermophilus* plasmids pTT8 and pVV8. IS1000 has imperfect terminal inverted repeats of six base pairs and no target site duplication. It has a long open reading frame which could code for a protein of 317 amino acids. The G:C content of the three codon positions closely matches the distribution found for other genes from *T. thermophilus* HB8. This protein has significant homology with open reading frames found in the transposable elements IS110 of *Streptomyces coelicolor* and IS492 of *Pseudomonas atlantica*. The homology between the deduced amino acid sequences for the open reading frames of IS110, IS492 and IS1000 suggests that since the host bacteria are unlikely to be related the three insertion sequences share a common ancestor. It may be possible to develop IS1000 as a genetic tool for gene transfer and mutagenesis of *Thermus* once a biological test for transposition has been developed.

2.2.8   *Promoter regions and other control sequences in* Thermus

There are now a number of sequences available for genes coding for enzymes and other proteins for *Thermus*. Many sequences have upstream and downstream sequences outside of the open reading frames. These data have been analysed for sequences that may represent promoters, ribosomal binding sites and transcriptional terminators (M.K. Ashby and P.L. Bergquist, unpublished). Some sequences which lacked the amino-terminal coding regions (for example, aqualysin I [41]) or where the genes were part of operons were excluded. The consensus promoter sequence was Gg–CCTC–

C—GG–CG–9–14 bp–CCtTTTa–, which is quite different from the *E. coli* consensus of tcTTGACat—t–9 bp–t–tg–TAtAaT– and different from the 23S/5S ribosomal operon of *Thermus thermophilus* described by Hartmann *et al.* [51].* The consensus ribosomal binding site was within 2–10 bp of the translational start site. Only five genes were suitable for the analysis of termination signals: 4 out of 5 of these genes contained inverted repeats located from 12–41 bp downstream of the stop codon. No doubt these consensus sequences will be further refined as more information becomes available.

2.2.9 *Genes and proteins from thermophilic strains of* Bacillus

Thermophilic bacilli have been the subject of attention for application in biotechnology because of their relative ease of isolation and culture and their nutritional diversity. Although there is still confusion over the taxonomy of many isolates, *Bacillus stearothermophilus* is the most frequently used culture. This organism is generally easy to grow, though some strains show instability in continuous culture [52] and other strains grow poorly on minimal medium.

The greatest interest in these organisms lies in their proteolytic activity— particularly at alkaline pH [53]—and saccharolytic activity, though claims have also been made for their use in ethanol production [54]. As yet, no enzymes derived from thermophilic bacilli have been used in substantial amounts in the starch hydrolysis industry, even though this process is dominated by the use of thermostable enzymes. Patents have been taken out on at least two enzymes derived from *B. stearothermophilus* (β-galactosidase and cyclodextrin glucotransferase). Production levels from the native organism do not match those of commerical strains, and cloning is viewed as the likely solution. There have been problems in developing this organism as a host-vector system, since some strains regenerate poorly while in other cases the plasmids are unstable at 60°C. Accordingly, most genes have been cloned in *E. coli* or *Bacillus subtilis*.

Earlier information on the isolation and cloning of genes from thermophilic bacilli was covered in a previous review [14]. The tryptophan synthetase genes of *B. stearothermophilus* have been cloned by complementation in *E. coli* (55). The 5′-proximal portion of the *trpA* gene was found to overlap the 3′-distal portion of the *trpB* gene by 20 nucleotides. The putative proteins from these open reading frames were 55–70% homologous with the same genes from *Bacillus subtilis* and 35–55% homologous with *E. coli trpA* and *B*. The genes were expressed efficiently in *E. coli* and their products constituted more than 20% of the soluble protein [55].

---

* Bold type, present in all sequences; upper case type, present in 75% or more sequences, lower case type, present in 50% or more sequences.

Kubo and Imanaka [56] have cloned and expressed a neutral protease from *B. stearothermophilus* MK232 in *B. subtilis*. This enzyme is more resistant to heat inactivation than thermolysin from *B. thermoproteolyticus* and has a higher specific activity, although it differs from thermolysin only by a substitution of Asp-37 by asparagine and Glu-119 by glutamine. Both substitutions result in the substitution of an uncharged polar amino acid and affect only one domain of the enzyme. Hence the changes may not affect the three-dimensional structure of the enzyme but may increase hydrogen bonding or enhance electrostatic interactions. The Asp → Asn change should enhance thermostability [57], and this example shows that enzymes with increased stability can be sought in nature as an alternative to *in-vitro* manipulation.

The lactate dehydrogenases from *Bacillus stearothermophilus* and *B. caldolyticus* have been cloned and sequenced [58]. The deduced amino acid sequences differ by 10 amino acids. The conservation of restriction enzyme sites between the two genes allowed the construction of hybrid enzymes in *Escherichia coli* which were examined for their activation by fructose 1,6-diphosphate and thermostability. Finer resolution was provided by site-directed mutagenesis [59]. Differences in the temperature stability of the two lactate dehydrogenases could be correlated with three amino acid exchanges in the middle portions of the enzymes, but the influence of the exchanged segments was not additive. Increases in thermostability were associated with hydrophobic amino acids replacing polar amino acids found in the same position in mesophiles.

The nucleotide sequence of the *B. stearothermophilus* glyceraldehyde-3-phosphate dehydrogenase gene and its flanking regions has been determined [60]. The gene is notable for a long stretch of DNA upstream from the start codon to the promoter sequence that is probably used in *E. coli* (more than a kilobase).

2.2.10   *α-Amylases from* Bacillus *spp.*

B. stearothermophilus can completely hydrolyse starch and starch-related polymers, but strains which only produce single enzyme activities can also be readily isolated [61]. The cloning and characterisation of starch-degrading genes from *Bacillus* has been an active research field, though it is again worth noting that increases in α-amylase production of over 11-fold were achieved in a culture of *B. caldolyticus* by simple selection procedures [62].

Ihara *et al.* [63] cloned an α-amylase from *B. stearothermophilus* DY-5 into *E. coli* and determined the nucleotide sequence of the gene. Comparison of the inferred amino acid sequence with other *Bacillus* α-amylases and with eukaryotic amylases showed three homologous sequences in regions known to be critical for enzyme activity. Homology comparisons will be discussed later in this review in connection with a pullulanase from an anaerobic thermophile. An α-amylase from a *B. stearothermophilus* plasmid was cloned and expressed in *E. coli* and in *B. subtilis* by Sen and Oriel [64]. The enzyme was overexpressed compared with the original strain, and was found both intracellularly and secreted into the medium. From the growth curves in their paper, the extracellular location of the enzyme may have resulted from cell death. Sohma *et al.* [65] have achieved the secretion of mature extracellular α-amylase of *B. stearothermophilus* by fusing the gene to a *B. subtilis* secretion vector, thus ensuring the correct processing from the preproenzyme.

Bacillus licheniformis is not regarded as a thermophile, but its α-amylase is extremely stable at 90°C and has a 100-fold longer half-life than the α-amylases from *B. amyloliquefaciens* and *B. stearothermophilus* [66]. There is significant homology between the three enzymes (65–80%) and their primary structures resemble each other. Suzuki *et al.* [67] constructed chimeric genes from the structural genes for the two enzymes and showed that two regions of the sequence were important in the determination of thermostability. A glutamine at position 178 (region I) and residues 255–270 (region II) determined the relative thermostability of the α-amylases. Deletion of arginine-176 and glycine-177 in region I and substitution of alanine for lysine-269 and aspartic acid for asparagine-266 in region II by site-directed oligonucleotide mutagenesis of the *B. amyloliquefaciens* α-amylase gene enhanced thermal stability. These changes affected thermostability independently and additively. Substitution of alanine for lysine is one of the preferred substitutions found in thermophilic proteins in the analysis of Argos *et al.* [57]. These changes differ from the amino acids predicted to increase the thermal stability of the *B. amyloliquefaciens* α-amylase from chemical studies by Tomazic and Klibanov [68].

Of considerable interest in this study was the finding that the mutant *B. amyloliquefaciens* gene product was as stable as the α-amylase from *B. licheniformis* with respect to irreversible denaturation at 90°C but nevertheless showed the same relatively low temperature optimum of the wild-type enzyme (65°C). Hence the mutant enzymes are still susceptible to reversible inactivation at temperatures above 65°C. Thermal inactivation of *Bacillus amyloliquefaciens* α-amylase involves two steps, reversible unfolding of the molecule and a conformationally irreversible step [66]. The deletion of Arg-176 and Gly-177 and the substitution of alanine for Lys-269 probably affect the unfolding step but not the equilibrium of the conformationally irreversible step.

2.2.11 *Pullulanases and related enzymes from thermophilic bacilli*

Pullulan is composed of α-1,6-linked maltotriose units and is degraded by at least four types of enzyme, each of which give characteristic end-products. The enzymatic degradation of starch is reviewed in detail in Vihinen and Mantsala [69], and readers should consult this paper for a detailed discussion of enzymes, substrates and end-products. The differences between the

α-amylase–pullulanase enzyme complex and a pullulanase with α-amylase activity (type II pullulanase, see later), as found in most thermophilic enzymes, has not been explained, although gene fusion under the selective pressure of temperature is a possible explanation. Only the genes for a few enzymes have been cloned, but they appear to fall into two distinct groups. First, those with products of high molecular weight are likely to be true pullulanases or pullulanase–amylase bifunctional proteins, whereas proteins that degrade pullulan but are of lower molecular weight (less than $M_r = 120\,000$ Da) are likely to be isopullulanases or neopullulanases.

Kuriki et al. [70] have cloned that gene coding for a pullulanase (debranching enzyme; end-product of pullulan degradation maltotriose) from B. stearothermophilus strain TRS128 into B. subtilis as a 4.2 kb HindIII fragment. The enzyme was very stable at 65°C but was rapidly inactivated at 70°C. The enzyme was found to be stabilised at 68°C by Ca^{2+} and the temperature optimum under these conditions was 75°C [71]. The nucleotide sequence of the gene (pulT) was determined and pulT was found to code for an open reading frame of 658 amino acid residues, $M_r = 75\,375$, in good agreement with sodium dodecyl sulphate–polyacrylamide gel electrophoretic determination of the molecular mass of the partially purified protein [71]. The enzyme showed homology at the protein level with the conserved amylase and pullulanase conserved regions (see later).

Another strain of B. stearothermophilus, TRS40, produced a pullulanase that converted pullulan to panose [61]. The gene for this enzyme, called neopullulanase, was also cloned in B. subtilis. This enzyme was stable at 60°C but was rapidly inactivated at temperatures above 70°C, especially in the presence of EDTA. The enzyme had an apparent molecular weight of 62 000 daltons compared with the observed molecular weight of approximately 76 000 daltons for the pullulanase from strain TRS128.

The nucleotide sequence of neopullulanase has been determined by Imanaka and Kuriki [72] and the translational start point determined by amino-terminal sequencing of the purified enzyme. The neopullulanase contained eight cysteines (generally considered to be undesirable for thermo-stability) but these were not associated with disulphide bonds. The deduced amino acid sequences showed significant homology in four areas with α-amylase [73], isoamylase (hydrolyses 1,6-α-D-glucosidic linkages of certain branched α-D-glucans, such as glycogen), pullulanase [74] and cyclodextrin glucanotransferase [75] but there was no homology in other regions between the several enzymes or with other glucanohydrolases such as cellulase [76] and glucoamylase [77].

2.2.12   Other genes and genetic systems in Bacillus stearothermophilus

The gene for a β-1,4-endoglucanase from the anaerobic thermophile Clostridium thermocellum, celA, was cloned into Bacillus subtilis in an E. coli–

B. subtilis shuttle vector using a promoter from pUB110 for expression [78]. Transformants expressing the antibiotic resistance of the vector yielded only deleted derivatives when *celA* was cloned using its own promoter. One plasmid stably transformed *B. stearothermophilus* to give carboxymethyl-cellulase activity at temperature up to 68°C, the maximal growth temperature for strain CU21.

Zhang *et al.* [79] have isolated a more readily transformable strain of *B. stearothermophilus* than CU21, which is unable to grow on minimal medium. Protoplasts of this strain, S1C1, could be transformed with both high- and low-copy-number plasmid vectors and the *amyT* (α-amylase) and *npoM* (neutral protease) genes were able to be cloned and expressed in it using the low-copy-number vector, pTB53. Wu and Welker [80] have developed an efficient protoplast transformation system for *B. stearothermophilus* NUB3621 by painstaking attention to protoplasting and regeneration conditions and showed that there was a direct relationship between temperature for efficient transformation and the minimum temperatures for growth.

2.3 Anaerobic eubacteria

There is a large diversity of anaerobes in thermal environments, particularly those associated with phototropic mats, which is possibly due to the diminishing solubility of oxygen at higher temperatures. Several isolates have been proposed as cultures suitable for thermophilic ethanol production. While fermentation at high temperatures appears to have many desirable attributes when volatile end-products are formed, no thermophilic fermentation yet matches that of yeast under optimum conditions. The potential in this area has been critically reviewed by Slapack *et al.* [4].

Interest in anaerobic thermophiles seems to have shifted in the meantime to either single-enzyme uses, either cloned or from the native organism, or the use of pure cultures in the breakdown of complex polysaccharides, especially cellulose. This new focus stems, in part, from the quite detailed studies which have been carried out on the cellulolytic complex from a moderately thermophilic anaerobe, *Clostridium thermocellum*. Although it is an ethanologenic fermenter, even under apparently optimum conditions and using hyperproducing ethanol mutants, there is still substantial production of organic acids, which precludes commercial application [81]. This organism can grow on crystalline cellulose as sole carbon source with hydrolysis yields comparable to those reported for *Trichoderma reesei* [3]. The cellulolytic complex has been investigated in the native organism [82] and by cloning the individual components [83]. The latter approach may well be desirable given the conflicting results that can arise from the synergistic action of minor impurities in apparently pure endoglucanase preparations [84]. The availability of unlimited amounts of each separate component of the cellulolytic complex by cloning should enable the mechanism of cellulose degradation to be unravelled. A quite diverse range of extremely thermophilic cellulolytic eubacteria has now been described with growth optima 10–15°C above that of *C. thermocellum*, including the only reported cellulolytic spirochaete [85].

2.3.1   *Cloning of genes involved in cellulose hydrolysis*

The isolation and description of genes encoding cellulases from various *Clostridium* species (primarily *C. thermocellum*) is one of the most thoroughly studied areas of thermophile molecular genetics. A comprehensive review of the molecular biology of cellulose degradation has been provided by Béguin [86].

The cloning and nucleotide sequences of *celA*, *B* and *D* genes from *C. thermocellum* were described in an earlier review [14], and the nucleotide sequence of *celC* and *celE* were reported recently [87, 88]. While *celE* showed some homology with the sequence of the other cellulase genes and retained the conserved, reiterated C-terminal domain, the *celC* gene showed no such homology, except for that with the *celA* gene product in a region which may be the active site of the enzyme. Endoglucanase C has an unusual substrate range compared with the other *Clostridium* endoglucanases and displays features shared by cellobiohydrolases in being able to cleave the agluconic bond of the aryl-β-glucosides. High-level expression of the *C. thermocellum* *celA* and *celZ* genes has been achieved in *E. coli* under the control of the λ_{PL} promoter [89]. Recombinant proteins made up 10–15% of the total cellular protein after temperature induction of the promoter. The production of large amounts of the *celA* product resulted in accumulation of the protein in the membrane and the killing of the host cells. Endoglucanase C (*celC* product) was retained in the cytoplasm and did not have a lethal effect on the host cell. The enzyme remained soluble and was not located in inclusion bodies. When the *C. thermocellum* *celA* gene was fused to the promoter and the prepro segment of the *Saccharomyces cerevisiae* α mating factor gene, secretion of endoglucanase A into the culture medium occurred when expressed from either integrating or replicating yeast plasmid vectors. The enzyme was highly glycosylated but retained its activity towards carboxymethylcellulose.

A comprehensive catalogue of *C. thermocellum* endoglucanases and xylanases has shown that this organism has at least 15 distinct endoglucanase genes, two xylanases and one β-glucosidase, which are apparently not clustered [83]. The reason for such a diversity of genes encoding enzymes that have similar activities is unknown. A number of the endoglucanases could degrade xylan, and some showed activity on 4-methylumbelliferylcellobioside (MUC), which has been (erroneously) regarded as specific for cellobiohydrolase activity. No cloned enzyme was able to degrade crystalline cellulose.

Sakka *et al.* [90] constructed a gene bank from a thermophilic cellulolytic strain of *Clostridium* and isolated 11 genes involved in cellulose degradation, as well as two xylanases. Some recombinant plasmids showed both endoglucanase and cellobiohydrolase activity on model substrates. A gene from '*Caldocellum saccharolyticum*' which is able to degrade carboxymethylcellulose and MUC but is a true avicellase as shown by its ability to degrade crystalline cellulose, has been described [91]. The gene, *celB*, has sequence homology with both exocellulases such as the *Cellulomonas cex* gene and endocellulases such as the endoglucanase from *B. subtilis* and with the *celB* endoglucanase from *C. thermocellum*. It has a central domain without enzymatic activity that is joined to the enzymatic domain by short stretches of polypeptides rich in proline and threonine (PT boxes) as seen in

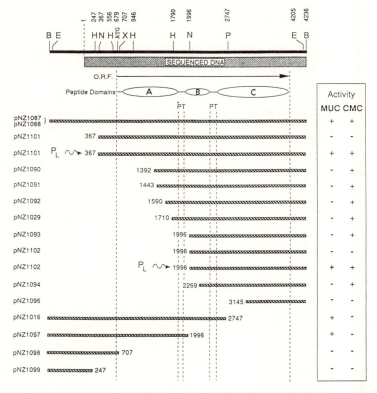

Figure 2.2 Deletion analysis of *celB* gene from *Caldocellum saccharolyticum* expressed in *Escherichia coli*. Diagrammatic representation of deletion mutations of the *celB* gene made by various techniques and assayed by CMCase and MUCase activity by plate assays. The numbers refer to the positions in the sequence at which the deletion commences. A diagrammatic representation of the peptide domain is included. Some recombinants in the expression vector pJLA602, using λP_L, are shown in the induced and uninduced situation (induced, with arrow). For further details, see ref. 91. (Reproduced with permission.)

Cellulomonas [92]. Deletion analysis has confirmed the location of the cellobiohydrolase activity in the amino-terminal domain and endocellulase activity in the carboxy-terminal domain (Figure 2.2). There are internal transcriptional and translational start sites within the gene, which confuse the analysis of the recombinant proteins produced in *Escherichia coli* [91].

The intact gene has been cloned in the temperature-inducible expression vector pJLA602 under the control of λ_{PL}. This recombinant plasmid allows overexpression of the gene product which was shown to be a protein of molecular mass 118 000–120 000. The *celB* gene has also been expressed at high levels in *Saccharomyces cerevisiae* in the yeast expression vector pUFα8, a secretion vector based on the α-pheromone (D. Saul, L. Williams and P.L. Bergquist, unpublished).

Schwarz *et al.* [93] have investigated the cellulase system of *Clostridium stercorarium*, an extreme thermophile capable of degrading cellulose. This cellulase system appears to be less complex than the multienzyme system found in *C. thermocellum*. They have cloned and characterised restriction fragments from a cosmid library that encode an endoglucanase (*celZ*), a β-glucosidase (*bglZ*) and a β-cellobiosidase with a broad substrate specificity (*celX*).

Hydrolysis of cellulose results in the accumulation of cellobiose, which has been shown to inhibit breakdown of cellulose by the *C. thermocellum* cellulase complex (reviewed in ref. 94). β-Glucosidase has been reported to stimulate cellulose degradation by the removal of cellobiose. The first β-glucosidase cloned was from the thermophile 'Caldocellum saccharolyticum' [95], and this enzyme has been shown to be remarkably heat-stable and to hydrolyse a variety of aryl-glucosides [96,97]. This enzyme has no homology with β-glucosidases from several fungi and with the *bglB* gene of *C. thermocellum* [98]. Kadam and Demain [99] have demonstrated that the degradation of crystalline cellulose is enhanced by the addition of the cloned enzyme to the cellulase complex from *C. thermocellum*.

2.3.2   *Hemicellulose hydrolysis*

Xylan is a major component of hemicellulose and is found in large amounts in straw and as a component of hardwoods and softwoods (Figure 2.3). The enzymatic hydrolysis of xylan, which is a heteropolymer of the pentose sugar xylose, is accomplished by the action of endo-β-1,4-xylanase and β-xylosidase. The first enzyme acts on xylan to generate small xylo-oligosaccharides, whilst the β-xylosidase hydrolyses dimers and trimers of xylose to the monomeric sugar [100].

Xylanases have been cloned and sequenced from two thermophilic micro-organisms, *C. thermocellum* and 'Caldocellum saccharolyticum' [102,103]. Grépinet *et al.* [104, 105] found that one of the *C. thermocellum* cloned products hydrolysed MUC, xylan and *p*-nitrophyl-β-D-xylobioside but not

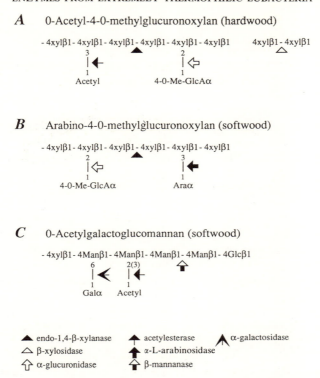

Figure 2.3   Structure of wood hemicelluloses and sites of attack of microbial hemicellulose activity [100, 101].

carboxymethylcellulose. The nucleotide sequence showed an open reading frame of 837 amino acids (2.51 kb) which contained a duplicated segment very similar to the conserved domain found at the carboxy-terminal ends of endoglucanases A, B and D from *C. thermocellum*. Xylanase activity was located towards the carboxy terminus by a series of gene fusions with the *E. coli lacZ* gene. The gene contained an internal translational reinitiation site 470 codons from the start, which gave xylanase activity.

Lüthi *et al.* [103] cloned and sequenced 6 kb of thermophilic DNA from '*Caldocellum saccharolyticum*' which contained five open reading frames, of which three could be identified with specific xylan-degrading enzymes. One open reading frame, ORF1, $M_r = 40\,455$ appears to code for a xylanase (XynA) which also acts on *o*-nitrophenyl-β-D-xylopyranoside. ORF5 (XynB, $M_r = 56\,365$) codes for a β-xylosidase (Figure 2.4). The *xynA* gene product showed significant homology with the xylanase from an alkophilic *Bacillus* (strain C125) and *C. thermocellum xynZ*, and there are further similarities with the catalytic domains for the exoglucanase of *Cellulomonas fimi*, the cellobiohydrolase of '*Caldocellum*' *celB* and the C-terminal domain of

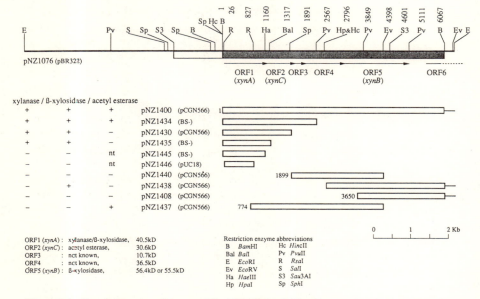

Figure 2.4 Restriction map of thermophilic DNA from *Caldocellum saccharolyticum* coding for xylan-degrading enzymes cloned in *Escherichia coli*. The shaded bar indicates the sequenced part of the cloned fragment and is represented as the pBR32 recombinant. The extent of thermophilic DNA remaining in derivatives of pNZ1076 by Bal31 exonuclease digestion or by subcloning of specific restriction enzyme fragments is shown by closed bars. Enzyme activities for the deletions are shown on the left-hand side of the figure [101]. (Reproduced with permission.)

Clostridium xynZ (which contains xylanase activity). As it has been suggested that *Bacillus*, *Cellulomonas* and *Trichoderma* glycosidases may have evolved from the shuffling of two catalytic domains and several binding domains [106], it is possible that the '*Caldocellum*' xylanase/β-xylosidase enzyme is the product of reshuffling of catalytic and binding domains, resulting in an enzyme with different substrate specificity.

The β-xylosidase has a higher temperature optimum and stability and does not show close homology with the xylanase. It has been cloned in à high-copy-number vector and expressed in *E. coli* [107]. The xylanase has also been overexpressed in *E. coli* comprising about 20% of the whole cell protein. This enzyme appears to be the only hemicellulase from an extreme thermophile which has been expressed in a regulated manner to high level in *E. coli* [23].

Depending on the source of pulp, xylan contains variable amounts of arabinosyl- and 4-*O*-methylglucuronic acid residues and acetyl groups [100]. A gene coding for an acetyl xylan esterase from '*Caldocellum*' (*xynC*, Figure 2.4), which releases acetic acid from acetylated xylan has been cloned and overexpressed [108].

Other softwoods contain mannose as a major component (Figure 2.3). A

β-mannanase from '*Caldocellum*' which is linked to the *celA* and *celB* genes on the genome has been isolated. The gene for this enzyme has been completely sequenced and has been overexpressed in *E. coli*. It shows virtually no sequence homology with *celB* or *xynA*, although it shows a small reiterated sequence homologous with a sequence in *celB* outside of the coding frame. The enzyme produced in *E. coli* completely hydrolyses guar gum (repeating mannan units) and shows limited hydrolysis of konjac gum and *Pinus radiata* glucogalactomannan, depending on the degree of glucose and galactose content [109]. The only other β-mannanase cloned and sequenced came from a mesophilic alkaline *Bacillus* [110] and had no significant sequence homology.

Other enzymes are thought to be involved in the breakdown of hemicellulose, and it is likely that activities such as β-mannosidase, α-arabinosidase, α-glucuronidase and α-galactosidase may be required for complete hydrolysis. We are currently endeavouring to isolate, clone and overexpress these enzymes from genomic libraries of '*Caldocellum*' and other anaerobic thermophiles. The temperature optima and the stability of the cellulolytic and hemicellulolytic enzymes produced by '*Caldocellum saccharolyticum*' exceed those reported for other cellulases and xylanases. These properties may become important as xylanases and other hemicellulases may play a role in the enzymatic bleaching of pulp in the manufacture of paper [111]. To be effective not only will the enzymes have to be produced cheaply and in large quantities, but a full spectrum of hydrolytic activities will be required, since it seems likely that commercial pulp bleaching will require a 'cocktail' of enzymes appropriate to the nature of the pulp.

2.3.3 Starch hydrolysis

Starch hydrolysis is a process already dependent on the use of thermostable enzymes. The main amylolytic enzymes used in saccharification are α-amylase, β-amylase, glucoamylase and pullulanase. An extremely thermostable α-amylase is available from the mesophile *B. licheniformis*, and glucoamylase and pullulanase with improved thermostability compared with the enzymes in use would be of value in this process. Because any new enzyme will have to fit into an existing industrial process, this consideration dictates a spectrum of other properties that a new enzyme must possess besides improved thermostability.

Starch utilisation is common among anaerobic thermophiles, which makes them a promising source of thermostable enzymes with the desired properties [112]. Although cloning holds the prospect of efficient production and purification of the enzymes, once again relatively simple alteration of growth conditions can achieve quite significant improvements in enzyme yield. Antranikian *et al.* [113] have reported a 100-fold increase in amylolytic enzyme production under starch limitation during growth, with the added benefit that over 90% of the enzymes were released into the culture broth. Further benefits for industrial application were reported by the high productivity of cells entrapped in alginate beads [7]. New strains of anaerobic thermophiles have also been isolated which produce large excesses of particular hydrolytic enzymes [114]; the further development of these strains may well compete with cloned enzyme production.

2.3.3.1 α-Amylases. α-Amylases from thermophilic anaerobes have been the subject of much interest for cloning and expression into mesophiles, with the genes from at least six different thermophilic species having been cloned into E. coli.

Dictyoglomus is notable for its extremely low G:C content (29%) and it produces several extracellular amylases. Three amylase genes have been cloned into E. coli [115]. The complete nucleotide sequence of an EcoRV–HindIII fragment containing the amyA gene was determined. The codon usage pattern of the open reading frame differed considerably from E. coli, reflecting the low G:C content of Dictyoglomus. This observation is common to the majority of genes cloned from anaerobes, which in general have low G:C contents (see also ref. 96). Removal of the non-coding region upstream of the translation start site gave increased yield of enzyme expressed from an E. coli promoter, but none of the enzyme was secreted into the medium, unlike the parent organism. No recognisable amino-terminal signal sequence was found in the deduced amino acid sequence of the amylase. The enzyme produced in E. coli had a temperature optimum of 90°C.

Two other amylases (amyB, amyC) have been cloned from genomic digests of Dictyoglomus. As with amyA, each had a ribosomal binding site but no obvious promoter-like sequences upstream of the translational start site. Both genes were sequenced and overexpressed under an E. coli consensus promoter or the tac promoter. Both enzymes produced in E. coli had lower temperature optima than the amyA gene product (amyB, 80°C; amyC, 70°C) and each gave different degradation products of starch to AmyA. AmyB and C were partially homologous to each other and to taka-amylase A at the protein level. AmyA showed no significant homology with other amylases or AmyB and C [116]. We have compared the three Dictyoglomus α-amylases to 'Caldocellum' pullulanase and other amylases (see later). AmyB and C are clearly true amylases in that they have the correct size and contain the 100, 200 and 300 regions in the correct positions (discussed in detail later). AmyA does not appear to be a true amylase because it is larger than other amylases and has none of the standard α-amylase regions, despite claims to the contrary [115]. It has no homology with any sequence in the database, even when retranslated to take into account frameshifts (G. Albertson, R.M. McHale and P.L. Bergquist, unpublished).

Kitamoto et al. [117] have cloned and sequenced a gene coding for a β-amylase from Clostridium thermosulfurogenes. It was concluded from the nucleotide sequence that the β-amylase was translated from a monocistronic mRNA that coded for a 32 amino acid signal peptide and that the gene could code for a protein of $M_r = 55\,000$. Homology comparisons with *B. polymyxa*, soyabean and barley β-amylase showed 12 regions that were relatively well conserved between the enzymes. Kitamoto *et al.* [117] suggest that the relatively high temperature stability of the *C. thermosulfurogenes* enzyme might reside in the presence of four additional cysteine residues and the low percentage of hydrophilic amino acids compared with the *Bacillus polymyxa* enzyme. Haeckel and Bahl [118] have cloned an α-amylase from the same organism into *E. coli*, where the gene was transcribed from its own promoter but was not secreted into the medium.

2.3.4   *Pullulanases*

Anaerobic thermophiles appear to be unique in their strategy for debranching starch molecules. In mesophilic bacteria this is accomplished by a pullulanase which specifically hydrolyses α-1,6-glycosidic linkages but has no effect on α-1,4-glycosidic bonds. In amylolytic thermophiles the production of an enzyme which hydrolyses both linkage types seems to be the rule, in addition to α-amylases which are specific for α-1,4-glycosidic linkages. This type of enzyme specificity has been termed a type II pullulanase, or amylopullulanase [119] to differentiate it from the type I pullulanase from mesophiles which has specificity for α-1,6-glycosidic linkages [7]. It now seems unequivocal that the active site of the type II pullulanase has dual specificity, which may make it useful for speciality corn syrup production. A strain of *Clostridium thermohydrosulfuricum* which secretes large quantities of this enzyme has been isolated [7].

The cloning of a thermostable pullulanase and neopullulanase from *B. stearothermophilus* was discussed earlier in this chapter. The cloning and expression of a type II pullulanase similar in specificity to the *Thermoanaerobium brockii* enzyme and the enzyme isolated from *Thermoanaerobium* spp. [120] has been reported by Melasniemi and Paloheimo [121] from *Clostridium thermohydrosulfuricum*. A variety of enzymatically active protein degradation products were produced in *E. coli*, and until the sequence was determined it was unclear whether or not the enzyme was produced from a single open reading frame and is cleaved at a later stage into more than one polypeptide or whether there were two coordinately regulated genes. The temperature optimum of the cloned enzyme, 80–85°C. was slightly lower than that of the enzyme purified from the wild-type bacterium.

The nucleotide sequence of the type II pullulanase (*apc*) gene has been determined by Melasniemi *et al.* (122) and an open reading frame of 4425 base pairs was found, which could code for a protein of $M_r = 165\,600$ daltons. The structural gene was preceded by sequences identical to *E. coli* and *B. subtilis* σ^{43} consensus sequences. The deduced amino acid sequence showed similarities to various amylolytic enzymes, particularly the neopullulanase of *Bacillus stearothermophilus*. The three conserved regions common to amylases were identified (see later), with two of the regions partly duplicated, suggesting that this gene may have arisen from the deletion and fusion of two separate genes, one an α-amylase and the other a pullulanase. The multiple protein species found in *E. coli* were presumably proteolytic degradation products or resulted from internal reinitiation events, as seen in *celB* expressed in *E. coli* [91].

We have cloned and sequenced a pullulanase from *Caldocellum saccharolyticum* which appears to be specific for α-1,6-glucosidic linkages and which degrades pullulan, limits dextrin and amylopectin to give maltotriose as the predominant product. The gene product produced in *E. coli* has a molecular mass of about 120 000 daltons, similar to the enzyme purified from the wild-type organism. The cloned enzyme has a temperature optimum of 85°C and a half-life of 80 minutes at 70°C when incubated in the absence of substrate. Pullulanase activity is confined to the carboxy terminus of the protein, as a series of gene fusions with the *lacZ* operon in a plasmid vector gave a molecular weight of 44 000 daltons for the smallest fusion with enzymatic activity against pullulan (Figure 2.5).

It is of interest to compare the predicted amino acid sequences of pullulanases, neopullulanase, isoamylase and α-amylases from selected organisms, particularly since the crystallographic structure is known for two α-amylases, one from pig pancreas and the other from the fungus, *Aspergillus oryzae*. Such comparisions show little overall conservation of homology except for three highly conserved regions of 5–8 amino acids at approximately amino acid positions 100, 200 and 300 (shown in Figure 2.6). There are also three weakly conserved regions around positions 50, 70 and 230. These homology relationships are conserved with the *amyB* and *amyC* gene products from *Dictyoglomus* described earlier in this chapter. However, the *amyA* product does not contain the conserved 100, 200 and 300 regions and may

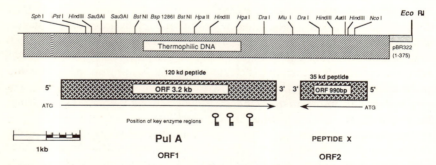

Figure 2.5   Restriction map and open reading frames in a fragment of thermophilic DNA from *Caldocellum saccharolyticum* encoding a heat-stable pullulanase. The entire fragment has been sequenced (G. Albertson, PhD thesis, University of Auckland).

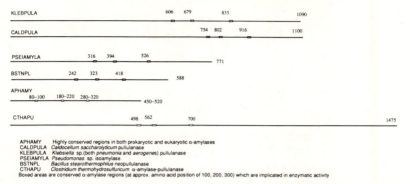

APHAMY   Highly conserved regions in both prokaryotic and eukaryotic α-amylases
CALDPULA *Caldocellum saccharolyticum* pullulanase
KLEBPULA *Klebsiella* sp.(both *pneumonia* and *aerogenes*) pullulanase
PSEIAMYLA *Pseudomonas* sp. isoamylase
BSTNPL *Bacillus stearothermophilus* neopullulanase
CTHAPU *Clostridium thermohydrosulfuricum* α-amylase-pullulanase
Boxed areas are conserved α-amylase regions (at approx. amino acid position of 100, 200, 300) which are implicated in enzymatic activity

Figure 2.6 Diagrammatic representation of the relative location of the three highly conserved regions of α-amylases extended to other glucosidases.

not be a true amylase as the protein is much larger than other amylases and close examination shows that none of the standard α-amylase regions is present. Furthermore, it has no protein sequence homology with any characterized enzymes in the GenBank database.

The three-dimensional structures of the pig and fungal α-amylases have similar structural elements. They have α-helices and β-sheets in a similar arrangement and position as well as an $(\alpha/\beta)_8$ barrel structure. In both cases the three conserved regions are associated around the active site cleft and/or the Ca^{2+}-binding site. The 100 and 200 regions consist of two parallel chains of adjacent amino acids running from the surface to the interior of the protein along one side of the active site cleft which is 'stapled' together by the binding of amino acids from the 100 region (asparagine, N) and the 200 region (histidine, H) to the essential Ca^{2+}. On the other side of the active cleft, two amino acids from the 300 region protrude from the interior of the protein to the surface of the active cleft. Substrate-binding studies have shown that two aspartic acid (D) side-chain residues, one from each of the 200 and 300 regions, are involved in catalytic activity. Other amino acids on the surface of the active cleft are involved with substrate binding, notably glutamic acid (E) at position 230 (reviewed in refs. 123–125).

Protein sequence comparisons of debranching enzymes such as pullulanase, isoamylase and neopullulanase show three conserved regions corresponding to the α-amylase 100, 200 and 300 regions. Although these regions are in the same order and have the same spacing (approximately 100 amino acids apart), the first region is found towards the middle of the amino acid sequence, in comparison to α-amylase, where it is found near the beginning of the sequence. The degree of conservation of the consensus α-amylase 100, 200 and 300 regions in the debranching enzymes is remarkable (Figure 2.7). Some sections are highly conserved, for example all of the amino acids of the 100 region, the 200 region catalytic site and its adjacent amino acids and the 300 region

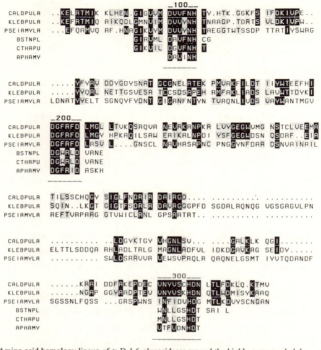

Amino acid homology lineup of α-D-1,6-glucosidases around the highly conserved alpha amylase regions (boxed areas). Also included are the same regions from other related glucosidases. Identical amino acids with respect to *C.saccharolyticum* pullulanase are shown in black reverse font; whereas similar amino acids are in grey background font.

APHAMY	Highly conserved regions in both prokaryotic and eukaryotic α-amylases
CALDPULA	*Caldocellum saccharolyticum* pullulanase
KLEBPULA	*Klebsiella* sp.(both *pneumoniae* and *aerogenes*) pullulanase
PSEIAMYLA	*Pseudomonas* sp. isoamylase
BSTNPL	*Bacillus stearothermophilus* neopullulanase
CTHAPU	*Clostridium thermohydrosulfuricum* α-amylase-pullulanase

Box areas are conserved α-amylase regions (at approx. amino acid position of 100, 200, 300) implicated in enzymatic activity

Figure 2.7 Amino acid sequence comparison of α-D-1, 6-glucosidase around the three highly conserved regions shown by α-amylases.

catalytic site amino acid. The 200 region Ca^{2+}-binding amino acids and the non-protruding amino acids adjacent to the 300 region catalytic site are not conserved and are different for each type of enzyme.

Since the three pullulanase sequences (*Klebsiella aerogenes* [73], *Caldocellum saccharolyticum* (G. Albertson, R.M. McHale and P.L. Bergquist, unpublished), *K. pneumoniae* [126] are identical in these regions and have a greater similarity to isoamylase than neopullulanase, we suggest that the non-conserved regions determine the nature of the enzyme activity—exactly how it cleaves pullulan to give maltotriose, panose or isopanose as end-products is not known. Alteration of the 200 region Ca^{2+}-binding region and the 300 region non-protruding region amino acids by cassette or PCR mutagenesis to produce a hybrid enzyme should allow the construction of an isoamylase with temperature stability and temperature optimum of the '*Caldocellum*' pullulanase—in short, a pullulanase that debranches glycogen. Our current experiments are designed to test this possibility using PCR mutagenesis.

2.4 Other enzymes from anaerobic thermophiles

The production of sweeteners from corn starch is an important application of enzyme technology in the food industry which has a requirement for thermostable enzymes. Of particular interest is the enzyme glucose isomerase, more thermostable variants of which would be desirable.

Zeikus *et al.* [127] have shown that this enzyme is common in amylolytic thermophiles and have developed a single-step process for the conversion of starch or lactose directly into fructose mixtures using cultures of *Thermoanaerobacter* pregrown under controlled conditions [128]. The same group has cloned the glucose isomerase of *C. thermosulfurogenes* into *E. coli* and then into a *B. subtilis* host. The enzyme was produced at high level from its own promoter in both organisms, and fructose production was observed at temperatures higher than those used in current industrial practice. This enzyme may have biotechnological promise because of its ability to be produced in a food-safe organism, its high temperature optimum allowing stability and an improved equilibrium towards a high-fructose equilibrium [127].

The alcohol dehydrogenases of some ethanologenic thermophiles have activity on a very diverse range of substrates, including primary and secondary alcohols, and linear and cyclic ketones. Bryant *et al.* [29] have reported two distinct forms of alcohol dehydrogenase in *Thermoanaerobacter ethanolicus*. While both are NADP-dependent, they are formed at different times in culture growth and have quite distinct substrate preferences. Zeikus *et al.* [27] have proposed that thermophilic alcohol dehydrogenases may have application as industrial ketone/aldehyde/alcohol oxidoreductases given their activity at high temperatures and substrate concentrations and their ability to form speciality chiral compounds.

2.4.1 Thermotoga

The Thermotogales represent a relatively new and exciting group of thermophilic eubacteria. *Thermotoga* was first isolated in 1986; several species have since been reported, but all have in common a growth temperature optimum of 80–85°C. This fact means that enzymes with even greater thermostability can be obtained [130]. It seems likely that the Thermotogales represent the oldest lineage of eubacterial descent [131]. Woese [132] has contended that where organisms live in an extreme and fluctuating environment, then enzyme specificity and control might be relaxed as a means of survival. If so, the enzymes of the Thermotogales may possess interesting and useful properties. *Thermotoga* produces extremely stable proteins with cellulase and xylanase activity yet is unable to grow or apparently metabolise these substrates [130]. Very little cloning of *Thermotoga* DNA has occurred. The first gene cloned from *Thermotoga maritima* was the 16S rRNA gene, and sequence comparisons with other eubacterial 16S rRNA sequences reinforced the unique characteristics of its cell wall and lipids in confirming that it is one of the more unusual eubacteria. The *Thermatoga* elongation factor Tu has been cloned and synthessed in *E. coli*. It was able to bind GDP only after heating to 65°C or more; presumably it is incorrectly folded when synthesised in *E. coli* at 37°C [133]. It might be expected that proteins such as the Tu factor would be expressed in alternative hosts because of its essential function in protein synthesis. Expression of other *Thermotoga* proteins may not be commonplace. Gene libraries of *Thermotoga* have been examined for the expression of cellulases, β-galactosidase and xylanase with negative results, and probes made from genes for these enzymes cloned from the thermophile *Caldocellum saccharolyticum* show no significant homology to *Thermatoga* genomic DNA in Southern blot experiments (R.M. McHale, R. Reeve and P.L. Bergquist, unpublished).

References

1. T.D. Brock, *Thermophilic Microorganisms and Life at High Temperatures*, Springer-Verlag (1978) 465 pp.
2. M.A. Innis, D.H. Gelfand, J.J. Sinsky and T.J. White (eds.), *PCR Protocols: A Guide to Methods and Applications*, Academic Press, San Diego (1990) 482 pp.
3. L.R. Lynd, H.E. Grethlen and R.H. Wolkin, *Appl. Environ. Microbiol.* **55** (1989) 3131.
4. G.E. Slapack, I. Russell and G.G. Stewart, *Thermophilic Microbes in Ethanol Production*, CRC Press, Florida (1987) 186 pp.
5. M. Oka, Y.-S. Yang, S. Nagata, N. Esaki, H. Tanaka and K. Soda, *Biotechnol. Appl. Biochem.* **11** (1989) 307.
6. M.L. Patchett, T.L. Neill, L.R. Schofield, R.C. Strange, R.M. Daniel and H.W. Morgan, *Enzyme Microb. Technol.* **11** (1989) 113.
7. M. Klingeberg, H. Hippe and G. Antranikian, *FEMS Microbiol, Lett.* **69** (1990) 145.
8. P.H. Janssen, H.W. Morgan and R.M. Daniel, *Appl. Microbiol. Biotechnol.* **34** (1991) 789.
9. M.S. Da Costa, J.C. Duarte and R.A.D. Williams, *FEMS Symposium No. 49*, Elsevier Applied Science, Amsterdam (1989) 429 pp.
10. G.J. Wiegel and L.G. Ljungdahl, *Crit. Rev. Biotechnol.* **3** (1986) 39.
11. T.D. Brock, in *Bergey's Manual of Systematic Bacteriology*, Vol. 1, ed. N.R. Kelly, Williams & Wilkins, Baltimore (1984) 333.
12. M.E. Bateson, K.J. Thibault and D.M. Ward, *System. Appl. Microbiol.* **13** (1990) 8.
13. R. Sharp, M. Munster, A. Vivian, S. Ahmad and T. Atkinson, in *Microbiology of Extreme Environments and its Potential for Biotechnology*, eds. M.S. Da Costa, J.C. Duarte and R.A.D. Williams, Elsevier Applied Science, Amsterdam (1989) 62.
14. P.L. Bergquist, D.R. Love, J.E. Croft, M.B. Streiff, R.M. Daniel and H.W. Morgan, *Biotech. Genet. Eng. Rev.* **5** (1987) 199.
15. K.B. Mullis and F. Faloona, *Methods Enzymol.* **155** (1987) 335.

16. E.Y. Loh, J.F. Elliot, S. Cwirla, L.L. Lanier and M.D. Davis, *Science* **243** (1989) 217.
17. F. Vallette, E. Mege, A. Reiss and M. Adesnik, *Nucleic Acids Res.* **17** (1989) 723.
18. T.A. Bechtereva, Y.I. Pavlov, V.I. Kramorov, B. Migunova and O.I. Kiselev, *Nucleic Acids Res.* **17** (1989) 10507.
19. F.C. Lawyer, S. Stoffel, R.K. Saiki, K. Myambo, R. Drummond and D.H. Gelfand, *J. Biol. Chem.* **264** (1989) 6427.
20. B.E. Slatko, J.S. Benner, T. Jager-Quinton, L.S. Moran, T.G. Simcox, E.M. Van Cott and G.G. Wilson, *Nucleic Acids Res.* **15** (1987) 9781.
21. J.E. Croft, D.R. Love and P.L. Bergquist, *Mol. Gen. Genet.* **210** (1987) 490.
22. K.R. Tindall and T.A. Kunkel, *Biochemistry* **27** (1988) 6008.
23. E. Lüthi, N.B. Jasmat and P.L. Bergquist, *Appl. Env. Microbiol.* **56** (1990) 2677.
24. D.H. Jones and B.H. Howard, *Biotechniques* **8** (1990) 178.
25. D.E. Kellog and S. Kwok, in *PCR Protocols: A Guide to Methods and Applications*, eds. M.A. Innis *et al.*, Academic Press, San Diego (1990) 337.
26. Y. Ting and M.M. Manos, in *PCR Protocols: A Guide to Methods and Applications*, eds. M.A. Innis *et al.*, Academic Press, San Diego (1990) 356.
27. S.-H. Park, in *PCR Protocols: A Guide to Methods and Applications*, eds. M.A. Innis *et al.*, Academic Press, San Diego (1990) 497.
28. D. Shibata, in *PCR Protocols: A Guide to Methods and Applications*, eds. M.A. Innis *et al.*, Academic Press, San Diego (1990) 368.
29. F. Barany, *Gene* **65** (1988) 167.
30. D.J. Nicholls, T.K. Sundaram, T. Atkinson and N.P. Minton, *Nucleic Acids Res.* **16** (1988) 9858.
31. L. Seidler, M. Peter, F. Meissner and M. Sprinzl, *Nucleic Acids Res.* **15** (1987) 9263.
32. A.V. Yakhnin, D.P. Vorozheykina and N.I. Matvienki, *Nucleic Acids Res.* **18** (1990) 3659.
33. K. Kunai, M. Machida, H. Matsuzawa, T. Miyazawa and T. Ohta, *Eur. J. Biochem.* **160** (1986) 433.
34. M. Ono, H. Matsuzawa and T. Ohta, *J. Biochem.* **107** (1990) 21.
35. N. Sashihara, N. Nakamura, H. Nagayama and K. Horikoshi, *FEMS Microbiol. Lett.* **49** (1988) 385.
36. D.A. Cowan, R.M. Daniel and H.W. Morgan, *FEMS Microbiol. Lett.* **43** (1987) 155.
37. D.A. Cowan, R.M. Daniel and H.W. Morgan, *Trends Biotechnol.* **3** (1985) 68.
38. C.W. Jones, H.W. Morgan and R.M. Daniel, *J. Gen. Microbiol.* **134** (1988) 191.
39. P. Kanasawud, S. Hjorleifsdotir, O. Holst and B. Mattiason, *Appl. Microbiol. Biotechnol.* **31** (1989) 228.
40. R.H. McHale, P.M. Stapleton and P.L. Bergquist, *Biotechniques* **10** (1991) 20.
41. S.-T. Kwon, I. Terada, H. Matsuzawa and T. Ohta, *Eur. J. Biochem.* **173** (1988) 491.
42. I. Terada, S.-T. Kwon, Y. Miyata, H. Matsuzawa and T. Ohta, *J. Biol. Chem.* **265** (1990) 6576.
43. H. Takagi, T. Takahashi, H. Momose, M. Inouye, Y. Moeda, H. Matsuzawa and T. Ohta, *J. Biol. Chem.* **265** (1990) 6874.
44. J.A. Wells and D.A. Estell, *Trends Biochem. Sci.* **13** (1988) 291.
45. B. Kroger, T. Specht, R. Lurz, N. Ulbrich and V.A. Erdmann, *FEMS Microbiol. Lett.* **50** (1988) 61.
46. Y. Koyama and K. Furukawa, *J. Bacteriol.* **172** (1990) 3490.
47. Y. Koyoma, S. Okamoto and K. Furukawa, *Appl. Environ. Microbiol.* **56** (1990) 2251.
48. Y. Koyama, Y. Arikawa and K. Furukawa, *FEMS Microbiol. Lett.* **72** (1990) 97.
49. S. Sen and P. Oriel, *FEMS Microbiol. Lett.* **67** (1990) 131.
50. M.K. Ashby and P.L. Bergquist, *Plasmid* **24** (1990) 1.
51. R.K. Hartmann, N. Ulbrich and V.A. Erdmann, *Biochimie* **69** (1987) 1097.
52. R.M. Burke and D.W. Tempest, *J. Gen. Microbiol.* **136** (1990) 1381.
53. T. Takami, T. Akiba and K. Horikoshi, *Appl. Microbiol. Biotechnol.* **30** (1989) 120.
54. B.S. Hartley and M.A. Payton, in *Biochem. Soc. Symp. 48*, eds. C.F. Phelps and P. Clarke, Biochemical Society (1983).
55. K.I. Ishiwata, S. Yoshino, S. Iwamori, T. Suzuki and N. Makiguchi, *Agric. Biol. Chem.* **53** (1989) 1941.
56. M. Kubo and T. Imanaka, *J. Gen. Microbiol.* **134** (1988) 1883.
57. P. Argos, M.G. Rossmann, U.M. Grau, H. Zuber, G. Frank and J.P. Tratschin, *Biochemistry* **18** (1979) 5698.

58. F. Zulli, H. Weber and H. Zuber, *Biol. Chem. Hoppe-Seyler* **368** (1987) 1167.
59. F. Zulli, H. Weber and H. Zuber, *Biol. Chem. Hoppe-Seyler* **371** (1990) 655.
60. C. Branlant, T. Oster and G. Branlant, *Gene* **75** (1989) 145.
61. T. Kuriki, J.-H. Park, S. Okada and T. Imanaka, *Appl. Environ. Microbiol.* **54** (1988) 2881.
62. C.Y. Cheng, I. Yabe and K. Toda, *Appl. Microbiol. Biotechnol.* **30** (1989) 125.
63. H. Ihara, T. Sasaki, A. Tsuboi, H. Yamagata, N. Tsukagoshi and S. Udaka, *J. Biochem.* **98** (1985) 95.
64. S. Sen and P. Oriel, *Biotechnol. Lett.* **11** (1989) 383.
65. A. Sohma, T. Fujita and K. Yamane, *J. Gen. Microbiol.* **133** (1987) 3271.
66. S.J. Tomazic and A.M. Klibanov, *J. Biol. Chem.* **263** (1988) 3086.
67. Y. Suzuki, N. Ito, T. Yuuki, H. Yamagata and S. Udaka, *J. Biol. Chem.* **264** (1989) 18933.
68. S.J. Tomazic and A.M. Klibanov, *J. Biol. Chem.* **263** (1988) 3092.
69. M. Vihinen and P. Mantsala, *Crit. Rev. Biochem. Mol. Biol.* **24** (1989) 329.
70. T. Kuriki, J.-H. Park and T. Imanaka, *J. Ferment. Bioeng.* **69** (1990) 204.
71. T. Kuriki, S. Okada and T. Imanaka, *J. Bacteriol.* **170** (1988) 1554.
72. T. Imanaka and T. Kuriki, *J. Bacteriol.* **17** (1989) 369.
73. R. Nakajima, T. Imanaka and S. Aiba, *J. Bacteriol.* **163** (1985) 401.
74. N. Katsuragi, N. Takizawa and Y. Murooka, *J. Bacteriol.* **169** (1987) 2301.
75. S. Sakai, M. Kutota, K. Yamamoto, T. Nakada, K. Torigoe, O. Ando and T. Sugimato, *J. Japanese Society of Starch Science* **34** (1987) 140.
76. P. Béguin, P. Cornet and J.-P. Aubert, *J. Bacteriol.* **162** (1985) 102.
77. I. Yamashita, K. Suzuki and S. Fukui, *J. Bacteriol.* **161** (1985) 567.
78. E. Soutschek-Bauer and W.L. Staudenbauer, *Mol. Gen. Genet.* **208** (1987) 537.
79. M. Zhang, H. Nakai and T. Imanaka, *Appl. Environ. Microbiol.* **54** (1988) 3162.
80. L. Wu and N.E. Welker, *J. Gen. Microbiol.* **135** (1989) 1315.
81. P. Tailliez, H. Girard, J. Millet and P. Béguin, *Appl. Environ. Microbiol.* **55** (1989) 207.
82. K. Hon-nami, M.P. Coughlan, H. Hon-nami and L.G. Ljungdahl, *Arch. Microbiol.* **145** (1985) 13.
83. G.P. Hazlewood, M.P.M. Romaniec, K. Davidson, O. Grépinet, P. Béguin, J. Mollet, O. Raynaud and J.-P. Aubert, *FEMS Microbiol. Lett.* **51** (1988) 231.
84. T.M. Wood, I.S. McCrae and K.M. Bhat, *J. Biochem.* **260** (1989) 37.
85. F.A. Rainey, P.H. Janssen, D.J.C. Wild and H.W. Morgan, *Arch. Microbiol.* **155** (1991) 396.
86. P. Béguin, *Ann. Rev. Microbiol.* **44** (1990) 219.
87. J. Hall, G.P. Hazlewood, P.J. Barker and H.J. Gilbert, *Gene* **69** (1988) 29.
88. W.H. Schwarz, S. Schimming, K.P. Rücknagel, S. Burgschwaiger, G. Kreil and W.L. Staudenbauer, *Gene* **63** (1988) 23.
89. W.H. Schwarz, S. Schimming and W.L. Staudenbauer, *Appl. Microbiol. Biotechnol.* **27** (1987) 50.
90. K. Sakka, S. Furuse and K. Shimada, *Agric. Biol. Chem.* **53** (1988) 905.
91. D. Saul, L. Williams, R. Grayling, L.W. Chamley, D.R. Love and P.L. Bergquist, *Appl. Env. Microbiol.* **56** (1990) 3117.
92. N.R. Gilkes, R.A.J. Warren, R.C. Miller Jr and D.G. Kilburn, *J. Biol. Chem.* **263** (1988) 10401.
93. W.H. Schwarz, S. Jauris, M. Kouba, K. Bronnenmeier and W.L. Staudenbauer, *Biotechnol. Lett.* **11** (1989) 461.
94. P. Béguin, N.R. Gilkes, D.G. Kilburn, R.C. Miller Jr., G.P. O'Neill and R.A.J. Warren, *CRC Crit. Rev. Biotechnol.* **2** (1987) 129.
95. D.R. Love and M.B. Streiff, *Biotechnology* **5** (1987) 384.
96. D.R. Love, R. Fisher and P.L. Bergquist, *Mol. Gen. Genet.* **213** (1988) 84.
97. A.R. Plant, R. Clemens, H.W. Morgan and R.M. Daniel, *Biochemical J.* **246** (1987) 537.
98. F. Gräbnitz, K.P. Rücknagel, M. Seib and W.L. Staudenbauer, *Mol. Gen. Genet.* **217** (1989) 70.
99. S.K. Kadam and A.L. Demain, *Biochem. Biophys. Res. Comm.* **161** (1989) 706.
100. P. Biely, *Trends Biotechnol.* **3** (1985) 286.
101. R.F.H. Dekker, in *Biosynthesis and Biodegradation of Wood Components*, ed. T. Higuchi, Academic Press, New York (1985) 505.
102. C.R. MacKenzie, R.C.A. Yang, G.B. Patel, D. Bilous and S.A. Narang, *Arch. Microbiol.* **152** (1989) 377.
103. E. Lüthi, D.R. Love, J. McAnulty, C. Wallace, P.A. Caughey, D. Saul and P.L. Bergquist, *Appl. Env. Microbiol.* **56** (1990) 1017.

104. O. Grépinet, M.-C. Chebrou and P. Béguin, *J. Bacteriol.* **170** (1988) 4582.
105. O. Grépinet, M.-C. Chebrou and P. Béguin, *J. Bacteriol.* **170** (1988) 4576.
106. C.A. West, A. Elzanowski, L.S. Yeh and W.C. Barber, *FEMS Microbiol. Lett.* **59** (1989) 167.
107. E. Lüthi and P.L. Bergquist, *FEMS Microbiol. Lett.* **67** (1990) 291.
108. E. Lüthi, N.B. Jasmat and P.L. Bergquist, *Appl. Microbiol. Biotechnol.* **34** (1990) 214.
109. E. Lüthi, N.B. Jasmat, R.A. Grayling, D.R. Love and P.L. Bergquist, *Appl. Env. Microbiol.* **57** (1991) 674.
110. T. Akino, C. Kato and K. Horikoshi, *Appl. Env. Microbiol.* **55** (1989) 3178.
111. A. Kantelinen, M. Ratto, J. Sunquist, M. Ranua, L. Vikari and M. Linko, in *International Bleaching Conference, Orlando, FL, TAPPI Proceedings*, ed. T.J. de Salvo, Technical Association of the Pulp and Paper Industry, Atlanta (1988) 1–4.
112. G. Antranikian, *FEMS Microbiol. Rev.* **75** (1990) 201.
113. G. Antranikian, P. Zablowski and G. Gottschalk, *Appl. Microbiol. Biotechnol.* **27** (1987) 75.
114. M. Klingeberg, K.D. Vorlop and G. Antranikian, *Appl. Microbiol. Biotechnol.* **33** (1990) 494.
115. S. Fukusumi, A. Kamizono, S. Horinouchi and T. Beppu, *Eur. J. Biochem.* **174** (1988) 15.
116. S. Horinuchi, S. Fukusumi, T. Ohshima and T. Beppu, *Eur. J. Biochem.* **176** (1988) 243.
117. N. Kitamoto, H. Yamagata, T. Kato, N. Tsukagoshi and S. Udaka, *J. Bacteriol.* **170** (1988) 5848.
118. K. Haeckel and H. Bahl, *FEMS Microbiol. Letts.* **60** (1989) 333.
119. H.H. Hyun and J.G. Zeikus, *Appl. Environ. Microbiol.* **49** (1985) 1168.
120. A.R. Plant, J.E. Oliver, M.L. Patchett, R.M. Daniel and H.W. Morgan, *Arch. Biochem. Biophys.* **262** (1988) 181.
121. H. Melasniemi and M. Paloheimo, *J. Gen. Microbiol.* **135** (1989) 1755.
122. H. Melasniemi, M. Paloheimo and L. Hennio, *J. Gen. Microbiol.* **136** (1990) 447.
123. G. Buisson, E. Duée, R. Haser and F. Payan, *EMBO J.* **6** (1987) 3909.
124. T. Kuriki and T. Imanaka, *J. Gen. Microbiol.* **135** (1989) 1521.
125. E.A. MacGregor and B. Svensson, *Biochem. J.* **259** (1989) 145.
126. M.G. Kornacker and A.P. Puglsey, *Molec. Microbiol.* **4** (1990) 73.
127. J.G. Zeikus, S.E. Lowe and B.C. Saha, in *Biocatalysis*, ed. D.A. Abramowicz, Catalysis Series, Van Nostrand Reinhold (1990) 243
128. C. Lee, L. Bhatnagar, B.C. Saha, Y.-E. Lee, M. Takago, T. Imanaki, M. Bagdasarian and J.G. Zeikus, *Appl. Environ. Microbiol.* **56** (1990) 2638.
129. F.O. Bryant, J. Wiegel and L.G. Ljungdahl, *Appl. Environ. Microbiol.* **54** (1988) 460.
130. R.M. Daniel, J. Bragger and H.W. Morgan, in *Biocatalysis*, ed. D.A. Abramowicz, Catalysis Series, Van Nostrand Reinhold (1990) 243.
131. L. Achenbach-Richter, R. Gupta, K.O. Stetter and C.R. Woese, *Syst. Appl. Microbiol.* **9** (1987) 34.
132. C.R. Woese, *Microbiol. Rev.* **51** (1987) 221.
133. O. Tiboni, A.M. Sanangelantoni, P. Cammarano, P. Cimino, G. Di Pasquale and S. Sora, *Syst. Appl. Microbiol.* **12** (1989) 127.

3 Biotechnological prospects for halophiles and halotolerant micro-organisms

E.A. GALINSKI and B.J. TINDALL

3.1 Introduction

The role of salt in our everyday lives is well established as a food flavouring and as a way of preventing microbial growth. However, little attention is usually paid to the wealth of micro-organisms which live at elevated salt concentrations.

Organisms which survive in elevated salt concentrations are either halo-tolerant or halophilic. Much attention has been paid to defining halotolerant or halophilic forms of life properly, and the various degrees, extreme, moderate, etc. [1–3]. To the authors' knowledge no satisfactory definition exists which adequately delineates all such groupings. It is not the purpose of this review to discuss at length these definitions, and the reader is referred to the articles by Larsen [1, 2], Kushner [3], and Rodriguez-Valera [4], where this is discussed in more depth. Suffice to say that an organism which does not grow in the absence of salt has an optimum at 3% (w/v) salt but tolerates 20% (w/v) salt may be described as both slightly or moderately halophilic and extremely halotolerant. In describing the exploitation of micro-organisms which survive in salt emphasis has been placed on processes for which the presence of salt is essential. As a consequence the majority of micro-organisms described are to some degree halophilic (slight, moderate, extreme) and halotolerant.

The concept of the occurrence of saline environments is largely coloured by knowledge of the current distribution of such environments. However, the geological record in various parts of the world shows quite clearly that hypersaline waters once covered vast areas. This is particularly evident from the large deposits of salt found over large areas of Europe and North America. While salt recovered by mining operations is referred to as 'rock salt', and that by the evaporation of salt water 'solar salt', this does not reflect the fact that the majority of salt deposits have arisen by evaporation of saline bodies of water. Thus modern-day evaporitic environments and geological evaporitic deposits have arisen by similar processes, not only indicating a geochemical continuity, but also suggesting that the presence of such environments over long periods of (geological) time has provided equal opportunity for the evolution of halotolerant/halophilic forms of life [5, 6]. Saline lakes are widely distributed, and the geochemical processes leading to the formation of particular brines has been reviewed by Eugster and Hardie [7] and Hardie and Eugster [8]. The commonest hypersaline bodies of water may be divided into three categories, those derived from marine sources, the alkaline saline waters (low in magnesium and calcium), and the magnesium-rich saline bodies of water. As such the classical view that the saline environment is limited to a neutral pH system with an ionic composition reflecting genesis from sea water does not reflect the true picture of the geochemical, and hence biological, diversity of these environments. In recent years a number of reviews have covered the topics of the physiology, biochemistry and taxonomy of halophiles and the ecology of saline lakes, however it is only recently that interest has turned to the exploitation of organisms from such environments for commercial purposes.

3.2 Micro-organisms in the food industry

3.2.1 Food spoilage

In the last two decades halophilic micro-organisms have attracted more attention, and have lost to some degree their label as exotic forms of life. It is, however, usually forgotten that halotolerant and halophilic micro-organisms were first investigated because of their importance in the food industry. Before the large-scale manufacture of modern refrigeration plants vast quantities of fish, meat and untreated hides (for the leather industry) were treated with salt in order to prolong their shelf life. Under adverse conditions (usually in hot weather when the fish, meat or hide was still damp) it was not uncommon to find areas of putrefication, which were often characterised by a distinctive red/pink colour as a result of the growth of aerobic, extremely halophilic archaeobacteria (for reviews see refs. 1 and 9). The major deleterious effect seems to have been due to the pronounced proteolytic activity of the organisms involved. The study of these organisms, which began at the end of the last century, centred largely on one of two problems, their potential involvement in cases of food poisoning and ways of preventing the growth of such halotolerant or halophilic micro-organisms. Although such problems have largely been overcome by the use of kiln-dried salt or by cold storage, in certain countries where solar salt is still used for salting meat or fish products these problems may still be encountered (ref. 10 and M. da Costa, personal communication).

Problems may continue to arise in certain areas of the food industry where meat products are still salted by immersion in brines or layering with salt. While the growth of members of the family Halobacteriaceae is less common, it appears that other halotolerant/halophilic organisms may also play a role. Members of the genera *Halomonas/Deleya* have been recovered from salting brines, and the ability of the strains recovered to reduce nitrate to nitrite may be a potential problem for the food industry. A variety of other organisms have been isolated from such brines, although it is unclear to what extent the various species may contribute to adverse effects. Certainly the growth of such organisms on the salted product is considered to be detrimental.

3.2.2   Fermentation products

In the western world the emphasis on the use of salt has generally been as a preserving agent to prevent the deterioration of the product. While the use of salt to prevent the development of adverse microbial growth (those producing unwanted flavours or pathogens) is also common in the Orient, a closer examination of salted foods typical of this region indicates the importance of halotolerant and halophilic micro-organisms in Oriental cuisine. The commonest examples are salted fermented vegetables, salted fermented sea food and, perhaps the best known, soy sauce. Although many of these foods or flavouring agents have been produced for centuries, first as a cottage industry, but more recently on an industrial scale, in all but a few cases very little is known about the types of organisms involved and their role and significance. In some cases studies have identified the organisms present, although caution should be exercised since the possibility that older studies may have used inadequate methods to identify the organisms involved cannot be excluded. An extensive review of the literature relating to the production of indigenous fermented foods has been published [11].

Often accepted as one of the national foods of Germany, sauerkraut, a salted cabbage product subjected to acid fermentation, appears to have its origins further east. A well studied process, sauerkraut is produced by adding approximately 2.5% (w/v) salt to shredded cabbage, which is pressed and fermented. Once a cottage industry and produced in special earthenware jars with a close-fitting stone weight, commercial production is far more significant today. Microbiological studies have shown that the succession of halotolerant bacteria *Leuconostoc mesenteroides–Pediococcus cerevisiae–Lactobacillus plantarum* is essential. While every household probably had its own recipe for sauerkraut, and the salt content and fermentation conditions gave rise to different flavours, we know that salt concentration is important in inhibiting the development of spoilage organisms. Equally, increasing the salt concentration prolongs or even prevents fermentation.

One of the oldest uses of halotolerant micro-organisms in the preparation of foods, which may be traced back thousand years, is the production of soy sauce [11]. Soy sauce production traditionally relies on the hydrolysis of soya beans by the fungus *Aspergillus oryzae*, followed by an anaerobic fermentation, which probably plays a role in flavour development. Although some modern soy sauces incorporate soya beans hydrolysed by chemical means, the flavour is considered to be inferior. In using the term soy sauce it should be remembered that the method of production and the subtle differences in flavour arising from variations in the method of preparation make the types of soy sauce almost as diverse as beer. An extensive review of the different types of soy sauce and the variations in the methods of production can be found in ref. [11]. In general the beans are cooked and then inoculated with a starter culture of *Aspergillus oryzae*. Depending on the method used, salt is added either before addition of the inoculum or afterwards. Once hydrolysis of the beans has begun, the second stage in the production of fermented soy sauce, the anaerobic fermentation by halophilic bacteria and yeasts, can begin. In some cases the organisms involved in the anaerobic fermentation are controlled by inoculation, while in other cases the fermentation proceeds by selection of suitable halophilic 'contaminants' in the vats. A detailed analysis of the composition of soy sauces may be found in ref. [11], and it may be assumed that a diversity of compounds contribute to the flavour and aroma of soy sauce. It is interesting that the modern food industry uses monosodium glutamate and nucleotides as flavouring agents, both of which play a role in the flavour of soy sauce. The salt concentration of soy sauces varies from 12 to 27% (w/v) depending on the type being produced.

Apart from being used to produce soy sauce, soya beans may also be used to produce soy pastes. As in the production of the sauces, soy pastes are produced by a combined hydrolysis of the beans using *Aspergillus oryzae*, followed by a bacterial/yeast fermentation. The addition of salt also plays a role in the selection of organisms capable of colonising the hydrolysed beans, since production on the cottage industry scale relies on 'natural contamination' to a certain extent. Although various halotolerant yeasts, including *Saccharomyces rouxii* and *Torulopsis versitilis*, and halotolerant bacteria like *Pediococcus halophilus*, *Enterococcus faecalis*, and *Lactobacillus delbruckii* have been reported, some caution should be exercised concerning the identification of certain organisms.

Apart from the production of salted soya bean and other vegetable products, other significant salted products in the oriental cuisine are the fish and shrimp pastes. The use of different fish and shrimps appears to be dependent not only on the area but also on the time of year. In general, these types of pastes are prepared from fish or shrimps which would otherwise not be easy to process because of their relatively small size. It is, therefore, not uncommon to find that whole fish or shrimps are added to the fermentation vats. One may speculate that the addition of salt in such cases was initially a way of inhibiting the growth of pathogens and controlling the development of strains suitable for fermentation of the fish/shrimps. Judging by the reports from detailed accounts of the manufacture of such fish/shrimp pastes (for a review see ref. 11), there appears to be no artificially controlled hydrolysis step, nor does there appear to be any active control of the micro-organisms involved. It seems that the two most important sources of the micro-organisms involved in the production of such pastes are the fish/shrimps themselves and the salt added to the vats. Although Steinkraus [11] quotes evidence that the microflora of the gut contributes largely to the fermentation of these pastes, it is not clear if the role of organisms in the solar salt had been taken into consideration. Microbiological investigations have shown that a wide variety of organisms may be recovered from the different products, including *Bacillus* spp., *Pediococcus* spp., *Lactobacillus* spp., *Micrococcus* spp., *Staphylococcus* spp., *Clostridium* spp., *Saccharomyces* spp., *Torulopsis* spp., *Halococcus* spp., *Halobacterium* spp., *Brevibacterium*-like organisms, *Flavobacterium*-like organisms, and *Corynebacterium*-like organisms [11]. It is interesting to note that members of the archaeobacterial family Halobacteriaceae have been recovered from such pastes. Considering that the salt concentration ranges from 10 to 25% (w/v), and that certain members of this family are strongly proteolytic (see also section 3.2.1) it is hardly surprising to find that they may play a significant role in the production of these foods. In general, the production of these types of fish/shrimp pastes is carried out in large vats to which the whole or ground fish/shrimps and salt are added and left for periods ranging from several weeks to several months. Considering the inferred potential significance of halophilic and halotolerant species in the fermentation process, together with our better understanding of the diversity of such bacteria, a close study of the microbiology of such fermentations may yield interesting additional information.

3.2.3   *Single-cell protein (SCP)*

3.2.3.1 *Prokaryotes.* The production of microbial biomass as a source of protein is still conceivable on both the industrial scale (consider the British developments in the 1980s on methanol-utilising bacteria) and on a low-skill level employing mass developments in natural ponds. The use of *Spirulina* as a supplementary food has a long tradition. At the time of the Spanish conquest dried *Spirulina* cakes were sold for human consumption in the region of Lake Texcoco near Mexico City [12], and even today these cyanobacteria occasionally serve as food for the natives of the Lake Chad area. As cyanobacterial cell walls, like those of other phototrophic bacteria, are much softer and easier to digest than those of green algal species, and because of *Spirulina*'s unusually high protein content (approximately 60% as compared with only 40% of soybean meal), research has focused on the use of *Spirulina* as a source for single-cell protein (SCP). The amino acid spectrum of *Spirulina* protein is similar to that of other micro-organisms (i.e. somewhat deficient in methionine, cysteine and lysine) [13, 14]. Although it is inferior to standard alimentary protein like meat or milk, it is certainly superior to all plant proteins, including that from legumes. Owing to uric acid production in the course of purine metabolism and impending pathological side-effects such as gout, a major concern lies with the total nucleic acid content ($< 5\%$ of dry weight), which is lower than that of bacteria and yeasts [15, 16].

Extensive investigations have been conducted in Israel on the possibility of large-scale production of *Spirulina platensis* using brackish water unsuitable for agricultural purposes [17]. The organism is to date produced and marketed as a health food by Ein Yahav Algal Products (Arava Valley, Israel). Another established plant for large-scale production of *Spirulina* operating all year is that of Lake Texcoco (Mexico), where a daily production of 2 tons (dry weight) has been reported [15], commercialised mostly as a feed additive for animals or as health food. The fact that the organism seems to grow optimally in alkaline lakes at a salt concentration ranging from 20 to 70 g/l is advantageous because the cyanobacterial population becomes practically monospecific under these conditions [18, 19]. Alkaline pH also ensures that carbon dioxide is retained in the system and (in combination with salinity) drastically reduces growth of other organisms. Furthermore the formation of gas vacuoles promotes floating bacterial mats, which may be easily harvested. An open-pond farm exploiting the natural productivity of alkaline saline lakes seems to be the obvious solution for arid areas of the tropics where malnutrition is often endemic. The productivity (on a protein basis) is at least 100 times higher than crop production and 50 times higher than fish farming. With respect to the problem of uric acid formation consumption of *Spirulina* can, however, only be supplementary, and a maximum dietary intake (daily) of 0.3 g per kg body weight is recommended [20].

3.2.3.2 *Eukaryotes.* Early trials on algal biomass production (feed and health food) concentrated on freshwater algae like *Chlorella* and *Scenedesmus*. Serious attempts to utilise mass culture systems for marine microalgae have been initiated within the last 10 years, prompted by investigations into the marine food chain biotechnology and by the search for new industrially interesting products [21]. For the marine and hypersaline thalassic environment halophilic *Dunaliella* strains are potential producers of a valuable protein containing feed material (requiring 6–12‰ salt for optimal growth) [22, 23]. *Dunaliella* shows a remarkable degree of adaptation and can grow in media containing 0.2 to 35% (w/v, saturated) salt. Typical examples of natural unialgal cultures of *Dunaliella* are the Dead Sea [24], Pink Lake in Australia [25] and the Great Salt Lake in Utah, USA [26]. The protein of *Dunaliella* is similar in composition to other common plant protein such as soy bean meal but has a relatively high lysine content [27]. It is, thus, suitable as a feed for aquaculture (shrimp, crab, shellfish) either directly or through the culture of zooplankton. Besides an increasing use of *Dunaliella* products as chicken feed [28], applications in the health food sector will certainly profit from recent findings that β-carotene can lower the risk of certain cancers in humans [29, 30]. With respect to biomass processing, which is critical to break up the cells, *Dunaliella* offers the advantage that the dried biomass (because of the absence of a cell wall) is easily and fully digestible by animals and humans.

3.2.4   Food colouring/flavouring

β-Carotene, which is used in the food industry as both provitamin A and a natural food colouring agent, has a high market value [22]. Pilot plants for the commercial cultivation of *Dunaliella* have been constructed by Koor-Foods, Israel (2 hectares of pond area), as well as by two American and two Australian companies [23, 31]. Of the many strains of the genus *Dunaliella* only *D. bardawil* and *D. salina* have been shown to produce large amounts of β-carotene. The reason for the unusual accumulation of β-carotene is not yet clear; it may be a form of protection against high intensities of irradiation, which would explain the predominance of the above two species over green strains of *Dunaliella* in solar salt lakes exposed to high light intensities. *Dunaliella bardawil* in 3 M sodium chloride contains about 30% (w/v) glycerol as a compatible solute (compare section 3.3.5), 29% protein and 8% carotene [32] and thus provides three valuable commercial products. The observed productivity of *Dunaliella* (maximised for β-carotene production) has been at most 0.5 g of β-carotene per m^2 per day in short-term experiments, but long-term yearly yields averaged around 0.25 g m^{-2} day^{-1}, corresponding to approximately 5 g algal dry weight (β-carotene content around 5%) [23, 31]. Another salt-tolerant microalga (up to 1 M sodium chloride) which shows biotechnological promise for the food industry is the unicellular red alga *Porphyridium*, a suitable producer of the red food colouring pigment phycoerythrin and of arachidonic acid, a human dietary supplement [33]. As regards the importance of water activity as a parameter influencing the physiological state of an organism, models have been proposed for microbial systems where decreased water activity results in new metabolic behaviour [34, 35], and it was finally demonstrated (at least for a fungus and a yeast) that water activity enhances to a certain extent aroma production and aroma release [36]. These findings may well become valuable for future developments in the food industry.

3.3   Production of commercially useful compounds

3.3.1   Biological fermentation processes at high salinities

3.3.1.1 *Material corrosion.* Biological halofermentations on an industrial scale would require adaptation of production processes to saline fermentation media. The major problems associated with processes at high salinity are the inherent technical difficulties such as increased material corrosion [37]. Natural resistance against corrosion of steel alloys is based on the formation of a thin passive layer (metal oxides) on the metal surface. The most important component to promote this passivation of stainless steel is chromium (e.g. a minimum of 17% for austenitic steel). However, this fair resistance of 'normal' austenitic chromium–nickel steel to neutral salt solutions applies to plane surfaces only. Localised corrosion, which gives rise to pitting, becomes *the* critical factor for passivated stainless steel used in saline media, because the chloride anion effectively destroys the protecting passive layer. This type of corrosion drastically increases above a critical temperature and therefore prevents the use of standard 'stainless steel' for halofermentations. A measure of resistance against pitting is the so-called pitting potential [38, 39], which displays a strong temperature dependence. Further obstacles to the use of steel in contact with salt water are intercrystalline corrosion, crevice corrosion and stress corrosion cracking. Intercrystalline corrosion takes place when chromium carbide is formed and deposited at welding zones leaving too little chromium to form a stable passive layer. Very low carbon contents of 0.03% or less are therefore essential to avoid this type of corrosion. Stress corrosion cracking may be experienced in salt solutions at higher temperatures (autoclaving) with material under tension. Increasing proportions of nickel largely relieve the dangers of this type of corrosion. Last but not least, unstirred or poorly aerated solutions lead to the formation of anaerobic zones (e.g. in crevices, under seals and encrustations) and thereby destroy the protecting metal oxide layer. Hence, fermenter design (avoiding niches and crevices), fermentation process (solely aerobic) as well as service and maintenance play an important role to safeguard materials.

In summary, large-scale fermentations of halophilic micro-organisms require special materials which affect both the stability of the passive layer and dissolution tendency of the metal. They are best described by the following specification: austenitic steel of *high* chromium, nickel and molybdenum but *low* carbon content.

Table 3.1 compares the chemical composition and associated pitting potential of different types of steel considered as fermenter materials. Whereas Avesta 17-12-2.5 closely resembles the V4A-fermenter material commonly supplied with high quality fermenters on the laboratory scale, the other types of steel represent materials which have been successfully used in halo-fermentation (904L) [37], or have been designed for salt-water applications (254SMO). The unusually high content of molybdenum in 254SMO typically increases the material's resistance to pitting at higher temperatures (compare pitting potentials). The copper content further confers resistance to a non-oxidising environment, and the proportion of nitrogen influences the mechanical characteristics of the material. The increasing resistance (from top to bottom in Table 3.1) is, therefore, explained by the elevated proportion of chromium, nickel and molybdenum. It can further be concluded from published data (Table 3.1), as well as from our own investigations, that chromium–nickel alloys with a composition similar to steel type 1.4539 (904L) may be used in fermentations up to 60°C, whereas at higher temperatures an alloy of the 254SMO type (very high molybdenum content) would be more suitable.

Table 3.1 Chemical composition of austenitic chromium–nickel alloys, used for fermentation processes and pitting potential, a measure for the material's resistance (information taken from Avesta 8559 and 8560, Avesta, S-77401 Avesta, Sweden). Values in parentheses represent measurements in 1 M NaCl.

Type of steel			Chemical composition (%)						Relative pitting potential (mV) in 3.5% (w/w) sodium chloride			
Avesta	DIN	ASTM	Cr	Ni	Mo	C	N	Cu	25°C	50°C	70°C	90°C
17-12-2.5	1.4436	316	17	11	2.7	0.05	—	—	550 (510)	(130)	—	—
18-13-3L	1.4438	317L	18.5	13.5	3.2	0.03	—	—	(930)	(260)	—	—
904L	1.4539	N08904	20	25	4.2	0.02	—	1.5	1000 (1000)	950 (650)	500	—
254SMO	—	S31254	20	18	6.2	0.02	0.02	Trace	1000	1000	1000	870

Similarly deleterious effects on submerged measuring and regulating devices must also be kept to a minimum. Thus, metal fermenter components like impeller, pH and oxygen electrode housings, aeration tube, sparge ring, etc. should be protected by inert layers of, for example, Teflon. However, long-term experience is still needed to prove the protective value of these coatings on submerged equipment. To avoid undue wear of the bearing assembly through contact with crystalline salts, overhead stirring is advisable [37]. It can therefore be concluded that technical problems of increased corrosion in halofermentation can be solved using appropriate materials and taking the necessary precautions. However, whether these halofermentation techniques (use of corrosion-resistant materials in combination with stringent corrosion precautions) will ever become applicable on an industrial scale remains doubtful. It will ultimately depend on the commercial value of the potential product. Given the present rapid rate of development in genetic engineering, transferring genes of interest from halophilies into well-known production organisms may sooner or later become a matter of routine. Thus, halofermentation may be confined to experimental research laboratories mainly investigating the valuable gene pool of halophiles, from which industry will draw its resources for biotechnological applications using non-halophilic production strains.

3.3.1.2 *Natural ponds.* These prospects will, however, appear in a different light if one looks at developing countries. Here the use of low-cost alternatives such as open-pond farming may be a more likely application in the near future. Experience with outdoor collection and cultivation techniques so far has largely been confined to certain Asian countries employing highly productive macroalgae [e.g. Phaeophyceae, *Macrocystis* (giant kelp), *Undaria* and *Laminaria*; Rhodophyceae, *Porphyra*; Chlorophyceae, *Enteromorpha*, *Monostroma* and *Ulva*] [40]. However, research on outdoor shallow pond techniques is a relatively young but expanding effort and has mainly been performed with highly productive microalgae and cyanobacteria (e.g. *Chlorella* and *Spirulina*). Experience has shown in these projects that within a few weeks large *Chlorella* ponds become contaminated with organisms, especially predatory protozoa, which feed on the *Chlorella* and thus dramatically reduce the yield [41]. Growing a particular species in outdoor ponds is therefore greatly hampered by contamination with other organisms that may dominate, depending upon environmental and media conditions. The ability to grow in an extreme environment therefore offers the opportunity to cultivate a single organism with less competition from other organisms. In addition, the only surfaces available for a new approach to open-air cultivation are arid lands, and the only water available in abundance is sea water.

Present experimental approaches towards an outdoor mass cultivation of micro-organisms centre on halotolerant cyanobacteria such as *Synechococcus elongatus* [41] or *Spirulina maxima* [42]. Among the halophilic, eukaryotic algae, *Dunaliella* is the test organism of choice [23, 43]. In general, these phototrophs are grown in oblong raceway channels following the basic design developed for growing *Chlorella* and *Spirulina* [44]. The ponds walls are generally constructed from brick, concrete or fibreglass. Special attention is required in case concrete is used to avoid deterioration of the metal reinforcing the concrete. Liners of plastic sheetings (UV-resistant polyvinylchloride or reinforced polyethylene) were found to be most suitable. The use of asphaltic concrete or asphalt is not recommended as salt penetration disintegrates the asphalt [23]. Units which do not require culture flow can also be built with natural clay lining, as used in many salt-harvesting plants. In order to prevent thermal stratification and to remove excess oxygen (photoinhibition) phototrophic cultures are usually mixed. This mixing process must take account of the organisms resistance to shearing forces and, in the case of fragile *Dunaliella* cells, is restricted to simple paddle-wheel devices. As sunlight is essentially fully absorbed within the top 5 cm of a dense phototrophic culture, the pond design ought to be as shallow as possible. However, owing to engineering restrictions most commercial ponds have a depth between 10 and 20 cm [23].

Calculations on the energy efficiency of phototrophic organisms in fields have shown that at most 1–3% of the energy available in solar irradiation is recovered in synthesised organic material. Theoretically, this means that in algal ponds no more that 25 g of organic matter can be produced per m^2 per day [23, 45]. In *Dunaliella* ponds optimized for β-carotene production (not biomass), long-term average yields were around 5 g m^{-2} day^{-1}. The mean annual yield of *Spirulina* biomass obtained on sea water plus urea in experimental ponds amounted to between 7 and 8 g m^{-2} day^{-1} (filtered, washed and sun-dried material), reaching a maximum of approximately 15 g m^{-2} day^{-1} in the summer months [42]. These values compare well with the maximum yield reported from the Sosa Texcoco company in Mexico (10 g dry matter m^{-2} day^{-1}) [40] and come reasonably close to the theoretical maximum. However, the economics and efficiency of outdoor cultivation techniques ultimately depend on cheap separation techniques. While *Spirulina* can be recovered by simple filtration, the majority of harvesting processes depend on centrifugation, flocculation or flotation. Centrifugation techniques are very efficient but require high initial investment, energy and labour cost for maintenance. Flocculation using aluminium sulphate (alum), ferrous and ferric ions and other commercially available polyelectrolytes excludes direct use for the food market unless the flocculant is safe or completely removed from the biomass. In addition, the cost of biomass dehydration, when added to the cost of growth and harvesting, may also be an economic barrier to the mass production for low-cost products. Although design, long-term maintenance and downstream processing techniques for outdoor ponds are at present still in an experimental stage, open-pond farming of halophiles makes use of the extreme environment so as to maintain quasimonospecific cultures. Mass production of micro-organisms is thus a very attractive alternative for cultivation on arid land where salty water is often available and the land not utilised for conventional crops.

3.3.2 Pharmaceutical compounds

3.3.2.1 *Antibiotics.* The majority of commercially useful strains of micro-organisms used in the pharmaceutical industry have been recruited from soils (e.g. *Actinomyces*). Recently the search for bioactive compounds has also focused on marine habitats as a source for technologically relevant producer strains [46, 47]. Unfortunately our knowledge of marine producers is still inferior to our understanding of bacterial freshwater systems. The marine and halophilic world therefore remains a profitable area for investigation. The majority of studied marine antibiotic producers comprise Gram-positive bacteria (bacilli, cocci and streptomycetes), which do not normally occur in large numbers in sea water [48, 49]. Novel antibiotics of relatively complex structure (istamycin, aplasmomycin, etc.) derived from these Gram-positive micro-organisms may become useful whenever the effectiveness of long-established products has diminished because of the presence of resistant strains. In addition, the investigation of strains from the marine environment has also led to the characterisation of a number of new and amazingly simple compounds, namely brominated pyrrole derivatives and quinolinol derivatives from Gram-negative bacteria (Figure 3.1) [50–54]. Quinolinol derivatives proved to be particularly effective against Gram-negative pathogens, which also failed to acquire plasmid-mediated resistance. Synthetic compounds marketed so far include Halquinol and Squibb [47]. The most marked inhibition, however, was shown for tetrabromopyrrole, which is antagonistic to *Escherichia coli*, *Pseudomonas aeruginosa*, *Staphylococcus aureus* and *Candida albicans* [52].

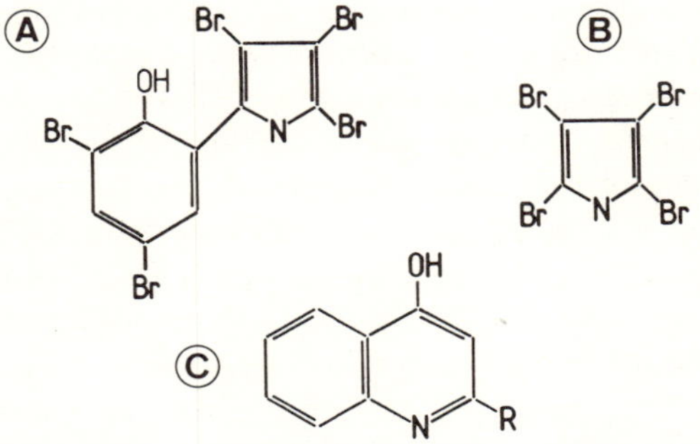

Figure 3.1 Novel antibiotics recovered from marine micro-organisms. (A) 2-(2′-Hydroxy-3′, 5′-dibromophenyl)-3,4,5-tribromopyrrole; (B) Tetrabromopyrrole (also dimer: hexabromo-2, 2′-bipyrrole); (C) 2-*n*-Alkyl-4-quinolinol, R = pentyl or heptyl. (After ref. 47.)

As there is no clear division between marine and more saline environments, the latter would also be worthy of the attention of biotechnologists. With a view to these studies on marine antibiotic-producing bacteria, it is therefore worth mentioning that extracts of microbial mats collected from hypersaline ponds also contain substances that are inhibitory to bacteria [55]. Studies on extremely halophilic archaeobacteria of the family Halobacteriaceae have also shown that they produce compounds with biocin-like properties, which have been called halocins [56]. Limited information is available on the diversity or specificity of these compounds, although their discovery provides evidence that organisms from such environments are also capable of producing antibiotic-type compounds. Halophilic and extremely halophilic producers of novel bioactive agents therefore still await exploitation. In addition it is widely accepted that viruses are inactivated in the marine environment [57]. This antiviral activity, at least against some enteroviruses, is associated with the presence of bacteria [58, 59]. Although the exact chemical composition of the inhibitory agents still needs to be resolved, these findings open up a wide field worthy of further investigation.

3.3.2.2 *Drug screening.* In the past drug screening has concentrated on functions specific for eukaryotes or prokaryotes. It is now clear that three lines of descent, the eukaryotes, the eubacteria and the archaeobacteria, may be distinguished. Irrespective of how one views the archaeobacteria, they provide an additional insight into the way biological systems function, and may provide alternative methods for screening drug activity to using eubacteria or higher animals. The plasmids pGRB-1 from *Halobacterium* strain GRB and plasmid pHV1 from *Halobacterium volcanii* can be used for the screening of new antibiotics and antitumour drugs affecting eukaryotic type II DNA topoisomerase or DNA gyrase (e.g. fluoroquinolones) or acting as DNA intercalators. The action of such compounds on archaeo-bacterial plasmids can be easily monitored by means of changed electro-phoretic mobility on agarose gels [60–63] and allows rapid evaluation of mode and degree of activity. The presence of high levels of magnesium (usually 0.5–2.0 M) in the medium may interfere with screening, and the use of alkaliphiles (which grow in the presence of 1–5 mM magnesium) with similar small plasmids may provide a useful alternative. In addition, an exopolysaccharide termed marinactan from the marine *Flavobacterium/ Cytophaga* group has been shown to display marked activity against sarcoma-180 solid tumour virus (S-180) in mice [64]. Clearly, other bacteria from the saline environment may also be likely producers of antitumour compounds, so an intensive search seems justified.

3.3.2.3 *Tumour detection.* An archaeobacterial 84-kDa protein obtained from *Halobacterium halobium* apparently shares some common epitopes with the human c-myc protein, which is an oncogene product in the serum of some cancer patients [65]. This archaeobacterial protein has been used for the detection of cancer specific antibodies by the western blot method. The conclusions obtained with the archaeobacterial antigen were consolidated by experiments in which the purified c-myc protein (obtained by genetic engineering) was used. Thus use of archaeobacterial antigens as a cancer probe seems to be rather promising at least for some types of cancer [66].

Naturally, one cannot predict the likelihood of detecting new physiologically active substances or the ultimate success of new drug-screening schemes. Screening for bioactive agents must, however, include as many microbial groups as possible, and phylogenetically distinguished clusters like halophiles of both archaeobacterial and eubacterial type should not be dismissed. Potential applications in medicine and the pharmaceutical industry are not always apparent, as has been shown with surfactin, which, apart from its uses as a surfactant (see section 3.3.3), has been patented as an agent for treating and preventing hypercholesterolaemia and counteracting thrombosis and embolism [67].

3.3.3 *Polymers*

3.3.3.1 *Emulsifiers*. Polymeric substances such as polysaccharides are widely used in industry and of immediate or prospective utility (e.g. for coatings, pharmaceutical formulations, emulsifying and gelling agents, food additives and thickeners, adhesives and solubilising agents, etc.) [68, 69]. These polymers generally resemble the chemical composition of capsule polysaccharides and do not reduce interfacial tension but may also prevent oil droplets coalescing. They are sometimes called 'biosurfactants', but in strict terminology should be referred to as biodispersants or bioemulsifiers [70, 71]. Such emulsifiers are, for example, emulsan from *Acinetobacter calcoaceticus*, which has already found an application as a cleaning agent for oil tankers [72–74], liposan from *Candida lipolytica* [75] and others from *Pseudomonas* and *Bacillus* species [76, 77]. Although screening for polysaccharides has so far largely concentrated on non-halophilic micro-organisms, the well-known producer of emulsan (*A. calcoaceticus* RAG-1) is a marine isolate. Other marine bacteria already well known for their production of xanthans and other viscous exopolymers have also come to scientific attention in the search for new bioemulsifiers [78, 79]. Similarly halophilic and halotolerant microbes, which have surprisingly been neglected, may also provide a valuable resource for biotechnological applications. An exopolysaccharide produced (up to 3 g/1) by *Haloferax mediterranei* is, in this respect, the first such product from an extreme halophile. Despite the wide phylogenetic gap separating archaeobacteria from eubacteria, the polymer of *H. mediterranei* (except for salt tolerance) resembles in many respects the most common exopolymers of eubacterial origin. It combines excellent rheological properties (high viscosity, pseudoplastic and thixotrophic behaviour), resistance to extremes of pH, temperature and, of course, salinity [80]. Chemically the polysaccharide of *Haloferax mediterranei* is a sulphated heterosaccharide containing mannose as a major component. The exact composition depends on the growth medium and growth conditions used [81]. Uronic acid, amino acids and inorganic sulphate are also present in the polymer, indicating a relationship to the polysaccharides of eukaryotic marine algae (e.g. Phaeophyceae and Rhodophyceae [82]. The most remarkable property, however, is the resistance of *H. mediterranei* polysaccharide to salt.

Significant oil deposits are associated with salt deposits and often elevated temperatures. Salt resistance is therefore of interest for potential use in microbially enhanced oil recovery (MEOR). Owing to regions of differing permeabilities within a reservoir, the oil entrapped in less permeable regions will be bypassed by injected fluid (secondary recovery). Hence the use of suitable polymers which may serve as mobility controllers and emulsifying agents has been discussed [83]. The properties needed for enhanced oil recovery include high viscosity at dilute concentrations and at elevated temperatures, pseudoplasticity, and higher resistance to shear, salt and thermal degradation than, for example, xanthan gum. An economical alternative to the industrial production of the polysaccharides would be *in-situ* production of microbial 'plugging' agents. In such a system the micro-organisms must grow and produce extracellular polymers under the environmental conditions that prevail in the oil reservoir. In addition to salinity and high temperatures, the anaerobic environment restricts the spectrum of suitable micro-organisms. Screening for (facultative) anaerobic, halo- and thermotolerant exopolymer producers has, however, produced a large number of potential candidates, probably *Bacillus* species, selected on a sucrose–mineral medium with 10% (w/v) salt (incubation at 50°C) and nitrate as an electron acceptor [83]. Bacteria with such properties should be able to grow in oil reservoir brines, if supplied with a carbon source and nitrate. While there are even indications that these bacteria may be recovered from oil reservoirs, it is not entirely clear if they are indigenous or enter during normal drilling operations. Thus the injection of bacteria into the reservoir may not be required, since the contaminating/indigenous populations may be stimulated by the addition of suitable nutrients. In addition to *Bacillus* species, the exopolysaccharide-producing species *Haloferax mediterranei* will also grow under these conditions, with the production of gas from nitrate (and hence cause pressure build-up), which may also be of benefit for oil recovery. Methanogenic bacteria have also been recovered from oil wells. However, when considering the commercial aspects of such a process, the choice of nutrients also plays a role, and the use of pure carbohydrates as a medium component is probably not economical.

3.3.3.2 *Surfactants*. Industry has great demand for synthetic surfactants as detergents for a wide variety of applications (e.g. washing liquid, etc.). Biological surfactants present a much wider spectrum of different types and properties than the synthetic products available. In addition they are usually biodegradable and, thus, avoid the potential danger of environmental pollution. Another use of biosurfactants which shows particular promise is in the cleaning up of oil spills both on water and on land. The release of bitumen from mixtures such as the Athabasca tar sands may also be an area worth further investigation.

A typical surfactant is characterised by the presence of both a hydrophilic and a lipophilic moiety within the molecule. Based on the character of the hydrophilic moiety they may be grouped into the following classes: glycolipids (lipopolysaccharides), lipopeptides, phospholipids, neutral lipids and fatty acids [71]. The most important of these groups are glycolipids (containing trehalose, rhamnose, mannose or sophorose as a sugar component) and lipopeptides. Many cyclic lipopeptides such as surfactin/subtilisin from *Bacillus subtilis* [84], viscosin from *Pseudomonas fluorescens* [85, 86] and cyclodepsipeptides from *Serratia marcescens* [87] have only recently been subjected to detailed investigations.

Reports on positive effects of biosurfactants in the process of crude oil degradation by marine bacteria [88] also ought to initiate ample research on biosurfactant producers among marine and halophilic micro-organisms. The natural occurrence of foam and rainbow-coloured soap bubbles associated with microbial mats in salt pans is a strong indication for biosurfactant-producing organisms (Figure 3.2); however isolation and characterisation of novel surfactants from the halophilic biotope remain open for future research.

Figure 3.2   Bacterial mat which is lifted from the sediment of a shallow saltern in Alicante (Spain). Abundance of soap bubbles and lowered surface tension indicate the presence of surfactant-producing halophilic micro-organisms.

Very little attention has been paid to the production of surface-active agents from halophilic bacteria, although it has been shown that halobacterial membrane lipids (phytanylglycerol diether) have surfactant properties with a hydrophilic/lipophilic balance (HLB) of 7–8, which falls within the optimum range for enhanced oil recovery [89]. Strain GSL-11, a Great Salt Lake isolate, displayed the greatest lowering of surface tension (41.0 mN/m) and was, therefore, used to extract bitumen from Utah tar sands [89]. Ether-linked lipids from archaeobacteria possess a very low melting point and are resistant to degradation by heat, acid and alkali and may, because of the ether linkage, even resist enzymatic degradation by other microbes. Whole cells as well as cell extracts appear to have a potential for enhanced oil recovery, as surfactants with such properties might be useful under conditions where other surfactants are unstable.

3.3.3.3 *Liposomes.*   Liposomes may be used in a variety of ways to carry certain compounds to target sites in the body. One popular use of liposomes is in the cosmetic industry. The disadvantage of liposomes based on fatty acid derivatives is that they are biodegradable, and as such their shelf life is limited. The use of ether-linked lipids from archaeobacterial halophiles may provide an interesting alternative, since they are not easily attacked by other bacteria.

3.3.3.4 *Bioplastics.*   Another group of polymers, the so-called bioplastics, will certainly become of increasing importance for industry. These poly-β-hydroxyalkanoates (PHA) can substitute for oil-derived thermoplastics [90] and, as a consequence of biological degradation, will help to reduce the amount of material dumped. The organism of choice for PHA production is *Alcaligenes eutrophus*, and the physiology and molecular genetics of its polymer synthesis have already been investigated [91–93]. At present a copolymer of β-hydroxybutyrate (PHB) and β-hydroxyvalerate (PHV) is produced by ICI (Imperial Chemical Industries) in Billingham, England, under the trade name Biopol. The production strain for the world's first marketed natural plastic is an *Alcaligenes eutrophus* mutant (NCIMB 11599), which accumulates up to 90% of its cell mass as poly-β-hydroxybutyrate/valerate (PHB-V). Although production costs today are calculated at approximately £10/kg (6–9 times more expensive than conventional plastic packing material), a cosmetic company has introduced a Biopol shampoo bottle onto the German retail market [94], the production costs of which are presently subsidised. The idea of marketing 'nature's plastic' seems to be rewarding, and as production volumes increase (short-term prospects 5000–10000 tons per annum), lowering production costs, biodegradable plastics will probably occupy an increasing proportion of the market.

The ability to produce poly-β-hydroxybutyrate is also found in the (halophilic) archaeobacterial world, the prime example being *Haloferax mediterranei* [95]. The pattern of PHB production in *Haloferax mediterranei* is very similar to that found in eubacteria such as *Alcaligenes eutrophus* [96], i.e. PHB synthesis is delayed with respect to biomass development and accumulation proceeds through the stationary phase until the carbon source is depleted [97]. Although the maximum yield obtained (0.33 g per g dry weight) equals four times the amount of protein, it is not as high as with commercially used *Alcaligenes eutrophus*, but still remarkable if one considers that PHB is not the sole biopolymer produced (see polysaccharides, above). Under most favourable conditions (2% carbon source, 0.00375% phosphate, 25% sea salt, 45°C, pH 7.2) a joint yield of 0.5 g per g dry weight was recorded [98]. Bearing in mind that the reported yields were obtained with the wild-type strain, there is certainly ample room for genetic improvement to increase strain productivity. In addition, the use of *Haloferax mediterranei* as a producer strain could offer a number of advantages. The requirement for sterile conditions is greatly reduced as very few organisms, if any, can grow as fast in hypersaline solution [4, 99]. Therefore, extremely simple production systems can be developed, such as open ponds. With a view to recent applications of recombinant DNA techniques to obtain PHB-producing organisms able to use cheap substrates [100], the ability of *H. mediterranei* to use starch as a cheap and abundant substrate becomes another important advantage (doubling time 6.9 h, the same as with glucose). In addition, genomic stability of the strain is a prerequisite for industrial processes, especially on a continuous or fed-batch basis. This is all the more vital as many halobacteria are known to have a high genomic instability. However, *H. mediterranei* complies to these demands and has been shown to remain stable over a period of 3 months in continuous culture [98].

Last but not least, technically useful plastics include various suitable copolymers such as PHB (poly-β-hydroxybutyrate) and PHV (poly-β-hydroxy-valerate), the biological production of which is achieved by co-metabolism of certain organic acids as precursors [90, 101–103]. Preliminary data indicate that polyhydroxyalkanoates of differing monomeric composition can also be obtained with *H. mediterranei* [98]. Furthermore, cell lysis and release of PHB, a rather critical step in conventional production processes, is easily achieved with archaeobacteria like *H. mediterranei*, which rapidly disintegrate upon treatment with low-salt solutions (e.g. water). Provided future developments present us with simple and economical ways of archaeo-bacterial biomass production and cell harvest (preferably in open ponds), *H. mediterranei* may well become the organism of choice for a low-skill bioplastic production mainly for countries with a suitable climate and access to salt.

3.3.4 *Enzymes*

3.3.4.1 *Degradation of polymers.* As a potential source for the production of enzymes, halophilic archaeobacteria offer the advantage that the enzymes produced have optimal activity at high salinities. This may be advantageous for harsh industrial processes where concentrated salt solutions may otherwise impede catalytic conversion. For halophilic eubacteria (with a low ionic cytoplasm) this only holds true for exoenzymes excreted into the saline environment. Thus halotolerant exoenzymes with polymer-degrading capacity may warrant commercial exploitation, for example amylases from *Micrococcus halobius* [104, 105] and *Halobacterium halobium* [106], nucleases from *Micrococcus varians* [107] and *Bacillus* spp. [108] as well as proteases from *Halobacterium salinarum* [109], *Bacillus* and *Pseudomonas* spp. [110, 111], to name only a few examples. A serine protease has been reported from an unidentified member of the family Halobacteriaceae [112]. This enzyme has been shown to be thermophilic and halophilic, showing high activity at 75–80°C in the presence of 25% (w/v) salt. In contrast, the serine protease from *Halobacterium halobium* reported by Izotova *et al.* [113] was unstable even at low temperature in the presence of 3 M sodium chloride. Degradation of RNA and DNA by an extracellular halophilic nuclease (nuclease H) from *Micrococcus varians* subsp. *halophilus* will produce the flavouring agents 5′-guanylic acid (GMP) and 5′-inosinic acid [114]. A bioreactor system of flocculated cells which preferentially adsorb nuclease H has already proved very useful for the production of 5′-GMP, provided competing 5′-nucleotidases were selectively inactivated by a number of techniques [115, 116]. Biotechnological applications of halophilic/halotolerant enzymes, therefore, appear to be within immediate reach.

3.3.4.2 *Biochemical tools.*   As a biochemical tool restriction endonucleases of unusual site specificity are well in demand and have already led to a patent on the production process for an enzyme from 'Halococcus acetoinfaciens' [117]. Restriction enzymes have also been isolated and characterised from the marine bacteria *Deleya marina* and 'Agrobacterium gelatinovorum' [118–120]. Other biochemical applications, mainly for genetic engineering purposes, are to be expected.

3.3.5   *Compatible solutes*

3.3.5.1 *Detection methods.*   To cope with the physiological stress of low water activity (high osmolality) of the surrounding medium, effective mechanisms of osmoadaptation have been evolved. With few exceptions, the halophilic and halotolerant eubacteria so far examined produce high concentrations of organic osmotica (1–3 mol per kg of cytoplasmic water), named compatible solutes. The term 'compatible solute' was coined by Brown [121] to describe a solute responsible for osmotic balance and compatible with the organism's metabolism. The production of compatible solutes is the typical property of halophiles closely associated with life at elevated salinities. Compatible solutes have attracted considerable attention, mainly because of their massive

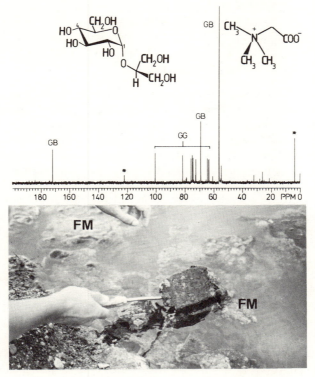

Figure 3.3 Floating microbial mat (FM) from a salina near Alicante (Spain) and ^{13}C-NMR spectrum of a crude cell extract (top). GB = glycine betaine, GG = glycerolglucoside. *acetonitrile reference (E.A. Galinski, unpublished).

occurrence and the detection of novel compounds with prospective application as stress protectants and stabilisers of enzymes. The identification of these osmolytes has been largely aided by natural abundance ^{13}C nuclear magnetic resonance (NMR) spectroscopy, which is especially suited for the detection of organic compounds accumulated to high cytoplasmic concentrations. In spite of the intrinsic insensitivity of the method (only 1% of the natural carbon is ^{13}C), the massive occurrence of solutes within halophiles (approximately 10–30% of the cells' dry weight) allows rapid detection from as little as 1 g of dried cell material. As can be judged from a sample of natural material collected from a floating microbial mat in a salina near Alicante (Spain) (Figure 3.3) the NMR spectrum is amazingly simple, as minor components such as intermediary metabolites do not contribute significantly. For screening purposes and rapid determination of the compatible solute 'cocktail' from as little as 10 mg of cell material, high-performance liquid chromatography (HPLC) methods have been developed for the most common organic osmolytes [122, 123].

3.3.5.2 *Diversity of microbial compatible solutes.*   The spectrum of compatible solutes (or osmolytes) found in nature has been reviewed before [124–128]. They can be grouped into the following classes of compounds: sugars, polyols, betaines, ectoines and some amino acids. The most common representatives are listed in Table 3.2 and discussed below.

(a) *Sugars.*   The disaccharides sucrose and trehalose often appear as part of the compatible solute 'cocktail' in a wide range of marine and halophilic/halotolerant micro-organisms. Only organisms of rather limited halotolerance rely solely on sugars as their osmolytes. There is increasing evidence that sugars at high concentration are less compatible with the organisms' metabolism [123, 129–131], and their possible role as general stress metabolites which ensure survival under adverse conditions is presently under discussion [132, 133]. Nevertheless the accumulation of the rare sugar trehalose may well become of technological importance as a cryoprotectant (see below).

(b) *Polyols* (*algae, yeast, fungi*).   Besides some euryhaline algae such as *Chlorella autotrophica* [134], moderately halophilic marine diatoms like *Cyclotella meneghiniana, Phaeodactylum tricornutum* [135, 136], *Navicula* spp. [137] and at least one oomycete, *Phytophthora cinnamoni* [138], which are known to accumulate proline, the majority of halophilic/xerotolerant algae, yeast and fungi principally produce polyols (chain length 3–7, including cyclitols and glycerol mannosides and galactosides) [137, 139–142].

However, the most halotolerant algae, which grow very well at high salinities, are generally glycerol producers, such as *Asteromonas* [143] and *Dunaliella*. Glycerol as a photosynthetic product in *Dunaliella* was first described by Craigie and McLachlan [144] and its role as a compatible solute was discovered in the early 1970s [145–147] and has since then been studied in great detail. Yeast and fungi also often respond to water stress by the accumulation of glycerol and, more rarely, other polyols [148, 149]. In *Debaromyces hansenii*, which uses glycerol and arabitol as the principal osmolytes, it was shown that glycerol, the predominant solute of the exponential growth phase, is also the typical inducible compatible solute in

Table 3.2 Main classes of organic osmolytes ('compatible solutes') frequently observed in halophilic, halotolerant and osmophilic micro-organisms.

Sugars	Polyols	Betaines	Ectoines	Amino acids
Sucrose	Glycerol	Glycine betaine	Ectoine	Glycine
Trehalose	Erythritol		Hydroxyectoine	Alanine
	Arabitol			Proline
	Mannitol			N-acetylated
	Sorbitol			diamino acids
				N-carbamoylated
				carboxamides
Glycerol glycosides				
(gluco-, manno-, galactoside)				

responding to increased salinity [150–153] underlining the importance of glycerol as the dominant osmotic regulator.

Although glycerol plays an outstanding role as the major osmolyte of eukaryotic micro-organisms, it is apparently not important for halophilic eubacteria; instead, a glycerol derivative glucosylglycerol (2-O-glycerol-α-D-glucopyranoside) is often found in both oxygenic and anoxygenic phototrophic bacteria of intermediate salt tolerance [122, 129, 154–156]. The presence of glucosylglycerol (GG) is clearly demonstrated by the NMR spectrum taken from a natural cyanobacterial mat (Figure 3.3), where this compound represents the minor solute next to glycine betaine.

(c) *Betaines and ectoines (eubacteria)*. The principal osmolytes produced by truly halophilic eubacteria are glycine betaine and the novel tetrahydro-pyrimidines ectoine and hydroxyectoine [130, 155, 157]. Earlier reports stating that glycine betaine was the main compatible solute of halophilic eubacteria [158, 159] have been put into perspective by showing that glycine betaine is easily transported and accumulated from the growth medium [157]. Common complex media supplements such as yeast extract apparently contain reasonable amounts of glycine betaine or suitable precursors (e.g. choline), which are readily used as osmolytes. It was further shown that the ability actually to synthesise glycine betaine is typical of extremely halophilic phototrophic bacteria, while the majority of aerobic chemoheterotrophic eubacteria (of pronounced halotolerance) produce the novel tetrahydro-pyrimidines, ectoine and hydroxyectoine, unusual cyclic amino acids of a betaine-like zwitterionic structure [128, 160, 161]. These novel cyclic amino acids of the ectoine type were first detected in and subsequently isolated from extremely halophilic species of the bacterial genus *Ectothiorhodospira*, hence the vernacular name [162]. Ectoine is chemically described as a 2-methylated hydrated pyrimidinecarboxylic acid (Figure 3.4) and biologically synthesised by cyclisation of N-acetylated diaminobutyric acid [163]. It is probably the simple biosynthetic pathway (offbranch of the reaction chain forming the aspartate family) which may account for the dominant role of ectoines within eubacterial haloadaptation.

(d) *Amino acids*. Organisms unable to synthesise and/or accumulate any of these widespread compounds (glycerol, glycine betaine or ectones) seem to achieve osmoadaptation using a number of amino acids [164]. Of the known polar α-amino acids, characteristically only those which have both an isoelectric point (pI) of between 6 and 7 and high solubility in their zwitterionic form seem to be important, e.g. glycine (pI 6.0; 3.3 mol/kg), alanine (pI 6.0; 1.9 mol/kg) and, most important, proline (pI 6.1; 14.1 mol/kg). The same applies for a few unusual diamino acid derivatives which have only recently come to attention: $N\delta$-acetylornithine (A. Wohlfarth and E.A. Galinski, unpublished) and $N\varepsilon$-acetyl-β-lysine [165]. In both cases acetylation of the amino group furthest from the carboxyl group converted a basic amino acid into a neutral polar molecule apparently more suitable for osmoregulatory

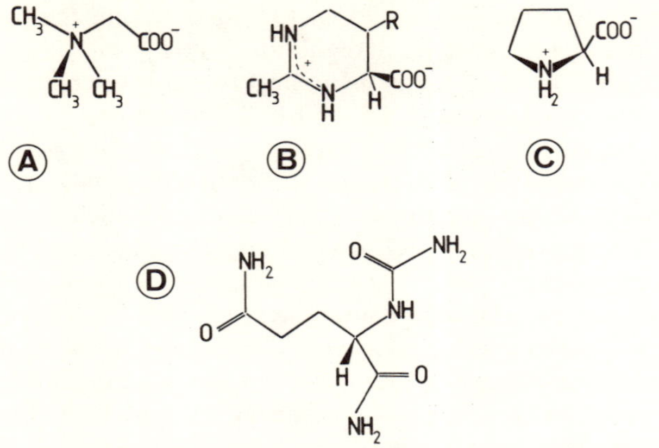

Figure 3.4   Common nitrogen-containing compatible solutes: glycine betaine (A), ectoines (B) R = H or OH, proline (C) and the novel carboxamide $N\alpha$-carbamoyl-glutamine amide (D).

purposes. Similarly the recently discovered $N\alpha$-carbamoylglutamine amide (CGA) (Figure 3.4) isolated from *Ectothiorhodospira marismortui* [166, 167] is derived from glutamate (not a true compatible solute) by transformation of anionic residues into polar non-ionic (mesomeric) carbonic acid amides and carbamoylation of the α-amino group, which also renders this moiety polar but uncharged. The resulting $N\alpha$-modified amino acid carboxamide therefore represents a novel class of compatible solutes for which the $CONH_2$ group is characteristic. It is from this compilation of compounds found in nature that the following principles have emerged to characterise effective organic osmolytes: the prospective compatible solute is small, carries polar functional groups, is uncharged at physiological pH and highly soluble in water.

3.3.5.3 *Biotechnological production of bulk chemicals.*

(a) *Glycerol.* The organisms' ability to produce and accumulate high concentrations of organic compounds (up to 30% of the cells' dry weight) makes them potentially useful for the biotechnological production of simple organic substances. Glycerol is used in drugs, cosmetics, in the food and beverages industries and in the production of urethane, cellophane and explosives. It is a typical commodity chemical, which is defined as a product of low monetary value and hence is sold primarily in bulk (market value $1.00/kg). So far glycerol is mainly produced by the petrochemical industry using propylene as a starting material. The price of glycerol therefore very much depends on the price of oil. As the intracellular concentration of glycerol can reach up to 3 M in *Dunaliella* cells (approximately 30% of dry weight), this organism appears to be the ideal candidate for a biotechnological process of glycerol production. Although a number of companies already use *Dunaliella* at pilot plant stage to produce glycerol, the process is still not economical.

This is mainly because of the fact that the price of oil has not increased as much as was predicted 10 years ago (predicted price of $50 as compared with actual price of $20 per barrel) [43]. In order to make this process economical, it must be coupled to other beneficial processes yielding high-value product (such as β-carotene) or organic solvents and acids, which could be obtained by microbial fermentation of glycerol supplied in the form of crude *Dunaliella* extracts. Although this prospect is highly speculative at the moment, first experimental results have been extremely encouraging for the conversion of glycerol into *n*-butanol, 1,3-propanediol and ethanol from *Dunaliella* biomass by fermentation with *Clostridium pasteurianum* [168]. Another approach to economise the process would be to pursue enhancement of glycerol production by use of overproducing or excreting mutants, such as *Dunaliella parva* 19/9, which leaks 90% of the glycerol produced [169]. Surprisingly the growth rate was not significantly less than that of the 'wild type'.

More unusual polyols such as erythritol, arabitol, mannitol and sorbitol and the glycerolglycosides face far less competition from preparative chemical processes; they are therefore all the more worthy of consideration for future applications such as dietary sugar replacements, provided of course the recovery from alternative sources like plant material proves uncompetitive.

(b) *Amino acids.* In contrast to cheap bulk chemicals, which will always meet severe competition from chemical production processes, prospects for a biotechnological production of more complex (mainly chiral) compounds seem to be more promising. Amino acids represent one such market, even though recent developments in stereochemical synthesis (using immobilised cells or enzymes for racemic separations) will certainly lower production costs even more in the future [170]. Unfortunately the amino acids used as compatible solutes (except proline) have a low market value, whereas those most in demand, for example as plant protein supplements (e.g. lysine, methionine, threonine and tryptophan), are apparently not produced for osmoregulatory purposes. The only exception is proline. It is not an essential amino acid but has found some application in the medical industries (e.g. infusion liquids) with an estimated annual output of between 50 and 100 tons/year in Japan [171, 172]. The original method of production from gelatin hydrolysate has been largely replaced by direct fermentation techniques using auxotrophic or deregulated mutants, mainly of the bacterial genera *Brevibacterium* and *Corynebacterium* [173, 174]. The maximum yield thus far was obtained with a mutant of *C. acetoacidophilum* reaching approximately 100 g/l on the laboratory scale [170, 174].

First reports displaying a correlation between proline production and osmotolerance and initiating attempts to use osmotic stress as a screening condition for proline-overproducing mutants were performed with non-halophiles like *Salmonella typhimurium* and *Serratia marcescens* [175, 176]. Further work is now in progress to explore the natural high production capacity of halophilic proline producers of the genus *Bacillus* [177]. This also includes the application of chemical mutagenesis under saline conditions in order to obtain compatible solute-overproducing eubacteria [178]. Although it is at present difficult to predict whether the use of halophilic proline producers may offer advantages over already highly productive conventional fermentation processes, the elucidation of possibly novel production and regulation mechanisms may well influence further developments. Considering the possibility of gene transfer into commercial production strains once the experimental stage has been passed, one is presently investigating the gene pool rather than establishing new production methods.

3.3.5.4   *Application as stabilising agents.*

(a) *Betaines and ectoines.*   Rather than providing mass chemicals as a starting material or as a supplement for industrial processes, the genuine cytoplasmic function of compatible solutes (as stabilisers and protective agents), presently under investigation, is more likely to lead to ultimate biotechnological application. If one considers that water freely equilibrates across the cell membrane it is obvious that the water activity or water potential within the cytoplasm is similarly low as in the surrounding medium (e.g. $a_w = 0.85$ for a saturated sodium chloride solution). Besides their function as an osmoticum, compatible solutes are therefore required to maintain the enzyme's hydration shell in an environment of low water activity and to counterbalance the effect of elevated ionic strength within the cytoplasm. Most abundant in nature (besides glycerol) are two zwitterionic classes of compounds, the betaines and the ectoines. Glycine betaine, almost always found in bacterial mats (composed of cyanobacteria and anoxygenic phototrophic bacteria) (Figure 3.3), is industrially produced as a cheap byproduct in the course of sugar beet extraction and, hence, of little interest as a biotechnological product. This is, however, not the case for ectoine and hydroxyectoine, two novel compounds only obtainable by biotechnological techniques; at least for hydroxyectoine, containing two chiral carbon atoms, a competitive chemical synthesis is out of the question.

The stability of an enzyme is a paramount consideration in its use, but surprisingly little conscientious work has been conducted in this area. Investigations into the enzyme-protective effect of compatible solutes have so far centred on salt, heat and cryoprotection as well as resistance against desiccation. Possibly some of these effects are based on a common physical phenomenon securing maintenance of a functional hydration shell. Salt-protective effects of compatible solutes have been reported in a number of cases where it has been shown that glycerol, betaine and proline are able fully or partially to relieve salt inhibition on certain enzymes [131, 179–184].

The majority of projects concerned with protection against freezing and thermal denaturation have been performed using labile model systems [phosphofructokinase (PFK), lactate dehydrogenase (LDH) and catalase]. It has been shown that the most common and generally available compatible solutes such as glycerol, proline, betaine and sugars are effective protectants [185–188], while only little information is presently available about the effect of ectoines [189, 190].

Although a number of theories have been put forward to explain the protective action of compatible solutes, the underlying principle is by no means clear. There is, however, an increasing body of evidence suggesting that compatible solutes are preferentially excluded from the hydration shell of proteins and exert an indirect effect on the surrounding water structure [186, 191–194]. This (hypothetical) model would provide an easy way of explaining why organic compounds of widely differing chemical structure have stabilising activities. However, the degree of protection seems to depend very much on the type of stress under investigation and the type of solute used. As an example, Figure 3.5 depicts the pattern of heat and freeze protection on LDH encountered with the most common bacterial osmolytes (glycine betaine, ectoine and hydroxyectoine). It can clearly be seen that the

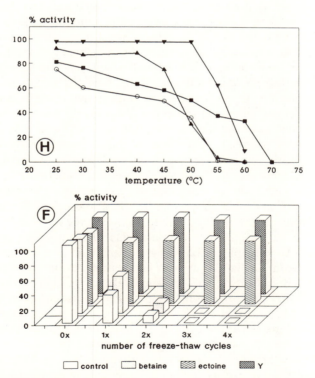

Figure 3.5 Stabilising effect of compatible solutes on lactate dehydrogenase against heat (H) and freeze–thaw treatment (F). ○, control; ▲, ectoine; ▼, hydroxyectoine (Y); ■, betaine. (From ref. 189 with permission.)

two ectoines confer the most pronounced stabilisation, whereas the protection of betaine (otherwise recognised as a good protectant) is only minor and—in the case of heat protection—seemingly of a different quality. Hence a lot of work remains to be done to resolve fully this apparently complex interrelationship between (labile) enzymes and appropriate protectants. A systematic study of this class of compounds and especially of the ectoines (at present only obtainable from halophilic producer strains) will therefore provide a significant contribution towards the understanding of enzyme action and stability. Advances in halofermentation, mutagenesis and genetic engineering of halophiles will hopefully soon provide virtually unlimited amounts of compatible solutes for further biotechnological research. Effective stabilisers which prolong lifetime and shelf life of otherwise labile enzymes will undoubtedly find many biotechnological applications.

(b) *Sugars*. Although the rare sugar trehalose has a relatively high market value, its market size is at present very small. This situation may, however, change very soon when potential applications against desiccation become more obvious. Starting from the observation that organisms capable of unhydrobiosis, as well as dried bakers' yeast, brine shrimp cysts (*Artemia salina*) and spores of many fungi contain large amounts of sugars, mainly trehalose (e.g. 20% of dry weight in *Artemia* cysts), it was subsequently shown that this sugar is correlated to survival in the dry state [195–199]. Trehalose has been shown to stabilise both phospholipid bilayers and soluble proteins against desiccation. Although understanding of the mechanism is still scarce, trehalose seems to replace water in the hydration shell of the lipid phosphate headgroup, thus preventing dehydration damage (e.g. fusion between bilayers, leakage, lateral separation). This is probably achieved by interaction of OH groups (hydrogen bonding) [200]. The most characteristic observation during dehydration in the presence of trehalose (and to a lesser degree of other sugars) is the depression of the phase transition temperature t_m observed without protectants. This phenomenon is explained by a lateral expansion of the bilayer due to the trehalose/polar headgroup complex [201, 202]. Thus, the hydrocarbon chains of phospholipids turn into the gel phase when dried without trehalose, but remain in the fluid phase in the presence of trehalose. This means that dried phospholipids are maintained in a liquid crystalline state (or, better, a new liquid phase) even though they are dry [203].

Similarly it has been shown that some sugars, especially trehalose, also serve to stabilise soluble proteins in the process of drying. Phosphofructokinase (PFK), when dehydrated in the presence of 100 mM trehalose, showed no loss of enzyme activity, while the compatible solutes glycerol and proline only protected until most of the water was removed, but did not stabilise against complete dehydration [204]. It was concluded that thermodynamic arguments (preferential exclusion of solutes from the protein surface) explaining protein stabilisation are no longer applicable when, during total desiccation, even the hydration shell is removed. It is further assumed that at some critical point direct bonding between OH groups and polar residues of the protein may be necessary for stabilisation. An infrared spectroscopic study has finally provided further evidence by showing that the frequency shifts observed for trehalose (when dried with lysozyme) mimic those of water on hydrated trehalose. This is a strong indication that hydrogen bonds occur between dried protein and trehalose [205].

Hence, freezing and drying are fundamentally different stress factors as the non-freezable water of the hydration shell (approximately 0.25 g per g dry weight) has to be replaced in the course of dehydration, probably involving hydrogen bonding between the stabilisers and polar residues of the protein. Until now only carbohydrates have been found to be effective, and of those tested trehalose is by far the most effective [206]. Therefore, trehalose may well find an application as a protectant for the freeze drying of biomolecules. On a whole-cell level, knowledge about the beneficial effects of trehalose as a membrane preservative may also become important for long-term conservation of micro-organisms. A search for novel and effective production techniques for trehalose (possibly exploiting halophilic producer strains) therefore seems to be worthwhile.

3.4 Future aspects

3.4.1 *Environmental biotechnology*

Environmental biotechnology may play an important role for the treatment of saline waste waters and the recovery and detoxification of soils. The biological treatment of liquefied wastes as an essential municipal activity has been a development of the last 100 years, initiated primarily in north-western Europe and the north-eastern United States, geographical areas characterised by low temperatures and low salinities. In recent years it has become evident that the methods applied worldwide (biofiltration and activated sludge) are not necessarily the best alternative for secondary treatment under conditions of high temperature and salinity prevalent in low-latitude regions, not to mention the fact that fresh water is far too valuable to be used as a flushing and carrying agent for domestic sewage where sea water would be more appropriate. In addition, conventional waste water treatment as described above involves many complex high-technology elements, which are difficult to apply in developing countries. A simple low-tech open-pond system would be much more appropriate [207]. It should further be mentioned that waste treatment ponds not only perform a disposal function (i.e. a destructive process) but also have a productive function in terms of algal biomass, which is a byproduct of the process. While the likelihood of sewage-grown algae being used as a direct source of food is at best remote, applications such as animal feed, starting material for chemical extraction and substrate for methane production are worth considering. In fact, some natural salt-harvesting plants already make use of the fertilising effect obtained from low doses of waste water. The intention is, however, of a totally different kind: promotion of algal and bacterial growth within a salt pan (and the concomitant development of colour) functions as a trap for solar radiation and hence speeds up the process of evaporation [208]. Nevertheless, these empirical techniques prove the point that microbial cultivation on sewage is feasible.

In addition, a considerable metal tolerance observed in halophilic bacteria may become of great importance for detoxification processes [209, 210], although it is at present not clear whether halophilic organisms generally express a higher metal tolerance. With a view to the denitrifying process, it has already been demonstrated that halophiles offer a valuable contribution. The moderately halophilic extremely halotolerant *Bacillus halodenitrificans* (ATTC 49067) enriched on 1 M sodium nitrate reduced nitrate to nitrous oxide during anaerobic growth. The production of nitrous oxide was not inhibited by high concentrations of nitrate (1 M sodium nitrate) and there was only a slight accumulation of nitrite. This *Bacillus* species proved resistant to nitrite up to a concentration of 0.6 M, and denitrification continued for several days after the culture had entered the stationary phase [211, 212]. By combination with the nitrous oxide reducing capacity of another denitrifier (*Pseudomonas stutzeri*, ATCC 17591) the authors succeeded in reducing nitrate to molecular nitrogen. Studies on the ability of certain members of the family Halobacteriaceae to reduce nitrate have also shown that some strains are also capable of producing nitrogen as the sole end-product [213, 214]. However, both growth conditions and the concentration of nitrate appear to have a bearing on the nature of the end-products of nitrate reduction.

All these varying facets of halophile activities indicate that environmental technology of the old world may also gain from new and powerful microbes recovered from the saline habitat.

3.4.2   Agricultural aspects

Availability of water is, in fact, the most serious limitation to crop productivity in large areas with high solar radiation. This limitation is also felt in areas where the rainfall is highly variable over a period of time. The arid and semiarid lands of the world constitute about 36% of the potentially arable land (mostly in the regions between 15° and 40° north and south of the equator) and are generally affected by elevated levels of salt. The concept of converting desert areas to suitable environments (by irrigation) for plants already used and understood in agriculture of the developed world seems to be the wrong approach. A logical alternative would be to make use of plants and organisms which not only tolerate but also thrive on salt water [215]. A further development of this approach would be to transfer salt tolerance to agriculturally important crops. The feasibility of adapting crops to seawater culture is enticing. The resources upon which this novel system would draw—sea and sand—are at present useless for this purpose and much thought has been given to developing an agriculture of the salt-tolerant kind [216–220]. Salinity is not *per se* incompatible with higher forms of life, and many different strategies have been evolved in desert plants to withstand the influence of salt in this extreme environment, e.g. shedding of salt-accumulating leaves, salt glands and salt bladders, but also the accumulation of organic osmotica like proline and betaines which can make up to 10–20% of the shoots' dry weight [221]. The accumulation of osmolytes appears to offer an easy target for genetic manipulation and has therefore attracted considerable interest, primarily at the level of well-characterised microbial systems. Knowledge of the relationship between compatible solute uptake and salt tolerance (including its regulation) is, naturally, most advanced with *Escherichia coli* [222–226]. Compatible solutes applied externally are taken up under osmotic stress and confer a moderate degree of salt resistance. This is illustrated in Figure 3.6. As little as 2 mmol of an externally supplied compatible solute (betaine, ectoine) is needed to enable the non-halotolerant *Escherichia coli* K 12 to grow on 5% w/v sodium chloride (E.A. Galinski, unpublished).

It was further shown that salt inhibition (0.65 M sodium chloride) of nitrogen fixation in *Klebsiella pneumoniae* was partially elevated in a proline-overproducing strain in the presence of glycine betaine and by other compatible solutes present in the medium [227–230]. Thus, the combination of two well-known compatible solutes (betaine, proline), also often accumulated during osmotic stress in higher plants (mainly of the family

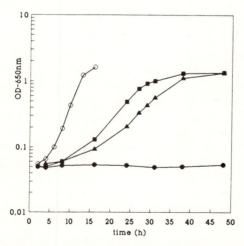

Figure 3.6 Influence of externally applied compatible solutes on the growth of *Escherichia coli* K 12 in 5% (w/v) sodium chloride. ○, control (no salt). Closed symbols represent 5% sodium chloride: ●, no compatible solutes added; ▲, ectoine (2 mM); ■, betaine (2 mM).

Chenopodiaceae), also proved a valuable tool for the engineering of stress tolerance in micro-organisms. The gene for proline overproduction (deregulated γ-glutamylkinase) was subsequently cloned and transferred to other micro-organisms [175, 231].

Subsequently research was focused on symbiotic nodule-forming bacteria of the genus *Rhizobium*, providing an imminent application in agriculture. *Rhizobium japonicum* is the most important nitrogen fixing organism in US agriculture today (inoculation of soya bean fields), and nitrogen fixation is extremely sensitive to osmotic stress (95% inhibition occurring at 0.5 M sodium chloride). Hence, this organism has attracted considerable interest as a promising object for genetic engineering and initiated work on the salt protection of nitrogen fixation by different compatible solutes, notably glycine betaine and proline betaine [232–235]. It was finally shown that these two compatible solutes can partially restore the symbiotic nitrogen-fixating ability of young nodulated alfalfa (*Medicago sativa*) subjected to salt stress [236, 237]. Thus, changing the salt tolerance of the symbiotic bacterial partner exerted an overall stimulatory effect on the crop plant. Although genes encoding for betaine production have not been located, let alone cloned, it is only a matter of time for genetic engineering to succeed. The state of the art is now at a point where micro-organisms can soon be manipulated for osmotic tolerance within certain limits. Engineering of compatible solutes production will, however, only do part of the job required, as true halotolerance probably also implies adaptations within the cell membrane and alterations at the level of ion regulation. Nevertheless one would expect that in the future genetic engineering technology may be applied for the construction of hardy varieties of crop plants.

3.4.3   Fuel from renewable sources (biofuel)

The 1970s marked a turning point as the oil crises of 1973 and 1979 clearly demonstrated the dependence on the fossil fuel oil and therefore initiated research into alternative sources of energy. Although because of a stabilisation in the oil market interest in energy autarchy has dwindled in the 1980s increasing concern about pollution and global atmospheric effects as a consequence of combustion of fossil fuels has revived the search for future energy carriers and alternative renewable resources. Besides conventional techniques (converting water, wind and direct solar energy) electrolytical and biological generation of hydrogen as well as the use of biomass for power generation are among the most promising proposals [238].

The ability of many photosynthetic organisms to produce hydrogen has been recognised for many years. The low efficiency of the process and the fact that hydrogen production is often sensitive to oxygen inhibition have largely precluded the exploitation of this process. However, recent experimental results indicate that there may be hope for resolving some of these problems and thereby enhancing future applications [239, 240]. Again marine (and halotolerant) micro-organisms play a dominant role, as for example *Oscillatoria* Miami BG7, which was shown to produce hydrogen over long periods of time and at high rates as a result of the absence of uptake hydrogenase activity. The rates were in the region of 8 μl of hydrogen per mg dry weight per hour ($=$ 1 ml of hydrogen per ml of reaction mixture within a 3-day illumination period) [241]. Without doubt this efficiency can certainly be improved in the future, and a technologically obtainable production rate of 1 l/day per litre of culture medium (at optimal illumination) is presently under discussion [240]. It should, however, not be overlooked that the running costs of biosolar hydrogen production plants are very high and that this process has to compete economically with electrolytically produced hydrogen. Hence, future costs of electrical energy will determine whether the biological process is viable. Estimates towards the next 20 years predict that the cost of electricity will drop, mainly because of increased production from renewable sources, so that the cost of electrolytically produced hydrogen of the same energy content may well approach the price of (heavily taxed) petrol [238].

The generation of fuel from biomass is presently recognised as a significant future energy source, especially if conversion into liquid or gaseous carriers is considered (e.g. ethanol, methane, oil, etc.). However, biomass 'energy farming' is limited because of severe competition with conventional food production. If one considers that the biomass needed to replace fossil oil totally is at least 2–3 times higher than the total annual crop production, the whole problem of production capacity becomes apparent. Concepts employing land and water resources not used or underutilised by conventional agriculture are the most attractive. Over one-third of the world's potentially arable land is found in semiarid regions, where severe salinity problems are a major factor in desertification. Apart from typical desert plants (succulents), potentially of interest for farming purposes, microscopic algae can be considered among those most useful for fuel production in a biosaline environment [242]. The main disadvantage of algal farming as opposed to desert plant cultures is its requirement of a pond from which water can evaporate freely. Present experience in outdoor microalgae mass culture concludes that yields of 15–20 g m^{-2} day^{-1} can be achieved over prolonged periods of time. The biomass yield (approximately 50 tons ha^{-1} year^{-1}) would, therefore, be at least 10 times higher than the desert plants produce [242]. In addition microalgae are capable of using saline waters unsuitable for even the most tolerant plants. As one would intend to establish specific (pure) algal populations selected for their specific properties, contamination under non-sterile conditions (outdoor ponds) may form a further obstacle. However, given the appropriate conditions (especially high salinity) it should be possible to obtain virtually oligo- or even monospecific cultures. The most expensive, and up till now most difficult (and power-consuming), part of microalgae cultivation systems is the harvesting. At present the only low-cost harvesting processes are screening (microstraining) or bioflocculation (which requires self-flocculating cultures). If these can be applied the bottleneck would be eliminated.

However, as with other plant materials, producing the biomass is only the first step; cheap processing technology is the key to actually obtaining a valuable product. As oil is the most densely stored and most readily available form of energy for mobile transport, conversion of autotrophic biomass into oil or oil-like fuel would be desirable. This would also resolve two problems of modern societies: firstly, the dependence on fossil oil would be largely alleviated and, secondly, global environmental consequences ('greenhouse effect') due to the combustion of fossil oil would be minimal since the carbon dioxide released is fully recycled as the carbon source for the autotrophically produced biomass, the starting material for the production of oil. A preliminary study has revealed that the conversion of algal biomass (*Dunaliella*) can be achieved by a catalytic process at 200–240°C (pyrolysis), which also recycles most of the nitrogen of the biomass. It was further shown that proteins rather than hydrocarbons are the most suitable source for the catalytic conversion [243]. As the overall process is exothermic, most of the thermal energy needed for pyrolysis may also be regained. Last but not least, the price will determine whether the process will become economically viable in the near future. First (optimistic) estimates rest at approximately $40 per barrel as compared with a world market price of approximately $20 per barrel for fossil oil [243]. The odds are, therefore, not too bad for at least partial replacement of fossil oil by liquidised renewable sources like biomass from saline ponds.

In considering the future development in the biotechnology of halotolerant and halophilic micro-organisms it is only possible to use current trends and concepts as guidelines. However, as in many areas of research, subsequent developments depend upon factors not necessarily determined at the laboratory level. It is obvious from this review that certain 'biotechnological' applications (production of fermented salted beans and fish) have been in use for centuries, while other ideas and concepts are being developed. However, it seems clear that a better understanding of the role played by halotolerant and halophilic micro-organisms may contribute to the future, whether it be in the Chinese restaurant, the biodegradable plastic carrier bag, or the use of enzymes stabilised by compatible solutes.

References

1. H. Larsen, in *The Bacteria* Vol. VI, eds. I.C. Gunsalus and R.Y. Stanier, Academic Press, New York (1969) 297.
2. H. Larsen, *FEMS Microbiol. Rev.* **39** (1986) 3.

3. D.J. Kushner, in *Microbial Life in Extreme Environments*, ed. D.J. Kushner, Academic Press, New York (1978) 317.
4. F. Rodriguez-Valera, in *Halophilic Bacteria* Vol. I, ed. F. Rodriguez-Valera, CRC Press, Boca Raton, FL (1988) 3.
5. F. MacIntyre, *Sci. Am.* **223** (1970) 104.
6. A.B. Ronov, *Sedimentology* **10** (1968) 25.
7. H.P. Eugster and L.A. Hardie, in *Lakes: Chemistry, Geology, and Physics*, ed. A. Lerman, Springer Verlag, New York (1978) 237.
8. L.A. Hardie and H.P. Eugster, *Miner. Soc. Am. Spec. Papers* **3** (1970) 273.
9. B.J. Tindall, in *The Prokaryotes, a Handbook on the Biology of Bacteria*, eds. A. Balows, H.G. Trüper, M. Dworkin, W. Harder and K.H. Schleifer, Springer Verlag, New York (1991), in press.
10. K. Sanderson, T.A. McMeekin, N. Indriati, A.M. Angawati and Y. Sudrajat, *Asian Food J.* **4** (1988) 31.
11. K.H. Steinkraus, *Handbook of Indigenous Fermented Foods*, Marcel Dekker, New York (1983).
12. O. Ciferri, *Microbiol, Rev.* **47** (1983) 551.
13. G. Clement, C. Giddey and R. Menzi, *J. Sci. Food Agric.* **18** (1967) 497.
14. S. Aaronson and Z. Dubinsky, *Experientia* **38** (1982) 36.
15. C. Santillan, *Experientia* **38** (1982) 40.
16. E.W. Becker and L.V. Venkataraman, in *Deutsche Gesellschaft für Technische Zusammenarbeit*, Eschborn (1982) 1.
17. A. Richmond and A. Vonshak, *Arch. Hydrobiol* **11** (1978) 274.
18. A. Iltis, *Cah. ORSTOM Ser. Hydrobiol.* **2** (1986) 119.
19. C.H. Tuite, *Freshwater Biol.* **11** (1981) 345.
20. P. Roth and K. Sattler, *Biologische Rundschau* **22** (1984) 159.
21. G. Parkinson, *Chem. Eng.* **94** (1987) 19.
22. A. Ben-Amotz and M. Avron, *Ann. Rev. Microbiol.* **37** (1983) 95.
23. A. Ben-Amotz and M. Avron, in *Algal and Cyanobacterial Biotechnology*, eds R.C. Cresswell, T.A.V. Rees and N. Shah, Longman Scientific and Technical Press (1989) 90.
24. A. Nissenbaum, *Microbiol. Ecol.* **2** (1975) 139.
25. L.J. Borowitzka, *Hydrobiologia* **81** (1981) 33.
26. F.J. Post, *Hydrobiologia* **81** (1981) 56.
27. A. Ben-Amotz and M. Avron, in *Algae Biomass: Production and Use*, eds. G. Shelef and C.J. Soeder, Elsevier, Amsterdam (1980) 603.
28. A. Ben-Amotz, S. Edelstein and M. Avron, *Br. Poultry Sci.* **846** (1986) 313.
29. K.F. Gey, G.B. Brubacher and H.B. Stähelin, *Am J. Clin. Nutr.* **45** (1987) 1368.
30. C.D. Jensen, T.W. Howes, G.A. Spiller, T.S. Pattison, J.H. Whittan and J. Scala, *Nutr. Rep. Int.* **35** (1987) 413.
31. A. Ben-Amotz and M. Avron, *Trends Biochem. Sci.* **6** (1981) 297.
32. A. Ben-Amotz, A. Katz and M. Avron, *J. Phycol.* **18** (1982) 529.
33. A. Vonshak, in *Micro-algal Biotechnology*, eds. L.J. Borowitzka and M.A. Borowitzka, Cambridge University Press (1988) 122.
34. B. Mattiasson and B. Hahn-Hägerdal, *Eur. J. Appl. Microbiol. Biotechnol.* **16** (1982) 52.
35. P. Gervais, W. Grajek, M. Bensoussan and P. Molin, *Biotechnol. Bioeng.* **31** (1988) 457.
36. P. Gervais, *Appl. Microbiol. Biotechnol.* **33** (1990) 72.
37. E.A. Galinski, in *Bioreactors and Biotransformations*, eds. G.W. Moody and P.B. Baker, Elsevier, London (1987) 201.
38. Z. Szklarska-Smialowska and M. Janik-Czachor, *Corrosion Science* **11** (1971) 901.
39. H. Böhni, *Werkstoffe und Korrosion* **25** (1974) 97.
40. A. Mitsui, in *The Biosaline Concept: An approach to the Utilization of Underexploited Resources*, eds. A. Hollaender, J.C. Aller, E. Epstein, A. San Pietro and O.R. Zaborsky, Plenum Press, New York (1979) 177.
41. K. Shinohara, Y. Zhao and G.H. Sato, in *Advances in Gene Technology: Molecular Genetics of Plants and Animals*, Academic Press, New York (1983) 327.
42. M.R. Tredici, T. Papuzzo and L. Tomaselli, *Appl. Microbiol. Biotechnol.* **24** (1986) 47.
43. D.J. Gilmour, in *Microbiology of Halophilic Microorganisms*, ed. C. Edwards, Open University Press, Milton Keynes (1990) 147.

44. A. Richmond, *Handbook of Microalgal Mass Culture*, CRC Press, Boca Raton, FL (1986).
45. M. Avron, in *Microbial Mats, Physiological Ecology of Benthic Microbial Communities*, eds. Y. Cohen and E. Rosenberg, American Society for Microbiology, Washington DC (1989) 387.
46. S. Nair and U. Simidu, *Appl. Environ. Microbiol.* **53** (1987) 2957.
47. B. Austin, *J. Appl. Bacteriol.* **67** (1989) 461.
48. R.B. Baarn, N.M. Gandhi and Y.M. Freitas, *Helgoländer Wissenschaftliche Meeresuntersuchungen* **13** (1966) 181.
49. Y. Okami, *Microb. Ecol.* **12** (1986) 65.
50. P.R. Burkholder, R.M. Pfister and F.H. Leitz, *Appl. Microbiol.* **14** (1966) 649.
51. F.M. Lovell, *J. Am. Chem. Soc.* **88** (1966) 4510.
52. R.J. Anderson, M.S. Wolfe and D.J. Faulkner, *Marine Biol.* **27** (1974) 281.
53. M.J. Gauthier and G.N. Flateau, *Can. J. Microbiol.* **22** (1976) 1612.
54. S.J. Wratten, M.S. Wolfe, R.J. Anderson and D.J. Faulkner, *Antimicrob. Agents and Chemother.* **11** (1977) 411.
55. B.J. Javor and J. Porta, in *General and Applied Aspects of Halophilic Microorganisms*, ed. F. Rodriguez-Valera, Plenum Press, New York (1991) 33.
56. F. Rodriguez-Valera, G. Juez and D.J. Kushner, *Can. J. Microbiol.* **28** (1982) 151.
57. R. LaBelle and C.P. Gerba, *Appl. Environ. Microbiol.* **39** (1980) 749.
58. R.S. Fujioka, P.C. Lok and L.S. Lau, *Appl. Environ. Microbiol.* **39** (1980) 1105.
59. A.E. Toranzo, J.L. Barja and F.M. Hetrick, *Can. J. Microbiol.* **28** (1982) 231.
60. M. Soud, G. Baldacci, P. Forterre and A.M. DeRecondo, *Nucleic Acids Res.* **15** (1987) 8217.
61. P. Forterre, in *Microbiology of Extreme Environments and its Potential for Biotechnology*, eds. M.S. da Costa, J.C. Durate and R.A.D. Williams, Elsevier, London (1989) 152.
62. P. Forterre, D. Gadelle, F. Charbonnier, C. Elie, M. Sioud and A. Hamal, *Can. J. Biochem.* **35** (1989) 228.
63. P. Forterre, in *General and Applied Aspects of Halophilic Microorganisms*, ed. F. Rodriguez-Valera, Plenum Press, New York (1991) 333.
64. Y. Okami, *Microb. Ecol.* **12** (1986) 65.
65. K. Ben-Mahrez, W. Sougakoff, M. Nakayama and M. Kohiyama, *Arch. Microbiol.* **149** (1988) 175.
66. K. Ben-Mahrez, I. Sorokine, D. Thierry, T. Kawasumi, S. Ishii, R. Salmon and M. Kohiyama, in *General and Applied Aspects of Halophilic Microorganisms*, ed. F. Rodriguez-Valera, Plenum Press, New York (1991) 367.
67. K. Arima, G. Tamura and A. Kakinuma, *US Patent* (1972) 3,687,926; U.S.Cl. 260/112.5, 195/80.
68. N. Kosaric, N.C.C. Gray and W.L. Cairns, in *Biotechnology* Vol. 3, eds. H.-J. Rehm and G. Reed, Verlag Chemie, Weinheim (1983) 575.
69. W. Crueger and A. Crueger, in *Biotechnology*, a Textbook for Industrial Microbiology, ed. T.D. Brock, Sinauer Associates, Sunderland, MA (1984) 288.
70. E. Rosenberg, C. Rubinovitz, A. Gottlieb, S. Rosenhak and E.Z. Ron, *Appl. Environ. Microbiol.* **54** (1988) 317.
71. R.K. Hommel, *Biodegradation* **1** (1990) 107.
72. Z. Zosin, D. Gutnick and E. Rosenberg, *Biotechnol. Bioeng.* **24** (1982) 281.
73. D.L. Gutnick and Y. Shabtai, in *Biosurfactants and Biotechnology* Vol. 25, eds. N. Kosaric, W.L. Cairns and N.C.C. Gray, Marcel Dekker, New York (1987) 211.
74. J.M. Foght, D.L. Gutnick and D.W.S. Westlake, *Appl. Environ. Microbiol.* **55** (1989) 36.
75. M.C. Cirigliano and G.M. Carman, *Appl. Environ. Microbiol.* **50** (1985) 846.
76. S. Banerjee, S. Duttagupta and A.M. Chakrabarty, *Arch. Microbiol.* **135** (1983) 110.
77. D.G. Cooper and B.G. Goldenberg, *Appl. Environ. Microbiol.* **53** (1987) 224.
78. C.D. Boyle and A.E. Reade, *Appl. Environ. Microbiol.* **46** (1983) 392.
79. R.M. Weiner, R.R. Colwell, R.N. Jarman, D.C. Stein, Ch. C. Somerville and D.B. Bonar, *Biotechnology* **3** (1985) 899.
80. J. Anton, I. Meseguer and F. Rodriguez-Valera, *Appl. Environ. Microbiol.* **54** (1988) 2381.
81. J. Anton Botella, I. Meseguer and F. Rodriguez-Valera, in *General and Applied Aspects of Halophilic Microorganisms*, FEMS-NATO Workshop, Alicante (1989) 114.
82. G.G. Geesey, *ASM News* **48** (1982) 9.
83. S.M. Pfiffner, M.J. MacInerney, G.E. Jenneman and R.M. Knapp, *Appl. Environ. Microbiol.* **51** (1986) 1224.

84. J. Vater, *Progr. Colloid Polymer Sci.* **72** (1986) 12.
85. R.M. Fakoussa, *Resources, Conservation and Recycling* **1** (1988) 251.
86. T.R. Neu, T. Härtner and K. Poralla, *Appl. Microbiol. Biotechnol.* **32** (1990) 5188.
87. T. Matsuyama, T. Murakami, M. Fujita, S. Fujita and I. Yano, *J. Gen. Microbiol.* **132** (1986) 865.
88. K. Poremba, W. Gunkel, S. Lang and F. Wagner, *Biologie in unserer Zeit* **19** (1989) 145.
89. F.J. Post and F.A. Al-Harjan, *Syst. Appl. Microbiol.* **11** (1988) 97.
90. P.A. Holmes, *Phys. Technol.* **16** (1985) 32.
91. A. Steinbüchel, P. Schubert, A. Timm and A. Pries, in *Novel Biodegradable Microbial Polymers*, ed. E.A. Dawes, Kluwer Academic Publishers, The Netherlands (1990) 143.
92. P. Schubert, N. Krüger and A. Steinbüchel, *J. Bacteriol.* **173** (1991) 168.
93. A. Steinbüchel and H.G. Schlegel, *Mol. Microbiol.* **5** (1991) in press.
94. Anon., *Bioengineering* **5** (1990) 11.
95. F. Rodriguez-Valera, J.A.G. Lillo, J. Anton and I. Meseguer, in *General and Applied Aspects of Halophilic Microorganisms*, ed. F. Rodriguez-Valera, Plenum Press, New York (1991) 373.
96. E. Heinzle and R.M. Lafferty, *Eur. J. Appl. Microbiol. Biotechnol.* **11** (1980) 8.
97. J.A. Garcia Lillo and F. Rodriguez-Valera, in *General and Applied Aspects of Halophilic Microorganisms*, FEMS-NATO Workshop, Alicante (1989) 130.
98. J. Garcia Lillo and F. Rodriguez-Valera, *Appl. Environ. Microbiol.* **56** (1990) 2517.
99. F. Rodriguez-Valera, G. Juez and D. J. Kushner, *Syst. Appl. Microbiol.* **4** (1983) 369.
100. A. Pries, A. Steinbüchel and H.G. Schlegel, *Appl. Microbiol. Biotechnol.* **33** (1990) 410.
101. A. Steinbüchel, *Forum Mikrobiologie* **10** (1989) 190.
102. R. Poole, *Science* **245** (1989) 1187.
103. A.J. Anderson and E.A. Dawes, *Microbiol.* **54** (1990) 450.
104. H. Onishi, *J. Bacteriol.* **109** (1970) 570.
105. H. Onishi, *Can. J. Microbiol.* **18** (1970) 1617.
106. V.A. Good and P.A. Hartman, *J. Bacteriol.* **104** (1970) 601.
107. M. Kamekura and H. Onishi, *J. Bacteriol.* **119** (1974) 339.
108. H. Onishi, T. Mori, S. Takeuchi, K. Tani, T. Kobayashi and M. Kamekura, *Appl. Environ. Microbiol.* **45** (1983) 24.
109. P. Norberg and B.V. Hofsten, *J. Gen. Microbiol.* **55** (1969) 251.
110. M. Kamekura and H. Onishi, *Appl. Microbiol.* **27** (1974) 809.
111. K. Makino, T. Koshikawa, T.H. Nishira, T. Ichikawa and M. Kondo, *Microbios* **31** (1981) 103.
112. M. Kamekura and Y. Seno, *Biochem. Cell Biol.* **68** (1990) 352.
113. L.S. Izotova, A.Y. Strongin, L.N. Chekulaeva, V.E. Sterkin, V.I. Ostoslavskaya, L.A. Lyublinskaya, E.A. Timokhina and V.M. Stephanov, *J. Bacteriol.* **155** (1983) 826.
114. M. Kamekura, T. Hamakawa and H. Onishi, *Appl. Environ. Microbiol.* **44** (1982) 994.
115. H. Yokoi and H. Onishi, *Agric. Biol. Chem.* **54** (1990) 2573.
116. H. Onishi, H. Yokoi and M. Kamekura, in *General and Applied Aspects of Halophilic Microorganisms*, ed. F. Rodriguez-Valera, Plenum Press, New York (1991) 341.
117. A. Obayashi, N. Hiraoka, K. Kita, H. Nakajima and T. Shuzo, *US Patent* **4** (1988) 724, 209, US Cl. 435/199.
118. Y. Yamada, H. Mizuno, T. Suzuki, M. Akagawa and K. Yamasato, *Agric. Biol. Chem.* **53**, (1989) 1747.
119. H. Mizuno, T. Suzuki, Y. Yamada, M. Akagawa and K. Yamasato, *Agric. Biol. Chem.* **54** (1990) 2863.
120. H. Mizuno, T. Suzuki, M. Akagawa, K. Yamasato and Y. Yamada, *Agric. Biol. Chem.* **54** (1990) 1797.
121. A.D. Brown, *Bact. Rev.* **40** (1976) 803.
122. E.A. Galinski, *PhD thesis*, University of Bonn (1986).
123. E.A. Galinski and R.M. Herzog, *Arch. Microbiol.* **153** (1990) 607.
124. P.H. Yancey, M.E. Clark, S.C. Hand, R.D. Bowlus and G.N. Somero, *Science* **217** (1982) 1214.
125. H.G. Trüper and E.A. Galinski, *Experientia* **42** (1986) 1182.
126. H.G. Trüper and E.A. Galinski, In *Microbial Mats—Ecological Physiology of Benthic Microbial Communities*, eds. Y. Cohen and E. Rosenberg, ASM Washington DC (1989) 342.
127. L.N. Csonka, *Microbiol. Rev.* **53** (1989) 121.

128. H.G. Trüper, J. Severin, A. Wohlfarth, E. Müller and E.A. Galinski, in *General and Applied Aspects of Halophilic Microorganisms*, ed. F. Rodriguez-Valera, Plenum Press, New York (1991) in press.
129. M.A. Mackay, R.S. Norton and L.J. Borowitzka, *Marine Biol.* **73** (1983) 301.
130. R.H. Reed, D.L. Richardson, S.R.C. Warr and W.D.P. Stewart, *J. Gen. Microbiol.* **130** (1984) 1.
131. S.R.C Warr, R.H. Reed and W.D.P. Stewart, *J. Gen. Microbiol.* **130** (1984) 2169.
132. A. van Laere, *FEMS Microbiol. Rev.* **63** (1989) 201.
133. A. Wiemken, *Antonie van Leeuwenhoek* **58** (1990) 209.
134. I. Ahmad and J.A. Hellebust, *Plant. Physiol.* **74** (1984) 1010.
135. B. Schobert, *Z. Pflanzenphysiol.* **74** (1974) 106.
136. B. Schobert, in *Plant Membrane Transport: Current Conceptual Issues*, eds. R.M. Spanswick, W.J. Lucas and J. Dainty, Elsevier/North-Holland Biomedical Press, Amsterdam (1980) 487.
137. A. Ben-Amotz and M. Avron, *Ann. Rev. Microbiol.* **37** (1983) 95.
138. E.J. Luard, *J. Gen. Microbiol.* **128** (1982) 2583.
139. H. Kauss, *Progress in Phytochemistry* **5** (1979) 1.
140. D.M.J. Dickson and G.O. Kirst, *Planta* **167** (1986) 536.
141. B.J.D. Meeuse, in *Physiology and Biochemistry of Algae*, ed. R.A. Lewin, Academic Press, New York (1962) 289.
142. J.A. Chudeck, R. Foster, I.R. Davison and R.H. Reed, *Phytochemistry* **23** (1984) 1081.
143. A. Ben-Amotz and T. Grunwald, *Plant Physiol.* **67** (1981) 613.
144. J.S. Craigie and J. MacLachlan, *Can. J. Bot.* **42** (1964) 777.
145. K. Wegmann, *Biochim. Biophys. Acta* **234** (1971) 317.
146. A. Ben-Amotz and M. Avron, *Plant Physiol.* **51** (1973) 875.
147. L.J. Borowitzka and A.D. Brown, *Arch. Microbiol.* **96** (1974) 37.
148. G.M. Gadd, J.A. Chudek, R. Foster and R.H. Reed, *J. Gen. Microbiol.* **130** (1984) 1969.
149. M.S.da Costa and M.F. Nobre, in *Microbiology of Extreme Environments and its Potential for Biotechnology*, eds. M.S.da Costa, J.C. Duarte and R.A.D. Williams, Elsevier, London (1989) 310.
150. M.F. Nobre and M.S.da Costa, *Can. J. Microbiol.* **31** (1985) 467.
151. M.F. Nobre and M.S.da Costa, in *Microbiology of Extreme Environments and its Potential for Biotechnology*, eds. M.S.da Costa, J.C. Duarte and R.A.D. Williams, Elsevier, London (1989) 424.
152. P.-A. Jovall, I. Tunblad-Johansson and L. Adler, *Arch. Microbiol.* **154** (1990) 209.
153. C. Larsson, C. Morales, L. Gustafsson and L. Adler, *J. Bacteriol.* **172** (1990) 1769.
154. L.J. Borowitzka, S. Demmerle, M.A. Mackay and R.S. Norton, *Science* **210** (1980) 650.
155. M.A. Mackay, R.S. Norton and L.J. Borowitzka, *J. Gen. Microbiol.* **130** (1984) 2177.
156. R.H. Reed, L.J. Borowitzka, M.A. Mackay, J.A. Chudek, R. Foster, S.R.C. Warr, D.J. Moore and W.D.P. Stewart, *FEMS Microbiol. Rev.* **39** (1986) 51.
157. A. Wohlfarth, J. Severin and E.A. Galinski, *J. Gen. Microbiol.* **136** (1990) 705.
158. J.F. Imhoff and F. Rodriguez-Valera, *J. Bacteriol.* **160** (1984) 478.
159. J.F. Imhoff, *FEMS Microbiol. Rev.* **39** (1986) 57.
160. W. Schuh, H. Puff, E.A. Galinski and H.G. Trüper, *Z. Naturforsch.* **40c** (1985) 780.
161. J. Severin, A. Wohlfarth and E.A. Galinski, *J. Gen. Microbiol.* (1991) submitted.
162. E.A. Galinski, H.P. Pfeiffer and H.G. Trüper, *Eur. J. Biochem.* **149** (1985) 135.
163. P.Peters, E.A. Galinski and H.G. Trüper, *FEMS Microbiol. Lett.* **71** (1990) 157.
164. J. Severin, A. Wohlfarth and E.A. Galinski, in *General and Applied Aspects of Halophilic Microorganisms*, FEMS-NATO Workshop, Alicante (1989) 132.
165. K.R. Sowers, D.E. Robertson, D. Noll, R.P. Gunsalus and M.F. Roberts, *Proc. Natl. Acad. Sci. USA* **87** (1990) 9083.
166. E.A. Galinski and A. Oren, *Eur. J. Biochem.* (1991) in press.
167. A. Oren, G. Simon and E.A. Galinski, *Arch. Microbiol.* (1991) 593.
168. J.P. Nakas, M. Schaedle, C.M. Parkinson, C.E. Cooney and S.W. Tannenbaum, *Appl. Environ. Microbiol.* **46** (1983) 1017.
169. B.C. Hard and D.J. Gilmour, in *General and Applied Aspects of Halophilic Microorganisms*, FEMS-NATO Workshop, Alicante (1989) 98.
170. W. Crueger and A. Crueger, *Biotechnologie— Lehrbuch der angewandten Mikrobiologie*, 3rd edition, Oldenbourg Verlag, Munich (1989), Kapitel 9.

171. Y. Hirose and H. Okada, in *Microbial Technology—Microbial Processes* 2nd edition Vol. I, eds. H.J. Peppler and D. Perlman, Academic Press, New York (1979) 211.
172. Y. Minoda, in *Biotechnology of Amino Acid Production*, eds. K. Aida, I. Chibata, K. Nakayama, K. Takinami and H. Yamada, Kodansha, Tokyo (1986) 51.
173. S. Okumura, in *The Microbial Production of Amino Acids*, eds. K. Yamada, S. Kinoshita, T. Tsunoda and K. Alida, Kodansha, Tokyo (1972) 473.
174. F. Yoshinaga, in *Biotechnology of Amino Acid Production, Progress in Industrial Microbiology* Vol. 24, eds. K. Alida, I. Chibata, K. Nakayama, K. Takinami and H. Yamada, Kodansha, Tokyo (1986) 117.
175. L.N. Csonka, *Mol. Gen. Genet.* **182** (1981) 82.
176. M. Sugiura and M. Kisumi, *Appl. Environ. Microbiol.* **49** (1985) 782.
177. E. Müller and H.G. Trüper, in *Microbiology of Extreme Environments and its Potential for Biotechnology*, M.S.da Costa, J.C. Duarte and R.A.D. Williams, Elsevier, London (1989) 422.
178. E. Müller and H.G. Trüper, in *General and Applied Aspects of Halophilic Microorganisms*, FEMS-NATO Workshop, Alicante (1989) 131.
179. A. Pollard and R.G. Wyn Jones, *Planta* **144** (1979) 291.
180. S.J. Coughlan and U. Heber, *Planta* **156** (1982) 62.
181. Y. Jolivet, F. Larher and J. Hamelin, *Plant Sci. Lett.* **25** (1982) 193.
182. K.A. Pavlicek and J.H. Yopp, *Plant. Physiol.* **69** (1982) 58 (suppl.).
183. E.J. Luard, *Arch. Microbiol.* **134** (1983) 233.
184. Y. Manetas, Y. Petropoulou and G. Karabourniotis, *Plant Cell Environ.* **9** (1986) 145.
185. T. Arakawa and S.N. Timasheff, *Arch. Biochem. Biophys.* **224** (1983) 169.
186. T. Arakawa and S.N. Timasheff, *Biophys. J.* **47** (1985) 411.
187. J.F. Carpenter, S.C. Hand, L.M. Crowe and J.H. Crowe, *Arch. Biochem. Biophys.* **250** (1986) 505.
188. J.F. Carpenter, L.M. Crowe and J.H. Crowe, *Biochim. Biophys. Acta* **923** (1987) 109.
189. E.A. Galinski and K. Lippert, in *General and Applied Aspects of Halophilic Microorganisms*, ed. F. Rodriguez-Valera, Plenum Press, New York (1991) 351.
190. K. Lippert, E.A. Galinski and H.G. Trüper, in *General and Applied Aspects of Halophilic Microorganisms*, FEMS-NATO Workshop, Alicante (1989) 10.
191. S.N. Timasheff, in *Biophysics of Water*, eds. F. Franks and S. Mathias, Wiley, New York (1982) 70.
192. P.S. Low, in *Transport Processes, Iono- and Osmoregulation*, eds. R. Gilles and M. Gilles-Baillien, Springer Verlag, Berlin (1985) 470.
193. J.F. Carpenter and J.H. Crowe, *Cryobiology* **25** (1988) 244.
194. P.M. Wiggins, *Microbiol. Rev.* **54** (1990) 432.
195. K.A.C. Madin and J.H. Crowe, *J. Exp. Zool.* **193** (1975) 335.
196. J.H. Crowe and J.S. Clegg, *Dry Biological Systems*, Academic Press, New York (1978).
197. C. Womersley, *Comp. Biochem. Physiol.* **B70** (1981) 669.
198. J.H. Crowe, L.M. Crowe and D. Chapman, *Science* **223** (1984) 701.
199. G.M. Gadd, K. Chalmers and R.H. Reed, *FEMS Microbiol. Letters* **48** (1987) 249.
200. J.H. Crowe, L.M. Crowe, J.F. Carpenter and C.A. Wistrom, *Biochem. J.* **242** (1987) 1.
201. B.P. Gaber, I. Chandrasekhar and M. Pattiabiraman, *Biophys. J.* **49** (1986) 435a.
202. I. Chandrasekhar and B.P. Gaber, *J. Biomol. Struct. Dyn.* **5** (1988) 1163.
203. C.W.B Lee, J.S. Waugh and R.G. Griffin, *Biochemistry* **25** (1986) 3737.
204. J.F. Carpenter and J.H. Crowe, *Cryobiology* **25** (1988) 459.
205. J.F. Carpenter and J.H. Crowe, *Biochemistry* **28** (1989) 3916.
206. J.H. Crowe, J.F. Carpenter, L.M. Crowe and T.J. Anchordoguy, *Cryobiology* **27** (1990) 219.
207. W.J. Oswald and J.R. Benemann, in *The Biosaline Concept: An Approach to the Utilization of Underexploited Resources*, eds. A. Hollaender, J.C. Aller, E. Epstein, A. San Pietro and O.R. Zaborsky, Plenum Press, New York (1979) 285.
208. A.G. Jones, C.M. Ewing and M.V. Melvin, *Hydrobiologia* **82** (1981) 391.
209. J.J Nieto, in *General and Applied Aspects of Halophilic Microorganisms*, ed. F. Rodriguez-Valera, Plenum Press, New York (1991) 173.
210. J.J. Nieto, R. Fernandez-Castillo, M.C. Marquez, A. Ventosa and F. Ruiz-Berraquero, in *General and Applied Aspects of Halophilic Microorganisms*, FEMS-NATO Workshop, Alicante (1989) 79.

211. G. Denariaz, W.J. Payne and J. LeGall, *Int. J. Syst. Bacteriol.* **39** (1989) 145.
212. G. Denariaz, W.J. Payne and J. LeGall, in *Microbiology of Extreme Environments and its Potential for Biotechnology*, eds. M.S.da Costa, J.C. Duarte and R.A.D. Williams, Elsevier, London (1989) 328.
213. L.I. Hochstein and G.A. Tomlinson, *FEMS Microbiol. Lett.* **27** (1985) 329.
214. R. Mancinelli and L.I. Hochstein, *FEMS Microbiol. Lett.* **35** (1986) 55.
215. J.C. Aller and O.R. Zaborsky, in *The Biosaline Concept: An Approach to the Utilization of Underexploited Resources*, eds. A. Hollaender, J.C. Aller, E. Epstein, A. San Pietro and O.R. Zaborsky, Plenum Press, New York (1979) 1.
216. D.W. Rush and E. Epstein, *Plant. Physiol.* **57** (1976) 162.
217. T.P. Croughan, D.W. Rains and S.J. Stavar, *Crop Science* **18** (1978) 959.
218. W.G. McGinnies, in *The Biosaline Concept: An Approach to the Utilization of Underexploited Resources*, eds. A. Hollaender, J.C. Aller, E. Epstein, A. San Pietro and O.R. Zaborsky, Plenum Press, New York (1979) 69.
219. E. Epstein, R.W. Kingsbury, J.D. Norlyn and D.W. Rush, in *The Biosaline Concept: An Approach to the Utilization of Underexploited Resources*, eds. A. Hollaender, J.C. Aller, E. Epstein, A. San Pietro and O.R. Zaborsky, Plenum Press, New York (1979) 77.
220. J.A. Bassham, in *The Biosaline Concept: An Approach to the Utilization of Underexploited Resources*, eds. A Hollaender, J.C. Aller, E. Epstein, A. San Pietro and O.R. Zaborsky, Plenum Press, New York (1979) 17.
221. D.W. Rains, in *The Biosaline Concept: An Approach to the Utilization of Underexploited Resources*, eds. A. Hollaender, J.C. Aller, E. Epstein, A. San Pietro and O.R. Zaborsky, Plenum Press, New York (1979) 47.
222. J. Cairney, I.R. Booth and C.F. Higgins, *J. Bacteriol.* **164** (1985) 1218.
223. J. Cairney, I.R. Booth and C.F. Higgins, *J. Bacteriol.* **164** (1985) 1224.
224. B. Perroud and D. LeRudulier, *J. Bacteriol.* **161** (1985) 393.
225. M.J. Gauthier and D. LeRudulier, *Appl. Environ. Microbiol.* **56** (1990) 2915.
226. I.R. Booth and C.F. Higgins, *FEMS Microbiology Reviews* **75** (1990) 239.
227. D. LeRudulier and R.C. Valentine, *Trends Biochem. Sci.* **7** (1982) 431.
228. D. LeRudulier and L. Bouillard, *Appl. Environ. Microbiol.* **46** (1983) 152.
229. D. LeRudulier, A.R. Strom. A.M. Dandekar, L.T. Smith and R.C. Valentine, *Science* **224** (1984) 1064.
230. D. LeRudulier, T. Bernard, G. Goas and J. Hamelin, *Can. J. Microbiol.* **30** (1984) 299.
231. A.R. Strom, D. LeRudulier, M.W. Jakowec, R.C. Bunnell and R.C. Valentine, in *Genetic Engineering of Plants*, eds. T. Kosuge, C.P. Meredith and A. Hollaender, Plenum Press, New York (1984) 39.
232. T. Bernard, J.A. Pocard, B. Perroud and D. LeRudulier, *Arch. Microiol.* **143** (1986) 359.
233. L.T. Smith, J.A. Pocard, T. Bernard and D. LeRudulier, *J. Bacteriol.* **170** (1988) 3142.
234. J.A. Pocard, T. Bernard, L.T. Smith and D. LeRudulier, *J. Bacteriol.* **171** (1989) 531.
235. F. Fougère and D. LeRudulier, *J. Gen. Microbiol.* **136** (1990) 157.
236. J.A. Pocard, T. Bernard, G. Goas and D. LeRudulier, *Comptes Rendus de l'Académie des Sciences* **298** (1984) 477.
237. J.A. Pocard, *Doctoral thesis*, University of Rennes I, France (1987).
238. C.J. Weinberg and R.H. Williams, *Spektrum der Wissenschaft* **11** (1990) 158.
239. A. Mitsui, in *The Biosaline Concept: An Approach to the Utilization of Underexploited Resources*, eds. A. Hollaender, J.C. Aller, E. Epstein, A. San Pietro and O.R. Zaborsky, Plenum Press, New York (1979) 177.
240. J.-H. Klemme, *Naturwissenschaftliche Rundschau* **44** (1991) 52.
241. A. Mitsui, in *Proceedings of Bio-Energy '80*, Bio-Energy Council, Washington DC (1980) 486.
242. J.R. Benemann, in *The Biosaline Concept: An Approach to the Utilization of Underexploited Resources*, eds. A. Hollaender, J.C. Aller, E. Epstein, A. San Pietro and O.R. Zaborsky, Plenum Press, New York (1979) 309.
243. B.Z. Ginzburg, in *General and Applied Aspects of Halophilic Microorganisms*, ed. F. Rodriguez-Valera, Plenum Press, New York (1991) 389.

4 Acidophilic bacteria: adaptations and applications

P.R. NORRIS and W.J. INGLEDEW

4.1 Introduction

Many species of micro-organisms are active at acid pH values. At pH 3, for example, a variety of algae, fungi and protozoa as well as bacteria can thrive and populate acidic environments such as mineral sulphide mine waters and leaching dumps, coal mines and their spoils, and some sulphidic geothermal springs and soils. Some isolates of the green alga *Chlamydomonas acidophila*, for example, grow and retain motility at pH 2, although most strains tend to be less acidophilic. The most extreme acidophiles appear to be bacteria which directly benefit from or create the acidity of their environment. The growth of iron-oxidising acidophiles is facilitated by the acidity which retards the spontaneous oxidation of their substrate, ferrous iron. Sulphuric acid is an end-product of the metabolism of the acidophilic, sulphur-oxidising bacteria, and sulphate appears to be the dominant anion in all naturally occurring acidic habitats which can sustain bacterial growth. An example of the reduction of diversity at extreme acidity in favour of such specialised acidophiles has been described: the proportion of iron-oxidising thiobacilli expressed as percentage biomass dry weight of the mixed microbial flora of mineral leach dump samples increased from approximately 70% in samples at pH 3.5–4 to over 90% at pH 2.4 [1]. The lower pH limit for growth of micro-organisms is not well defined, with final pH values in batch cultures of some mineral sulphide- and sulphur-oxidising bacteria reaching well below pH 1, but with the extreme acidity probably resulting from abiotic and biochemical oxidations which are not coupled to growth. It is these extremely acidophilic bacteria that occupy most of the following discussion, which will concentrate on areas where acidophily is a key feature of the biology or biotechnology in question. For example, many interesting features of acidophilic, thermophilic archaeobacteria which derive largely from aspects of their thermophily, phylogeny or metabolism and which are not directly related to their acidophily are not considered. However, some physiological and bioenergetic consequences of acidophily for chemiosmosis are discussed in detail. In addition, the application of acidophiles in metal extraction where a strongly acid environment is both a requirement for and a consequence of bacterial mineral sulphide oxidation is discussed in relation to the diversity of the potentially useful bacteria. Finally, and again with reference to the problems posed by the organisms' acidic environment, the progress in molecular genetic studies that could provide a foundation for strain improvement in regard to the industrial application of these bacteria is described.

4.2   Constraints on growth at acid pH

Some of the major questions that arise when considering acidophily concern the survival and functioning of biomolecules exposed to extreme acidity; the conditions pertaining in the periplasm, where present; the pH of the cytoplasm and, moreover, if this is close to neutrality, the functioning of the chemiosmotic mechanism with regard to a large pH difference across the cytoplasmic membrane. The chemiosmotic consequences of acidophily have been explained and reviewed extensively [2–5]. In this chapter, these consequences are briefly summarised and discussed further where controversy remains or to provide some new perspectives.

The enzymes and metabolic processes found in acidophiles are generally the same as those found in neutrophiles, as they are designed to function at neutrality and do not have exceptional acid tolerance, i.e. most would be destroyed or impaired if exposed to the acidity at which the cells grow [6–8]. Thus, acidophiles must maintain their cytoplasmic pH close to neutrality. The cytoplasmic components of a cell can be maintained at a different pH to that of the surrounding medium because the cytoplasmic membrane is an osmotic barrier between the two environments. The composition of the cytoplasmic membranes of acidophilic bacteria has been investigated and found to contain no distinguishing features when compared with other bacteria [9]. The cell envelopes of acidophilic eubacteria have the same basic structural features as other Gram-negative and Gram-positive bacteria [10, 11]. However, the region exposed to the acidity is not only structural but also the site of important biochemical processes such as the periplasmic, respiratory reactions of Gram-negative bacteria.

4.2.1   Chemiosmotic considerations

4.2.1.1 *Measurement of the transmembrane pH difference (ΔpH) and the membrane potential ($\Delta\Psi$).* The chemiosmotic theory adequately describes energy conservation in acidophiles. When a proton is pumped from the inside to the outside of a cell the chemical entity of the proton is moved and charge is moved. These two parameters are the chemical potential (represented as ΔpH) and the electrical potential (the membrane potential). Both these parameters apply force to the distribution of a proton and can be summated to give the transmembrane proton–electrochemical potential (Δp) [$\Delta p = \Delta\Psi - 59\,\Delta$pH]. The value of 59 (mV) used in the equation is a conversion constant ($2.303\,RT/nF$, where R = gas constant, T = absolute temperature

(K), n = equivalence and F = Faraday's constant) to convert the concentration term into mV. The negative sign in the equation stems from the sign convention of ΔpH (pH$_{out}$ − pH$_{in}$). These forces can be estimated by measuring the distribution of membrane-permeable probes which are affected by the $\Delta\Psi$ (anions and cations) or the ΔpH (weak acids and bases).

Many reports have established that under normal conditions the cytoplasmic pH of acidophiles is close to neutrality. This was first indicated indirectly by demonstrating that putative cytoplasmic processes were acid-labile and later by direct calculation from the distribution of radiolabelled probes [2–5, 12–14]. More recently, magnetic resonance techniques have been applied to this problem [13–15]. The methods used have been similar to those developed for the study of neutrophile and mitochondrial bioenergetics except that care is required in the choice of probes in order to obtain good accumulation ratios to facilitate measurement of the required parameters [2, 16, 17]. A number of reviews have collated measurements of ΔpH and $\Delta\Psi$ from different bacteria and plotted these as a function of the pH of the suspending media [2, 3, 5]. For ease of illustration, this type of plot is shown in Figure 4.1. Because the external pH is acid and the cytoplasm neutral, the transmembrane pH difference (ΔpH) is very large in acidophiles. The ΔpH term of the chemiosmotic relationship shows little variation because it is the difference between the imposed external pH and the nearly neutral cytosol. Figure 4.1a shows measurements of pH$_{in}$, plotted as a function of pH$_{out}$, which have been determined for a number of different acidophiles. The methods used have to be employed with care and are prone to errors, which may explain the scatter. In addition, strain variations and experimental conditions employed may result in measurements away from the norm. Ignoring possible systematic error in all these measurements, the internal pH in energised (respiring) cells appears to be around 6.5 and independent of external pH (Figure 4.1a). When the cells are de-energised the pH$_{in}$ does not fall significantly (except for some measurements made under inappropriate conditions). Further discussion of the maintenance of pH$_{in}$ follows later. Thus it is the $\Delta\Psi$ which must vary with the energy status of the cell. This constraint is not unique to acidophiles: in all cells tested the ΔpH is constrained and it is therefore the $\Delta\Psi$ parameter which reflects the energy state. The relationship between $\Delta\Psi$ and pH$_{out}$ is illustrated in Figure 4.1b. The plot clearly shows the dependence of $\Delta\Psi$ on pH$_{out}$. The lines drawn are the theoretical ones obtained by assuming pH$_{in}$ is 6.5 and Δp is 240 mV (energised) or pH$_{in}$ is 6.0 and Δp is 0 mV (de-energised) throughout. It is apparent from these data that $\Delta\Psi$ is altered by the energy state of the cell, becoming large and relatively positive inside, to balance the ΔpH when Δp tends to zero.

4.2.1.2 *The chemiosmotic consequences of acidophily.*   The consequences of a large ΔpH, acidic outside, for the cell and for the magnitude and polarity of the $\Delta\Psi$ have recently been considered in quantitative detail [3, 17]. Here

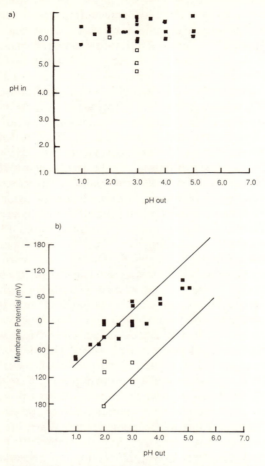

Figure 4.1   The relationships between the pH of the suspending media and chemiosmotic parameters. (a) A plot of the cytoplasmic pH (pH_{in}) of energised (solid symbols) and de-energised cells (open symbols) against the pH of the suspending media. The data points come from a number of investigations on different bacteria (*Thiobacillus ferrooxidans*, *Bacillus coagulans*, *Thermoplasma acidophilum* and *Bacillus acidocaldarius*; refs. 4, 5, 16, and 20). (b) A plot of membrane potential against pH of the suspending media. The solid lines are explained in the text. The data points come from a number of investigations on different bacteria (refs. 4, 5, 16 and 20). The measurements were taken on respiring cells (solid symbols) and de-energised cells (open symbols).

the salient points will be summarised and the reader is referred to these articles. The large transmembrane pH difference imposes the problem that this concentration difference can result in the non-specific accumulation of weak acids. As a consequence, acidophiles are very sensitive to simple organic acids as these can be concentrated in the cytoplasm. The mechanism by which an acid distributes across a membrane in response to the ΔpH has been discussed elsewhere [3, 18]. The premise is that the protonated form of the acid is membrane-permeable and the dissociated form is not. The permeable form (HA) will equilibrate across the membrane, giving the same concentration on each side, i.e $[HA_{in}] = [HA_{out}]$. To illustrate the consequences of this for the distribution of the dissociated form it is instructive to set up Henderson–Hasselbach equations, $pH = pK_a + [\log A^-/HA]$, one for the external pH and one for the internal pH and substract them. Given a pH_{out} of 2 and a pH_{in} of 7, and a weak acid with a pK_a of 2.5, the equilibrium distribution of the combination of both forms $(HA + A^-)$ is 1:24 033 out to in. In practice, if the amount of weak acid added externally is low, the cytoplasmic pH can be maintained, but if the concentration of the acid is significant the internal pH will decrease in response to the entry and dissociation of the acid [3]. This phenomenon explains the toxicity of simple organic acids to acidophiles.

From Figure 4.1, it can be seen that in an acid medium (for example, pH 2), the $\Delta\Psi$ is relatively small in the energised state and large and positive in the de-energised state $(pH_{out} = 2)$ [2, 4, 5, 16, 17]. The consequences of these values of $\Delta\Psi$ have been discussed in detail [3]. A permeant ion can distribute according to the $\Delta\Psi$. This distribution is determined by the Nernst equation; for example, for an anion (A), $\Delta\Psi = 2.303\ RT/nF \log_{10}[A^{n-}_{in}]/[A^{n-}_{out}]$. With a relatively positive $\Delta\Psi$ inside, as would be the case in the de-energised state in an acid medium, permeant anions would tend to accumulate and permeant cations would tend to be excluded. This is why the lipophilic anion thiocyanate can be used as a probe for $\Delta\Psi$ as well as a lipophilic cation [16]. The mechanism of accumulation is different to that affecting acids: it is the charged form which enters, and the driving force is $\Delta\Psi$. The end result of permeant ions in the medium may be accumulation and poisoning of the cytoplasm. This has been used to explain in part the unusual sensitivity of acidophiles to anions, with the exception of relatively impermeable sulphate ions, and the potentiation of this sensitivity by de-energisation [17]. Conversely, acidophiles are tolerant of many cations. Extreme acidophiles require sulphate as the predominant anion; only the chemically similar selenate can effectively substitute for sulphate [19]. Low concentrations of thiocyanate, nitrate and iodide are deleterious, as are higher concentrations of bromide and chloride. Studies on *Thiobacillus ferrooxidans* have shown that these ions are particularly toxic when the cells are de-energised, i.e. when $\Delta p = 0$ and thus $\Delta\Psi$ is large and positive inside [17]. In general, the toxicity of the anions follows their expected non-specific permeability, with the possible exception of chloride, which appears to have a higher than expected permeability that may be explained by the existence of porter systems [13, 14, 20].

4.2.1.3 *Uncoupling and pH_{in} homeostasis.* As discussed earlier, de-energised cells retains pH_{in} homeostasis in the absence of permeable anions. In most of the work on de-energised cells, de-energisation was achieved by making the membrane permeable to protons, i.e. uncoupling. In *T. ferrooxidans* in the presence of an uncoupler (dinitrophenol, DNP) or the respiratory poison azide the ΔpH is largely maintained, but the $\Delta \Psi$ offsets this and as a consequence is large and positive inside [2, 16, 17]. Similar results have been obtained with *Thermoplasma acidophilum, Bacillus acidocaldarius* and other bacteria (refs. 12–14 and Figure 4.1). Two points arise from these observations which were discussed earlier [2, 16, 21] but which seem not to have been fully appreciated. These are (i) how uncouplers work in acidophiles and (ii) how the ΔpH is maintained in the presence of uncouplers. The action of DNP in acidophiles has been considered in some detail by Ingledew [2]. DNP is only one type of uncoupler, an acidic uncoupler which works by being permeable in both its anionic and protonated forms. DNP is particularly effective in acidophiles because of its pK_a (4.0) between the pH_{out} and pH_{in}. It dissipates the Δp by entering the cell in the protonated form (which would predominate at pH 2) and dissociating in the cytosol, the anionic form predominating at pH 6. Initially, the DNP^- anion will tend to leave the cell to be reprotonated externally, completing the uncoupling cycle. As the uncoupling process proceeds, a $\Delta \Psi$ positive inside develops [16], and this will tend to cause accumulation of the anion in the cytoplasm (see section on anion accumulation). Thus the DNP will distribute both as an acid and as an anion (note that when the fully uncoupled state is reached, the $\Delta \Psi$ is equal and opposite to the ΔpH (mV), Δp is zero and the extent of DNP accumulation by both mechanisms is equal). In the process of uncoupling (before $\Delta p = 0$) the accumulation of DNP by the acidic mechanism will exceed the maximum accumulation supportable by the $\Delta \Psi$, thus the anion will leak out, completing the uncoupling cycle. The distribution of DNP when $\Delta p = 0$ will be a dynamic equilibrium, and any perturbation by which $\Delta p = 0$ will re-establish net cycling. For example, if the cell has a pH_{in} maintained at pH 6, with pH_{out} at 2, then DNP with a pK_a of 4 and membrane-permeable in the anionic and protonated forms will reach an equilibrium distribution of 1 $[DNP^-]$ and 100 $[DNPH]$ outside to 1000 $[DNP^-]$ and 100 $[DNPH]$ inside, when $\Delta p = 0$.

Most studies have found that the maintenance of the pH_{in} is uncoupler-insensitive. However, there is one detailed report which claimed that uncoupler concentrations generally used are too low [12, 22] and that when higher concentrations are used the internal pH does fall. These authors used extremely high concentrations of DNP in some experiments, three orders of magnitude greater than required for uncoupling. The flaws in this work have been summarised by Matin [4]. An additional serious consequence would result simply from the accumulation of the DNP as a weak acid, causing acidification of the cytoplasm as predicted by Alexander *et al.* [17]. To maintain pH_{in} at 6 with a pH_{out} of 2 in the presence of 1 mM DNP, as Michels and Bakker did [12, 22], would mean an internal DNP concentration of approximately 0.1 M! Clearly, the cell could not cope with this and, not surprisingly, the internal pH does come down.

A basic uncoupler would need to enter the cell in the cationic form and leave in the neutral form to establish an uncoupling cycle. In the later stages of uncoupling, the entry of the cationic form would be against the $\Delta\Psi$. The concentration of the uncoupler inside the cell will be lower than outside but the same arguments apply as above.

It is necessary at this point to attempt to define the resting state and discuss the nature of the $\Delta\Psi$ in this state. In starved cells, the pH_{in} has been shown to be stable over a long time period [23, 24]. Matin and colleagues have shown that everted vesicles made from *Bacillus coagulans*, internally buffered (β-alanine) at pH 5 and suspended in phosphate buffer at pH 7.2, maintained a ΔpH of over 1 unit and a $\Delta\Psi$ of approximately 50 mV. This was stable for the 19 hours' duration of the experiment, although collapse resulted from ionophore addition (Matin, described in ref. 4). During experiments with starved cells, Matin and colleagues noted that pH_{in} homeostasis was maintained long after cell viability was lost, cells of *Thiobacillus acidophilus* maintaining a ΔpH of about 2 units for several weeks [25]. In the ground state ($\Delta p = 0$) there is no energy available to drive any process. The pH_{in} is maintained against a concentration difference of protons however, because there is no net force on a proton to enter, the ΔpH in one direction is equal and opposite to the $\Delta\Psi$ in the other direction. The $\Delta\Psi$ arises simply because as the energised state relaxes protons enter the cytoplasm—the buffering capacity of the cytoplasm minimises their impact on pH_{in} but the charge they bring builds up and eventually halts further proton entry. The question that has arisen is what to call this $\Delta\Psi$, a Donnan potential or a proton diffusion potential [4, 5]. As the process is probably correctly perceived in its operation, it would seem unnecessary to dwell on this point were it not for possible confusion over what is meant by the terms used. A Donnan potential arises in cells because of the fixed negative charges of the internal proteins and lipids. When cells are suspended in a medium of low ionic strength, the more permeable ions redistribute and may give a net chemical potential offset by a membrane potential. The observed membrane potential in acidophiles is not a Donnan potential in the classic sense: although the change of charge on internal groups clearly plays a major role, the net cytoplasmic fixed charge is unlikely to change sign on de-energisation. A proton diffusion potential is similar to other ion diffusion potentials: they are created by having a pre-existing concentration difference and then making the membrane permeable to that species. That the membrane potential is created by proton permeability in the presence of a proton concentration difference is not disputed. However, the protons which enter must bind (the time fraction average number of protons as hydronium ions in a bacterial cytoplasm at pH 7 is approximately one proton per cell). Therefore, the $\Delta\Psi$ is created by a buffered proton diffusion potential.

4.2.1.4 The enigma of pH_{in} homeostasis and the putative need to convert $\Delta\Psi$ into ΔpH. In an acidophile, the membrane potential may be small and the Δ_pH large, but when a proton is pumped out both charge and chemical entity are moved, as noted earlier. Thus it has been argued that there must be an ion transport system with the role of converting the $\Delta\Psi$ into a ΔpH and in doing so regulating the pH_{in} [4, 5, 26]. Two questions arise: (i) is this hypothesis correct? (i.e. is there a need for systems to convert $\Delta\Psi$ into ΔpH as part of the mechanism of pH_{in} homeostasis?) and (ii), if this is so, are the proposed mechanisms capable of this function?

If the proposed mechanisms by which the $\Delta\Psi$ is converted into a ΔpH and the pH_{in} regulated is considered first, then in the case of an acidophile the proposal [5, 26] is that the charge movement resulting from (primary) respiratory proton pumping is compensated for by counterion movement, such as that of K^+ moving in via a uniporter (electrophoretically). This would achieve the desired aim of lowering the $\Delta\Psi$ and allowing the development of a large ΔpH, but it could only work transiently until the transmembrane distribution of K^+ was equal and opposite to the $\Delta\Psi$ (which with $\Delta\Psi = 0$ would mean no accumulation of K^+). Thus in the steady state K^+ cycling is required and herein lies the problem: how to efflux K^+? It has been proposed that K^- leaves electroneutrally in exchange for a proton. Refinements might be proposed but under steady-state conditions all ion movements must balance as the net result of such mechanisms is that charge compensation for proton extrusion is by proton re-entry. This is merely an uncoupling cycle. A much better mechanism allowing protons to re-enter would be via a H^+-ATPase. Analogous cycles do exist (e.g. the Ca^{2+} cycle in mitochondria, see later). They also consume proton current, but this is acceptable assuming the primary aim of the cycle is to regulate the internal concentration of the ion and the proton current consumed is only a small part of the total. However, such mechanisms are clearly not acceptable as a means of charge compensating the primary pumping of protons, as this would need to be stoichiometric as envisaged by the proponents of these mechanisms (in the steady state $\Delta\Psi$ may be close to zero). Also, how can cells use K^+ (or chloride) pumps to regulate the balance of $\Delta\Psi$ and ΔpH when they need to regulate the cytoplasmic concentrations of these ions for metabolic purposes? On the basis of the requirement of *Thiobacillus ferrooxidans* for sulphate ions, Bakker [5] has proposed that sulphate plays a role in pH homeostasis and the maintenance of $\Delta\Psi$ and ΔpH in this bacterium by carrying out a transmembrane transport cycle composed of sulphuric acid uptake and SO_4^{2-} extrusion. However, the cell gains nothing by allowing protons to enter the cell by such a mechanism. The contention is also refuted by the observation of Alexander *et al.* [17], who demonstrated the impermeability of sulphate in these cells.

This leads to a consideration of whether it is necessary to convert the $\Delta\Psi$ into ΔpH. If the starting position is taken as the de-energised state ($\Delta p = 0$), then there may be a $\Delta\Psi$ of approximately 240 mV, positive inside, and a ΔpH of -4 units. When cells start pumping protons, the membrane potential does change by the full extent expected. It is the confusion of a starting $\Delta\Psi$ of up to 240 mV rather than $\Delta\Psi = 0$ which has led to a lot of confused discussion and the proposal of Na^+, K^+, Cl^- and even SO_4^{2-} cycles to charge compensate proton movement. It is the resting state (which is a stable ground state in the absence of permeant ions) which again needs some consideration as these misconceptions arise from a failure to recognise the ground state as the starting condition. A $\Delta\Psi$ is generated fully; it is not converted into ΔpH. On the transition from the resting state to the energised state, the $\Delta\Psi$ goes, for example, from 240 mV to approximately 0 mV, a span of 240 mV. It has been argued that the energy-dependent ΔpH is small and the energy-dependent $\Delta\Psi$ in acidophiles is large, as it is in all other micro-organisms. The ground state pH_{in} and $\Delta\Psi$ are stable in the de-energised cell in the absence of permeant anions and can last for days even in cells starved for long periods and not viable [25]. Those authorities proposing the extensive ion transport system as a pH_{in}-regulating system seem to have looked at the absolute size of these parameters in the energised state and concluded that $\Delta\Psi$ is small and ΔpH is large therefore the $\Delta\Psi$ must be converted into a ΔpH.

So there are three arguments against such mechanisms predominating: (i) the need for such a system is not as great as implied by some; (ii) the proposed regulating ions are often ones which need to be regulated themselves (independently); (iii) the use of these mechanisms as proposed for acidophiles is almost stoichiometric with proton extrusion and hence highly uncoupling. Some of these cycles will exist, but to regulate the cytoplasmic concentrations of these anions and cations and possibly to fine-tune the pH_{in}, not to cause the wholesale conversion of the $\Delta\Psi$ into ΔpH. The pH_{in} can be regulated principally by the respiratory systems combined with the internal buffering capacity and perhaps by some fine-tuning by other porter systems. For instance, if, from the resting state, respiratory-driven proton extrusion commences, the $\Delta\Psi$ becomes relatively much more positive and the internal pH rises slightly. Once the Δp is large enough, protons will re-enter via the H^+-ATPase, generating ATP. There are resting states in which permeant anions have been present and in which the pH_{in} may be a little lower than in the normal resting state, but provided this is only marginal the cells can recover [17]. In this case, during establishment of a Δp some of the proton current is used to pump out the anions, but this ought to be transient.

4.2.1.5 *Ion transport in acidophiles.* Acidophilic bacteria, like all cells, need to control the ionic constituents of their cytoplasm, in particular chloride, sodium and potassium ions. The chemiosmotic considerations outlined in the preceding section give the general principles of ion permeability and distribution in acidophiles, but the role of porter systems must also be considered. A chloride porter has been suggested to have a role in the acid tolerance of the acidophile *Bacillus coagulans* [20]. Furthermore, although it has been demonstrated that chloride can distribute according to the $\Delta\Psi$ (i.e electrophoretically [17]), this cannot be the complete story where there are regulatory mechanisms designed to maintain the cytoplasmic chloride concentration within a particular range. A single porter system cannot do this in an environment where the external concentration of the ion is variable. Extreme acidophiles generally do not grow in the presence of high chloride concentrations, and the concentrations in their natural environments are sometimes low. Thus, cells must have the ability to accumulate chloride into respiring cells against a concentration gradient when the $\Delta\Psi$ is small or even opposing. Such a mechanism is possible using the ΔpH as the driving force. An electroneutral chloride porter in *B. coagulans* has also been suggested by Matin and colleagues [15, 20]. This must be a proton symport or hydroxide antiport system. A chloride uniporter (or finite ion permeability) would distribute chloride in response to the $\Delta\Psi$, the neutral porter in response to the ΔpH (Figure 4.2a). The two-porter system will be a dynamic mechanism

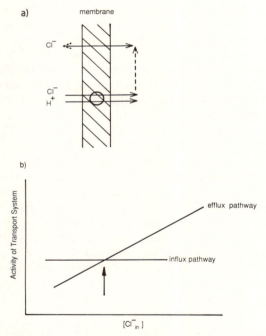

Figure 4.2   Chloride transport systems in acidophilic bacteria. (a) An illustration of a chloride uniporter (top) and an electroneutral chloride porter (bottom). (b) A model of a regulatory system which will maintain a cytoplasmic chloride concentration at a designated level. The activity of each porter is shown as a function of the internal chloride concentration. The activity of the efflux pathway (the uniporter) is affected by the internal chloride concentration. The activity of the influx pathway is shown as independent of the cytoplasmic chloride concentration over the required range. The two-porter model will regulate cytoplasmic chloride concentrations as described in the text.

for controlling the intracellular chloride concentration and has parallels with other systems, such as the calcium ion porter systems of the mitochondrion [18]. The distribution of chloride via non-specific permeability or a uniporter system will be given by $\Delta\Psi = 59 \log_{10} [Cl^-_{in}]/[Cl^-_{out}]$. The distribution of chloride by the electroneutral antiporter will be governed by $\Delta pH = \log_{10} [Cl^-_{in}]/[Cl^-_{out}]$. Thus, except in complete de-energisation, the electroneutral porter can accumulate chloride to an extent unsustainable by the electrophoretic pathway and thus the uniporter will be driven backwards. These are thermodynamic relationships; in the dynamic two-porter system they only set limits on the accumulations each porter can sustain. The activities of the porters must also be controlled. These relationships do show, however, that the electroneutral porter can accumulate chloride to a ratio which exceeds the capacity of the electrophoretic uniporter; thus a cyclic system is developed in which chloride entering on the electroneutral porter can exit on the uniporter. Such a cycle consumes some proton current but has the advantage of control. The required $[Cl^-_{in}]$ can then be obtained by regulating the activities of the two carriers. The two porters and how their activities can be controlled are illustrated in Figure 4.2b. The simplest version is if the rate of uptake is independent of the internal concentration and the rate of the exit (or leak) is dependent on the internal concentration; then the internal concentration of the chloride will be regulated at the intersection of the two lines. For example, if the internal chloride concentration falls below the intersect concentration, then the uptake pathway is the more active (as required), and if it is higher than the intersect concentration then the exit pathway activity is the greater.

The above mechanism is consistent with the observed toxicity and accumulation of chloride ions in the fully de-energised state [17]. The $\Delta\Psi$ and the ΔpH are approximately equal and opposite but each is trying to accumulate chloride to the same extent via the operating porters.

The internal K^+ ion concentration will probably be regulated in a comparable way to that suggested for chloride. Such mechanisms have been proposed but to explain pH regulation rather than control of the K^+ concentration [4, 5, 26].

4.2.2 *Considerations of the conditions in the periplasm and the implications for its processes*

Gram-negative bacteria have the periplasmic space between cell wall and the osmotic barrier of the cytoplasmic membrane. Although difficult to measure directly, we believe that the pH of the periplasm is close to that of the suspending medium. There much dispute over the volume and form of the periplasm [27, 28] but agreement that it contains a high concentration of enzymes. The periplasmic proteins are separated from the external medium by a cell wall which is perforated by porins, allowing channels of approximately

15 Å diameter [29]. These channels will allow free passage of small ions but not of the proteins of the periplasm. It is important to define a starting position to consider this further, and for the purpose of argument a cell at neutrality will be considered and acidification of its suspending medium further postulated. As the external medium equilibrates with the periplasmic space the entering protons will titrate the carboxyl groups of the proteins, leaving the protein with an overall positive charge. The anions (for example sulphate) will accompany these charges. Thus the sulphate concentration in the periplasm will tend to be that associated with the fixed positive charges plus the same concentration as outside. Thus the sulphate concentration in the periplasm will be higher than in the medium. However, the proton (and cation) concentration should be the same in the periplasm as outside (this would not be the case if the fixed charges were negative). This is because the extra protons, compensating for the extra sulphates, are responsible for the fixed positive charge. The compensation of the positive charges of the periplasmic proteins by sulphate anions is probably responsible for the reported binding of sulphate by *Thiobacillus ferrooxidans* [30].

4.2.2.1 *Protein stability and protein–protein interaction at acid pH*.　There is no indication of what confers acid tolerance on a protein. Many proteins are denatured at pH 2, but those that function at such acidity are not noticeably unusual. For example, rusticyanin, a copper-containing, periplasmic, respiratory chain protein of *T. ferrooxidans* has been purified and sequenced (R.P. Ambler, W.J. Ingledew and J.C. Cox, unpublished data and refs. 31 and 32). This protein belongs to a large class of blue copper proteins from a range of bacterial and eukaryotic sources. It shows no sequence peculiarities and nuclear magnetic resonance (NMR) studies have shown that it remains folded and stable at pH 1 (Hill and Ingledew, unpublished data).

As with neutrophilic Gram-negative bacteria, the periplasm of some acidophiles appears to contain such a high concentration of protein that it may be misleading to think of it in terms of solution chemistry. However, the attainment of protein–protein interactions can be considered, bearing in mind that the proteins will be positively charged in an acidic periplasm. Clearly, anionic bridging can be invoked, but is it necessary to consider this beyond the extent of its involvement in reactions at neutral pH? At pH 2, the exposed carboxylic acid groups of a protein will spend approximately 1% of their time in a dissociated state in a very rapid equilibration. Each proton–carboxylic association/dissociation can be considered as an individual event, but the protein–protein interaction cannot be; there are more bonds being formed and the process can be considered synergistic in terms of proton displacement, i.e. the forces of protein–protein interaction being composed of many associations are substantially greater than those of individual proton binding in an aqueous medium. Thus the positive charges on the interacting proteins can replace the protons, enabling the interactions to proceed in the normal way. An additional problem which is an extension of this is how proteins (and co-factors) can assemble in the periplasm. This, however, is a developing area of microbial biochemistry, very poorly understood in neutrophiles, and likely to involve the same processes in acidophiles although within the additional constraints of an acidic environment as previously discussed.

4.3 The diversity of the extreme acidophiles

The fundamental adaptations to high acidity and its consequences, such as the large transmembrane pH gradient, are probably universal in acidophiles. The characterisation of previously undescribed strains with some unique features may nevertheless provide useful model systems for examining further some of the problems posed by acidophilic growth. For example, the isolation of the halotolerant bacterium *Thiobacillus prosperus* [33] may encourage further consideration of the previously discussed control of intracellular chloride concentrations in acidophiles. The perceived diversity of the extreme acidophiles is steadily increasing, particularly through studies directed towards examining the role and potential applications of bacteria in the oxidation of mineral sulphides.

4.3.1 *Iron- and sulphur-oxidising acidophiles*

The assessment of the phylogeny of these acidophiles on the basis of molecular data is far from complete, but it is possible to begin to consider the distribution of some biochemical or physiological characteristics of particular groups in the context of their phylogeny. The acidophiles which have received most study in the context of mineral leaching (Table 4.1) can be grouped in relation to their optimum growth temperatures in a scheme which also reflects some major taxonomic divisions. The characteristics of many of these bacteria have been reviewed (refs. 34–36 and 37 for *Sulfobacillus thermosulfidooxidans*) and *T. prosperus* [34], *T. cuprinus* [38] and *Metallosphoria sedula* [39] have been described by Stetter and colleagues. The mesophiles with optimum activity at about 30°C all appear to be Gram-negative, but the types within this group are not closely related to each other and different species have some unique characteristics. There is also considerable genomic diversity among isolates of *T. ferrooxidans* [40], and some may merit different binomials after further inspection. '*T. ferrooxidans*' strain m-1 for example [35] is not closely related to *T. ferrooxidans* on the basis of 5S rRNA sequence comparisons [41]. Two mesophilic acidophiles have received considerable study but are not noted in Table 4.1 as they do not oxidise iron and are of uncertain or little significance in mineral sulphide dissolution. *Thiobacillus acidophilus* [42] has been utilised in the study of the mechanisms

Table 4.1   Major groups of acidophilic iron- and sulphur-oxidising bacteria.

Bacteria	DNA mol% GC	Growth on		
		Iron	Sulphur	Yeast extract
Gram-negative mesophiles				
Thiobacillus ferrooxidans	58	+	+	−
Thiobacillus thiooxidans	52	−	+	−
Leptospirillum ferrooxidans	51–55	+	−	−
Thiobacillus prosperus	64	+	+	−
Thiobacillus cuprinus	66–69	−	+	+
'Thiobacillus ferrooxidans' m-1	65	+	−	−
Gram-negative thermophile				
strain BC13	61	−	+	−
Gram-positive thermophiles				
Sulfobacillus thermosulfidooxidans	45–49	+	+	+
strain TH1/BC1	50	+	+	+
strain ALV	57	+	+	+
strain LM2	60	+	+	+
strain TH3	69	+	+	+
Extreme thermophiles (archaeobacteria)				
Sulfolobus acidocaldarius	37	+	+	+
Sulfolobus strain BC	ND	+	+	+
Sulfolobus strain B6-2	ND	−	+	+
Acidianus brierleyi	31	+	+	+
Metallospaaera sedula	45	+	+	+

ND: not determined

of acidophily as discussed earlier (e.g. ref. 25) and is able to grow hetero-trophically in the absence of sulphur compounds as well as autotrophically on elemental sulphur, thiosulphate and tetrathionate [43]. The obligately heterotrophic *Acidiphilium* has been studied in relation to possible interactions in mixed culture with *T. ferrooxidans* (see ref. 44) and in relation to the development of genetic systems for acidophiles (see below).

Strain BC13 (Table 4.1) is an example of the thermotolerant or moderately thermophilic sulphur-oxidising acidophiles [43] which grow at higher temperatures than *Thiobacillus thiooxidans* but otherwise resemble it and likewise do not oxidise iron. These bacteria can be readily isolated from acid mine waters and hot springs. The only moderately thermophilic iron-oxidising acidophiles which have been well described so far all appear to be Gram-positive (Table 4.1). The presence of key signature nucleotides in their 16S rRNA has confirmed their affiliation with the Gram-positive bacteria, but their precise relationship to other bacteria in the phylum is not yet certain [45]. Recent isolations have increased the number of types in pure culture to about 12, and the spread of mole percentage G + C values of the most studied strains (Table 4.1) indicates considerable variety in the group.

The upper temperature limit for the moderately thermophilic iron-oxidising eubacteria appears to be about 60°C, above which the bacterial oxidation of iron and sulphur at low pH is associated with archaeobacterial acidophiles. These bacteria superficially resemble *Sulfolobus acidocaldarius* [46], the first such isolate to be described in detail, but, as with the moderate thermophiles, the perceived diversity of types is overdue for revision. Some of the type-cultures of *S. acidocaldarius* and *Sulfolobus solfataricus* which have been available from culture collections have been unable to oxidise sulphur [47]. These cultures appear to be enrichments of acidophilic obligate heterotrophs from mixed type-cultures rather than recently facultatively autotrophic bacteria which have lost the sulphur oxidation capacity. The phenotypic variation of *Sulfolobus* type-cultures has been described [48]. *Sulfolobus brierleyi* has been renamed *Acidianus brierleyi* to join *A. infernus* in a new genus [49]; other isolates have been designated *Desulfurolobus ambivalens* [50], *Sulfolobus shibatae* [51], *Sulfurococcus mirabilis* [52] and *Metallosphaera sedula* [39]. Several other iron- and mineral sulphide-oxidising extreme thermophiles which can be readily distinguished from the mineral-oxidising *Sulfolobus* strain BC, *A. brierleyi* and *M. sedula* have also been isolated but require further characterisation (P.R. Norris, unpublished work).

4.3.2   *Phylum- and group-specific traits?*

In addition to the different growth temperature ranges of the phylogenetically distinct iron-oxidising acidophiles, the distribution of some other phenotypic traits may also be aligned with particular taxonomic groupings. The capacity for autotrophic growth has been confirmed in all the iron- and sulphur-oxidising acidophiles noted in Table 4.1 with the exception of the moderate thermophile strain TH3. However, within this generally shared capacity there are some group-specific features. The Gram-negative iron- and sulphur-oxidising mesophilic acidophiles have generally been regarded as obligate autotrophs. However, the recently described *T. cuprinus* can grow heterotrophically, albeit at less acidity than the obligate autotrophs [38]. In contrast, all of the iron-oxidising thermophiles show considerable nutritional versatility, with most capable of mixotrophic and heterotrophic growth as well as autotrophy. The key enzymes of the Calvin cycle have been found in all of the acidophilic, autotrophic eubacteria in which they have been sought, whereas carbon dioxide fixation in the archaeobacterium *Sulfolobus* appears to involve a reductive citric acid cycle [53] and possibly a key role for an acetyl CoA carboxylase [54].

There is so far little information on the mechanisms of iron and sulphur oxidation by acidophiles with the exception of *Thiobacillus ferrooxidans*. This precludes much discussion of the distribution of these key processes in relation to the phylogeny of these bacteria. However, with iron oxidation occurring at the cell surface and with the surfaces of Gram-negative bacteria,

Gram-positive bacteria and archaeobacteria having distinctive chemical composition and cell wall architecture, some fundamental differences in the interaction with ferrous iron are likely. A variety of respiratory chains for the subsequent transfer of electrons to oxygen exist [55, 56]. The process in *Leptospirillum ferrooxidans* resembles that in *T. ferrooxidans* in that electron transfer from ferrous iron in these Gram-negative bacteria involves small, soluble, acid-stable, periplasmic proteins with high mid-point potentials ($+ 680\,mV$ at pH 3.5). These proteins are quite distinct in the species so far examined with a blue copper protein, rusticyanin, and an acid-stable cytochrome c in *T. ferrooxidans* [2, 31, 32] but a novel cytochrome which appears to contain zinc as well as iron in *L. ferrooxidans* [57]. Among the iron-oxidising archaeobacteria, a conformity rather than diversity of respiratory chains is suggested by the presence, only during growth on ferrous iron, of a cytochrome with alpha-peak absorption at 573 nm in oxidised minus reduced difference spectra of species of *Sulfolobus*, *Acidianus* and *Metallosphaera* [55]. The moderately thermophilic, iron-oxidising, Gram-positive acidophiles have received insufficient study for an assessment of any variation in their substrate oxidation systems. Optical spectroscopy of some isolates (strains BC1, ALV, TH3 and LM2) has revealed that the most conspicuous components of the respiratory chain in each case are probably *b* and *aa*₃-type cytochromes [55]. As might be expected in the absence of a true periplasm, and in contrast to the examples of the Gram-negative acidophiles so far examined, conspicuous, abundant, soluble electron carriers were not found. For all these groups of acidophiles, the initial interaction of the cells with ferrous iron requires further investigation. The demonstration of the induction of iron oxidation-specific outer membrane proteins in *T. ferrooxidans* [58] and the cloning of the rusticyanin gene [59, 60] have marked the development of molecular biological studies of the iron oxidation systems.

Several important details of the mechanisms and pathways of inorganic sulphur oxidation by acidophiles remain to be discovered (see ref. 61). Pathways have been resolved in a number of bacteria for the oxidation of a range of sulphur oxy-anions, but little is known about the mechanism of oxidation of 'native substrates', sulphur in pyrite or of sulphur itself (in its octet form, S_8^o). A range of bacteria (all thiobacilli and several other species) can derive energy from the oxidation of reduced sulphur compounds to (ultimately) sulphuric acid, and many of these are also acidophiles. The pathways of sulphur compound catabolism (S^{2-}, S^o, etc. through to S^{6+}) in the thiobacilli are complex and varied and they have been extensively reviewed (see ref. 61). In the complete oxidations of S^{2-} from pyrite and of S^o the chemistry involves the removal of $8e^-$ and $6e^-$ respectively to give SO_4. The terminal oxidant is oxygen ($E_{m7} + 800\,mV$, $E_{m2} + 1100\,mV$), except in *Thiobacillus denitrificans*, which can use oxides of nitrogen.

The first step, S^{2-} oxidation, occurs either chemically or, in the absence of ferric iron (from pyrite for example), via a sulphide oxidase. The reducing equivalents are fed into the respiratory chain and S° is thought to be produced in a bound reactive form (work with *T. thiooxidans*, see ref. 61). This may then be attacked by a sulphur oxygenase or an oxidase. One pathway for S° oxidation (to SO_3^{2-}) has been demonstrated in some thiobacilli and other organisms (e.g. *Sulfolobus (Acidianus) brierleyi* [62]); this is the S° oxygenase pathway:

$$S° + O_2 + H_2O \rightarrow H_2SO_3$$

This route, however, cannot be the sole route for bound S° oxidation as this process can occur at equivalent rates in the absence of oxygen when oxides of nitrogen (*T. denitrificans*) or ferric iron (*T. ferrooxidans*) are provided as alternative respiratory oxidants. Therefore an oxidase system must also be present in these bacteria:

$$S° + 3H_2O \rightarrow H_2SO_3 + 4H^+ + 4e^-$$

or

$$S° + 4H_2O \rightarrow H_2SO_4 + 6H^+ + 6e^-$$

These reactions would be linked to the respiratory chain and hence to ATP synthesis. When the pathway proceeds via the oxygenase reaction, this would not yield energy for the cell. Finally, SO_3^{2-}, when and if produced as a free intermediate, is oxidised to SO_4^{2-}. Two pathways have been described: the adenosine 5'-phosphosulphate (APS) pathway and the sulphite oxidase pathway. SO_3^{2-} will also react directly with ferric iron. The APS pathway:

$$SO_3^{2-} + AMP \rightarrow APS + 2e^- (E_m - 60\,mV)$$

$$APS + PO_4^{3-} \rightarrow ADP + SO_4^{2-}$$

leads to substrate-level phosphorylation as well as the possibility of oxidative phosphorylation. The alternative pathway:

$$H_2SO_3 + H_2 \rightarrow H_2SO_4 + 2H^+ + 2e^- (E_{m7} - 280\,mV)$$

has a lower potential but does not involve substrate-level ATP synthesis.

The next step, S° oxidation, is more problematical when the substrate is the native form of S°, the octet ring. An initial problem is that this is hydrophobic and insoluble and must be opened for enzymic attack. This has been found in *T. ferrooxidans* to involve an initial reduction of the octet ring to open it followed by sequential oxidation of the linear polysulphide in the periplasm [63] with the reducing equivalents passing through the complex II and complex III regions of the respiratory chain to oxygen, generating a Δp and enabling ATP synthesis [3]. Alternative models for sulphur oxidation have been proposed whereby reduced glutathione would be utilised to release hydrogen sulphide, which would be oxidised by ferric iron, which would then be reoxidised by the respiratory chain (see ref. 64). This mechanism seems wasteful, unlikely in its use of periplasmic glutathione, and only applicable to *T. ferrooxidans*.

The commercial potential of the various groups of mineral sulphide-oxidising acidophiles will be discussed in relation to the influences of environmental factors on their activity with regard to some of the phylum- or group-specific traits. These factors can influence processes directly involved in the mineral oxidation, such as iron and sulphur oxidation, and indirectly affect the mineral-oxidising capacity of the bacteria through influencing growth, as in the cases of nutrient availability and inhibition by toxic compounds. The following overview of the nature of the industrial processes in which particular bacteria could be utilised provides a useful background to this discussion.

4.4 The bacterial extraction of metals from mineral sulphides

The capacity of iron- and sulphur-oxidising, acidophilic bacteria to enhance the extraction of metals from ores is utilised in the large-scale dump and heap leaching of copper ores, the *in-situ* leaching of uranium ore, and in the treatment of gold-bearing mineral sulphide concentrates in reactors. The microbiology of the mineral sulphide oxidation and the scope for controlling or improving the activity of the micro-organisms may depend on the nature and design of the industrial processes.

The recovery of copper by dump leaching of low-grade ores and waste from open-pit mining has been practised for many years in the major copper-producing countries. An acid solution is percolated through dumps that may contain millions of tonnes of ore and the copper is stripped from the effluent solution. A mixed population of bacteria would be expected to countribute to iron, sulphur and mineral sulphide oxidation *in situ* where heterogeneous microenvironments and gradations of temperature and oxygen concentration occur. There are few reports (e.g. ref. 1) of attempts to influence metal extraction by inoculation of heaps of low-grade ore with particular strains. Although this could shorten a lag phase before the establishment of an active indigenous microbial population in freshly prepared heaps, improvements in design to extend oxygenation and the interface of ore–solution contact rather than direct manipulation of the microbiology of the systems are likely to be easier routes to improved metal extraction. The activity and composition of the microbial population can more readily be studied in dump leaching simulations which could create conditions similar to those encountered in full-scale operations [65], although there have also been studies of the microbial ecology of mineral sulphide wastes dumps (for example ref. 66).

Bacterial leaching of uranium ore in a Canadian mine is producing about 500 tonnes of uranium per year, about 10–15% of the total mine production, more cheaply than the conventional sulphuric acid leaching process [67]. Bacterial leaching allows metal recovery from low-grade ore and potentially from ore of unsuitable composition for conventional process technology. Studies of the microbiology of the mine have concentrated on the nutrient requirements of the bacteria (supplementation of the leach solution with phosphate is required) and on the influence of temperature on the bacterial oxidation of iron. Isolates of *Thiobacillus ferrooxidans* from colder parts of the mine were found to be most adapted to growth at low temperatures and are recommended as inocula for *in-situ* leaching in which the solution temperature averages about 12°C and limits bacterial activity.

Mineral sulphide concentrates rather than low-grade material, with the possible exception of some gold-bearing ores, are the potential substrates for bacterial oxidation in reactors. An auriferous pyrite/arsenopyrite concentrate from one mine has been treated successfully for several years by acidophilic, mesophilic bacteria in stirred-tank reactors [68], the bacteria oxidising iron and sulphide to release the gold from the mineral sulphide matrix before recovery of the precious metal via cyanidation. There have been several pilot plant and some larger scale demonstrations (e.g. ref. 69) of the bioleaching of refractory gold ores, including one pilot plant utilising moderately thermophilic acidophiles [70], and the introduction of further commercial plants seems likely.

4.4.1 *Factors influencing the selection of bacteria for mineral-leaching processes*

The variety of process designs reported in the previous section, from massive dumps and *in-situ* processes to the relatively easily controlled reactor systems, means that the bacteria involved can be exposed to a range of conditions and stresses. This range is widened by the variety of ore types and potentially toxic metals that are solubilised. It is unlikely, therefore, that any one type of organism will be ideal for all commercial processes. This is particularly evident with reference to the temperature at sites of mineral oxidation, which may vary from those at which psychrotolerant strains are required (uranium leaching, see above) to those at which only thermophiles could be active. Exothermic mineral oxidations can lead to temperatures of at least 60°C inside dumps (see ref. 71), and the use of thermophiles in reactors would allow increased rates of mineral dissolution as well as a reduction in the cooling costs that, particularly in hot climates, would be incurred with processes utilising mesophiles. Beyond the obvious influence of temperature selection on the active bacteria in a process, different types of mineral-oxidising bacteria can respond differently to other factors influencing growth, and in some cases at least the possession of phylum- or group-specific traits underlies the different responses. The various types of bacteria can all be inhibited by extremes of, for example, acidity or carbon dioxide limitation, but the relative significance of potentially limiting factors can vary with the organism. This is illustrated in an overview (Figure 4.3) based on observations of the

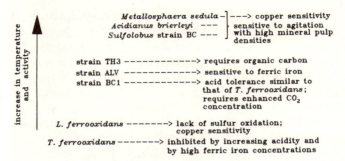

Figure 4.3 A summary of factors likely to limit the growth of iron-oxidising acidophiles on mineral sulphides (see text).

performance of bacteria in laboratory reactor leaching of pyrite and chalco-pyrite concentrates [72–76]: the importance of the limiting factors may relate to both or only one of these minerals, as with carbon dioxide limitation and sensitivity to copper respectively.

Thiobacillus ferrooxidans, as might be expected, is used in most bacterial leaching process development since it is in most respects an effective agent of mineral sulphide dissolution, being a proficient oxidiser of both iron and sulphur, generally robust in the context of agitation in the presence of minerals, and tolerant of high concentrations of many potentially toxic metal ions. However, it appears to be less acid-tolerant than *Leptospirillum ferrooxidans* [72, 77] and more susceptible to end-product ferric iron inhibition of ferrous iron oxidation [73]. *L. ferrooxidans* appears to have a higher affinity for ferrous iron than other iron-oxidising acidophiles which probably accounts for the virtual elimination of *T. ferrooxidans* from ferrous iron-limited chemostats which initially contained both species [73]. Several factors may contribute to the capacity of *L. ferrooxidans* to become the major iron-oxidising acidophile in the microflora of serial, mine water-enrichment cultures containing pyrite. Against a background of a decreasing ferrous–ferric iron ratio and increasing acidity in the cultures, some significance could be attached to its high affinity for ferrous iron, to its high tolerances of ferric iron and acidity, and even to a greater tendency than *T. ferrooxidans* to attach to the mineral particles [73] which would ensure its presence in the region of the highest concentration of the growth-limiting substrate, assuming this was ferrous iron being released from the solid. In turn, however, *L. ferrooxidans* suffers in comparison with *T. ferrooxidans* through its apparent inability to oxidise sulphur. This severely restricts its degradation of some mineral sulphides, including chalcopyrite, on which a sulphur-rich layer may accumulate during dissolution unless a sulphur-oxidising acidophile such as *Thiobacillus thiooxidans* is also present [72, 78]. *L. ferrooxidans* is also more sensitive than *T. ferrooxidans* to some metal ions, including copper [73].

The thermophilic bacteria have the merit of being active at temperatures which allow more rapid degradation of mineral sulphides when the overall rate-limiting steps in dissolution are temperature-dependent reactions and diffusion rates of ions at the solid–liquid interface. The thermophilic iron-oxidising acidophiles do not generally have the capacity for more rapid autotrophic growth than *T. ferrooxidans* when the substrates are ferrous iron or sulphur compounds in solution. The factors that primarily limit the potential application of the Gram-positive, moderate thermophiles are in many cases strain-dependent (Figure 4.3). The strains so far examined in pure culture, such as BC1, are relatively inefficient in their utilisation of carbon dioxide from air in comparison with the mesophilic and archaeobacterial mineral-oxidising acidophiles [76]. This would require an increase in process costs to enhance the carbon dioxide concentration in reactor aeration in order to utilise fully the mineral-oxidising capacity of such strains.

The potential application of the extremely thermophilic acidophiles lies in their capacity to produce the most rapid rates of bacterial mineral dissolution of finely ground mineral sulphides in reactors (for example, see ref. 76). These bacteria also produce the highest yields of target metals in solution when this yield is dependent on the temperature, as it is with copper release from chalcopyrite [47, 75]. So far, few of the iron- and mineral-oxidising archaeo-bacterial acidophiles now available have been studied in this context, but a potential problem appears to be their sensitivity to agitation in the presence of high concentrations (> 10–15% w/v) of mineral sulphides [74].

4.5 Molecular genetic studies of acidophiles

Studies of gene structure, organisation and expression in *Sulfolobus* have been reviewed in the context of this organism as a member of the Archaeobacteria [79]. Studies of the molecular genetics of the other extreme acidophiles, and particularly of their plasmids as potential cloning vectors, have mostly been initiated with a long-term view to the development of improved strains for bacterial leaching. Generally, the structure, expression and regulation of the genes so far examined in *Thiobacillus ferrooxidans* (in nitrogen metabolism [80] and sulphate assimilation [81] for example) are typical of Gram-negative bacteria. However, some interesting features of the molecular genetics of *T. ferrooxidans* which could have some bearing on its growth in an industrial environment have been revealed.

In mercury-resistant *T. ferrooxidans*, the *merA* gene which encodes the mercuric reductase enzyme responsible for reducing mercuric ions to the element is located on the chromosome rather than being plasmid- or transposon-borne [82, 83]. This *merA* gene has 78% homology with Tn*501* and 76% homology with plasmid R100 *merA* genes [82]. Comparison of the codon usage frequency for the *merA* gene with other genes in *T. ferrooxidans* and other *merA* genes has suggested that the *T. ferrooxidans* gene may have originated from the transposon or plasmid, or from a common ancestral gene, rather than in *T. ferrooxidans* itself [82].

Two different families of mobile, repeated DNA sequences, each of about 1 kb and 20–30 copies per *T. ferrooxidans* genome, have been described [84]. It has been speculated [85] that the mobility of these insertion sequences might cause phenotypic variation which is manifested in a reversible loss of iron-oxidising activity (revealed as colourless, spreading colonies on thiosulphate–iron agar). Positional changes in the family 1 repeats during the culture of *T. ferrooxidans* in the presence of a copper-containing industrial waste have been observed and raised the question of cause or effect in relation to strain adaptation [86].

4.5.1   The development of genetic systems for acidophiles

The prospect of strain improvement in the context of metal extraction from mineral sulphides encompasses both natural selection of adapted or mutant acidophiles and potentially the controlled manipulation of certain desired characteristics. The latter requires an understanding of the molecular biology of the key bacteria and the development of genetic systems for them.

The development of plasmid cloning vectors has involved strains of *T. ferrooxidans*; species of the acidophilic heterotroph *Acidiphilium* which have been seen as possible donors of genes to *T. ferrooxidans* and as models for developing genetic techniques for acidophiles; and more recently the Gram-positive iron-oxidising bacteria. A 15.5-kb plasmid and an associated virus-like particle have been found in *Sulfolobus* strain B12 [87], now designated *Sulfolobus shibatae* [51], but there is no information on gene transfer in the extremely thermophilic acidophiles. The plasmids which have received most attention in studies with the acidophilic mesophiles are noted in Table 4.2.

Plasmids have been found in the majority but not all of the *T. ferrooxidans* isolates that have been screened. An apparent coincidence of uranium resistance and the presence of the 13-MDa plasmid in some isolates has been

Table 4.2 Plasmids of mesophilic acidophiles.

Organism	Plasmids	References
T. ferrooxidans ATCC 33020	6.7, 16, > 23 kb	88, 89
T. ferrooxidans FC2	12.4, 20 kb	90
T. ferrooxidans	6, 19, 24 MDa	91
T. acidophilus	12, 18 MDa	92
T. acidophilus	12, 18, 54, > 60 MDa	93
Acidiphilium organovorum	3.3, 3.8, > 30 kb	94
Acidiphilium spp.	1.5 − 7.9 kb	95

observed [96], and the loss of resistance to mercury and silver in other isolates appears to follow the loss of a 19-kDa plasmid [91]. However, evidence for metal resistance genes or any other selectable markers on natural plasmids of any *T. ferrooxidans* strain has not been obtained. As noted above, the mercury resistance genes located so far are chromosomal.

The 6.7- and 12.4-kb plasmids of *T. ferrooxidans* have received particular study from initial restriction mapping following cloning in *Escherichia coli* [88, 90] to the location and sequencing of genes involved in their mobilisation and transfer [89, 97]. These plasmids share several similarities with the IncQ incompatibility group plasmids RSF1010 and R1162. They have a broad host range (a *T. ferrooxidans* plasmid is able to replicate in *E. coli*, *Klebsiella pneumoniae*, *Pseudomonas aeruginosa* and *Thiobacillus neapolitanus*) and comparable copy numbers, 12–15 per chromosome, in *E. coli* [97]. The predicted amino acid sequences of 42.6- and 11.4-kDa proteins encoded by *mob* genes of the 6.7-kb plasmid of *T. ferrooxidans* are similar, in certain regions, to those of mobilisation proteins of RSF1010 [97]. Differences between the *T. ferrooxidans* and *E. coli* plasmid gene arrangements include the minimum size of the replication regions, located on a 2-kb fragment of the 12.4-kb plasmid of *T. ferrooxidans* but comprising about 5.5-kb of RSF1010 [97]. The *oriT* sequences of RSF1010 and the 6.7-kb plasmid of *T. ferrooxidans*, despite sequence conservation, are not interchangeable [89].

Small and large plasmids have been detected in strains of the Gram-positive, moderately thermophilic iron-oxidising bacteria. A plasmid of about 2.6-kb from strain BC1 has been cloned in *E. coli* to facilitate mapping and further study (F.E. Gibson, J.C. Murrell and P.R. Norris, unpublished work). Sequence information has indicated that it is of the type which replicates via a single-stranded intermediate and is found in many Gram-positive bacteria. It contains an open reading frame which is predicted to encode a 42-kDa protein that would exhibit significant homology to the rep proteins encoded on plasmids pC194 [98] and pUB110 [99] from *Staphylococcus aureus*. The putative rep protein encoded by the BC1 plasmid was expressed using an *E. coli in-vitro* transcription/translation system. However, the development of a genetic system for the moderate thermophiles could benefit from the utilisation of potentially compatible marker genes and vectors originating from other Gram-positive bacteria.

4.5.2 *Gene transfer*

As noted above, potentially useful shuttle vectors for *T. ferrooxidans* and a moderately thermophilic acidophile are available, but descriptions of successful conjugation or transformation are still awaited. Gene transfer from *E. coli* to *T. ferrooxidans* by conjugation has not been described, at least in part reflecting the problem of devising a suitable medium for the mating of bacteria with such different pH optima for growth. The use of non-acidophilic but somewhat acid-tolerant thiobacilli such as *Thiobacillus neapolitanus* as intermediates in the gene transfer has been considered. Arsenic resistance has been expressed from a *T. ferrooxidans/E. coli* recombinant plasmid in *E. coli* strains, among which it was efficiently mobilised and then transferred by conjugation to *Thiobacillus novellus* [100]. However, mating between *T. novellus*, in which a small increase in arsenite resistance was observed, and *T. ferrooxidans* has not been achieved. Plasmids have been transferred from *E. coli* to *Thiobacillus acidophilus* via *T. novellus*, and the *T. acidophilus* transconjugants were able to transfer resistance markers back to *E. coli* indirectly via *Thiobacillus perometabolis* [93]. Although most difficulty in crosses was experienced with the acidophile among these thiobacilli, further evidence that acidophily need not present insuperable problems for gene transfer has been obtained with the acidophilic heterotroph *Acidiphilium facilis*. Broad host range cloning vectors and *E. coli/A. facilis* recombinant plasmids were introduced into *A. facilis* by conjugation with *E. coli* and by electroporation [95]. The efficiency of gene transfer was found to be greatly strain-dependent, at least partly reflecting different tolerances of the neutral pH at which conjugation and electroporation were performed. A bacteriophage infecting some species of *Acidiphilium* has been described [101]. This raises the possibility of developing a gene transfer system for the acidophiles by transduction, but any phages of the mesophilic and moderately thermophilic iron- and sulphur-oxidising acidophiles remain to be described. The transfer of recombinant plasmids bearing the mercury resistance genes of *T. ferrooxidans* into mercury-sensitive strains of this iron-oxidising acidophile has recently been achieved (T. Kusano, personal communication), indicating that electroporation would seem to offer the most immediate route to iron-oxidising acidophile strain manipulation. The transformants appeared stable with the capacity for Hg^{2+}-dependent NADPH oxidation.

4.6 Concluding comments: diversity, identification and applied molecular biology

Obtaining single colonies of some of the acidophilic iron-oxidising eubacteria and archaeobacterial acidophiles on solid media can be difficult, with success often depending on the type of solidifying gel (many standard agars being toxic to the iron-oxidising autotrophs). Some confusions in the descriptions of strain characteristics have arisen through the lack of a simple and reliable method of ensuring culture purity: these include the assignment of heterotrophy to *T. ferrooxidans* after the examination of cultures of the iron-oxidiser which also contained acidophilic heterotrophs (see ref. 44) and, presumably, the mixture of phenotypes associated with *S. acidocaldarius* noted earlier. This problem has also restricted the ease and accuracy of identification of the active strains in the mixed cultures of mineral-leaching environments and reactors. The application of immunological detection techniques and nucleic acid probe technology should facilitate progress in this important area. The use of an immunofluorescence assay has shown that uncharacterised acidophiles may effectively compete with *T. ferrooxidans* in the removal of pyrite from coal [102]. The use of genomic DNA probes for the enumeration and identification of *T. ferrooxidans* has been considered [103]. Of two 5S rRNA types extracted from the mixed population of a copper leach dump pond sediment, one was from *T. ferrooxidans* while the other did not appear closely related to others in the sequence catalogue [40].

The certain identification of the extremely thermophilic acidophiles is also desirable, with activities other than mineral oxidation also being examined for industrial potential. It would perhaps be unlikely that the same 'strain' of *S. acidocaldarius* would be most active in the removal of both inorganic (pyritic) sulphur [104] and organic (e.g. thiophenic) sulphur [105] from coal. The activity of a thermostable amylase of *S. solfataricus* has been described [106]. However, the obligately heterotrophic archaeobacterial acidophile which was isolated from a culture designated *S. solfataricus* (see earlier) has been shown to be more active in the degradation of starch and of thiophenes than the *Sulfolobus*-like bacteria, which can oxidise sulphur (M. Constanti, J. Giralt and P. Norris unpublished work). Molecular genetic studies of acidophiles have so far added little to the understanding of the mechanisms of iron and sulphur oxidation but are providing the foundation for the potentially useful manipulation of some commercial strains. An increase in arsenic resistance, for example, is a realistic goal which could enhance the industrial potential of some of the mineral-oxidising acidophiles which can be used to release gold from refractory arsenopyrite ores.

References

1. S.N. Groudev, F.N. Genchev and S.S. Gaidarjiev, in *Metallurgical Applications and Related Microbiological Phenomena*, eds. L.E. Murr, A.E. Torma and J.A. Brierley, Academic Press, New York (1978) 253.
2. W.J. Ingledew, *Biochim. Biophys. Acta* **683** (1982) 89.
3. W.J. Ingledew, in *Microbiology of Extreme Environments*, ed. C. Edwards, Open University Press, Milton Keynes (1990) 33.
4. A. Matin, *FEMS Rev.* **75** (1990) 307.
5. E. Bakker, *FEMS Rev.* **75** (1990) 319.
6. N.L. Gale and J.V. Beck, *J. Bacteriol.* **94** (1967) 1054.
7. C. Adapoe and M. Silver, *Can. J. Microbiol.* **21** (1975) 1.
8. R. Tabita, M. Silver and D.G. Lundgren, *Can. J. Biochem.* **47** (1969) 1141.
9. A.S. Short, D.C. White and M.I.H. Aleem, *J. Bacteriol.* **99** (1969) 142.
10. T. Oshima, H. Arakawa and M. Baba, *J. Bacteriol.* **81** (1977) 1107.
11. C. Remsen and D.G. Lundgren, *J. Bacteriol.* **92** (1966) 1765.
12. M. Michels and E.P. Bakker, *J. Bacteriol.* **161** (1985) 231.
13. A. Matin, *Biochim. Biophys. Acta* **1018** (1990) 267.
14. A. Matin, in *Microbial Growth on C₁ compounds*, ed. R.L. Crawford, Wiley, New York (1984) 62.

15. D. McLaggen, S. Belkin, L. Packer and A. Matin, *Arch. Biochem. Biophys.* **273** (1989) 206.
16. J.C. Cox, D.G. Nicholls and W.J. Ingledew, *Biochem. J.* **178** (1979) 195.
17. B. Alexander, S. Leach and W.J. Ingledew, *J. Gen. Microbiol.* **133** (1987) 1171.
18. D.G. Nicholls, *Bioenergetics: an Introduction to Chemiosmotic Theory*, Academic Press, London (1982).
19. N. Lazaroff, *J. Gen. Microbiol.* **101** (1977) 85.
20. D. McLaggan, K. Keyhan and A. Matin, *J. Bacteriol.* **172** (1990) 1485.
21. W.J. Ingledew, J.C. Cox and P.J. Halling, *FEMS Microbiol. Lett.* **2** (1977) 193.
22. M. Michels and E.P. Bakker, *J. Bacteriol.* **169** (1987) 4342.
23. E. Zychlinski and A. Matin, *J. Bacteriol.* **153** (1983) 371.
24. E. Goulborne, M. Matin, E. Zychlinski and A. Matin, *J. Bacteriol.* **166** (1986) 59.
25. E. Zychlinski and A. Matin, *J. Bacteriol.* **156** (1983) 1352.
26. I.R. Booth, *Microbiological Rev.* **49** (195) 359.
27. J.A. Hobot, E. Carelemalm, W. Villeger and E. Kellenberger, *J. Bacteriol.* **160** (1984) 143.
28. T.J. Beveridge, *Int. Rev. Cytol.* **72** (1981) 229.
29. R. Benz, *Ann. Rev. Microbiol.* **42** (1988) 359.
30. K. Imai, T. Sugio, T. Tsuchida and T. Tano, *Agric. Biol. Chem.* **51** (1975) 1349.
31. J.C. Cox and D.H. Boxer, *Biochem. J.* **174** (1978) 497.
32. W.J. Ingledew, in *Workshop on Biotechnology for the Mining, Metal-Refining and Fossil Fuel Processing Industries, Biotech. Bioeng. Symp. No. 16*, eds. H.L. Ehrlich and D.S. Holmes, Wiley. New York (1986) 23.
33. H. Huber and K.O. Stetter, *Arch. Microbiol.* **151** (1989) 479.
34. S.R. Hutchins, M.S. Davidson, J.A. Brierley and C.L. Brierley, *Ann. Rev. Microbiol.* **40** (1986) 311.
35. A.P. Harrison Jr, *Biotechnol. Appl. Biochem.* **8** (1986) 249
36. P.R. Norris, in *Microbial Mineral Recovery*, eds. H.L. Ehrlich and C.L. Brierley, McGraw-Hill, New York (1990) 3.
37. G.I. Karavaiko, R.S. Golovacheva, T.A. Pivovarova, I.A. Tzaplina and N.S. Vartanjan, in *Biohydrometallurgy, Proc. Int. Symp.*, eds. P.R. Norris and D.P. Kelly, Science and Technology Letters, Kew (1988) 29.
38. H. Huber and K.O. Stetter, *Appl. Environ. Microbiol.* **56** (1990) 315.
39. G. Huber, C. Spinnler, A. Gambacorta and K.O. Stetter, *Syst. Appl. Microbiol.* **12** (1989) 38.
40. A.P. Harrison Jr, *Arch. Microbiol.* **131** (1982) 68.
41. D.J. Lane, D.A. Stahl, G.J. Olsen, D.J. Heller and N.R. Pace, *J. Bacteriol.* **163** (1985) 75.
42. R. Guay and M. Silver, *Can. J. Microbiol.* **21** (1975) 281.
43. P.R. Norris, R.M. Marsh and E.B. Lindström, *Biotechnol. Appl. Biochem.* **8** (1986) 318.
44. A.P. Harrison Jr, *Ann. Rev. Microbiol.* **38** (1984) 265.
45. D. Lane, A.P. Harrison Jr, D. Stahl, B. Pace, S. Giovanni, G.J. Olsen and N.R. Pace, *Abstracts 8th Annual Meeting ASM* (1988) 240.
46. T.D. Brock, K.M. Brock, R.T. Belly and R.L. Weiss, *Arch. Microbiol.* **84** (1972) 54.
47. R.M. Marsh, P.R. Norris and N.W. Le Roux, in *Recent Progress in Biohydrometallurgy*, eds. G Rossi and A.E. Torma, Associazione Mineraria Sarda, Iglesias (1983) 71.
48. D.W. Grogan, *J. Bacteriol.* **171** (1989) 6710.
49. A. Segerer, A. Neuner, J. Kristjansson, K.O. Stetter, *Int. J. System. Bacteriol.* **36** (1986) 559.
50. W. Zillig, S. Yeats, I. Holz, A. Böck, M. Rettenberger, F. Gropp and G. Simon, *System. Appl. Microbiol.* **8** (1986) 197.
51. D. Grogan, P. Palm and W. Zillig, *Arch. Microbiol.* **154** (1990) 594.
52. R.S. Golovacheva, K.M. Val'ekho-Roman and A.V. Troitskii, *Microbiology* **56** (1987) 84.
53. G. Fuchs and E. Stupperich, in *Archaeobacteria '85*, eds. O. Kandler and W. Zillig, Fischer Verlag, Stuttgart (1986) 364.
54. P.R. Norris, A. Nixon and A. Hart, in *Microbiology of Extreme Environments and Its Potential for Biotechnology*, eds, M.S.da Costa, J.C. Duarte and R.A.D. Williams, Elsevier, London (1989) 24.
55. D.W. Barr, W.J. Ingledew and P.R. Norris, *FEMS Microbiol. Lett.* **70** (1990) 85.
56. R.C. Blake, E.A. Shute and K.J. White, in *Biohydrometallurgy 1989*, eds. J. Salley, R.G.L. McCready and P.L. Wichlacz, Canmet, Ottawa (1989) 391.
57. A. Hart, J.C. Murrell, R.K. Poole and P.R. Norris, *FEMS Microbiol. Lett.* **81** (1991) 89.

ACIDOPHILIC BACTERIA: ADAPTATIONS AND APPLICATIONS 141

58. N. Mjoli and C.F. Kulpa, in *Biohydrometallurgy, Proc. Int. Symp.*, eds. P.R. Norris and D.P. Kelly, Science and Technology Letters, Kew (1988) 89.
59. C.F. Kulpa, M.T. Roskey and N. Mjoli, *Biotechnol. Appl. Biochem.* **8** (1986) 330.
60. E. Jedlicki, R. Reyes, X. Jordana, O. Mercereau-Puijalon and J.E. Allende, *Biotechnol. Appl. Biochem.* **8** (1986) 342.
61. D.P. Kelly, in *Autotrophic Bacteria*, eds. H.G. Schlegel and B. Bowien, Science Technical Publishers/Springer Verlag, Madison/Berlin (1989) 193.
62. T. Emmel, W. Sand, W.A. König and E. Bock, *J. Gen. Microbiol.* **132** (1986) 3415.
63. M. Bacon and W.J. Ingledew, *FEMS Microbiol. Lett.* **58** (1989) 189.
64. T. Sugio, T. Katagiri, K. Inagaki and T. Tano, *Biochim. Biophys. Acta* **973** (1989) 250.
65. L.E. Murr and J.A. Brierley, in *Metallurgical Applications of Bacterial Leaching and Related Microbiological Phenomena*, eds. L.E. Murr, A.E. Torma and J.A. Brierley, Academic Press, New York (1978) 491.
66. A.E. Goodman, A.M. Khalid and B.J. Ralph, *Australian Atomic Energy Commission Report*, AAEC/E531, Sutherland (1981).
67. R.G.L. McCready and W.D. Gould, in *Biohydrometallurgy 89*, eds. J. Salley, R.G.L. McCready and P.L. Wichlacz, Canmet, Ottawa (1989) 477.
68. P.C. van Aswegen, A.K. Haines and H.J. Marais, *Randol Perth Gold* (1988) 124.
69. R.P. Hackl, F.R. Wright and L.S. Gormely, in *Biohydrometallurgy 89*, eds. J. Salley, R.G.L. McCready and P.L. Wichlacz, Canmet, Ottawa (1989) 533.
70. P.A. Spencer, J.R. Budden and R. Sneyd, in *Biohydrometallurgy 89*, eds. J. Salley, R.G.L. McCready and P.L. Wichlacz, Canmet, Ottawa (1989) 231.
71. J.A. Brierley and C.L. Brierley, in *Thermophiles: General, Molecular, and Applied Microbiology*, ed. T.D. Brock, Wiley, New York (1986) 279.
72. P.R. Norris, in *Recent Progress in Biohydrometallurgy*, eds. G. Rossi and A.E. Torma, Associazione Mineraria Sarda, Iglesias (1983) 83.
73. P.R. Norris, D.W. Barr and D. Hinson, in *Biohydrometallurgy, Proc. Int. Symp.*, eds. P.R. Norris and D.P. Kelly, Science and Technology Letters, Kew (1988) 43.
74. P.R. Norris and D.W. Barr, in *Biohydrometallurgy, Proc. Int. Symp.*, eds. P.R. Norris and D.P. Kelly, Science and Technology Letters, Kew (1988) 532.
75. P.R. Norris, in *8th Int. Biotech. Symp.*, eds. G. Durand, L. Bobichon and J. Florent, Société Francaise de Microbiologie, Paris (1988) 1119.
76. P.R. Norris, in *Biohydrometallurgy '89*, eds. J. Salley, R.G.L. McCready and P.L. Wichlacz, Canmet, Ottawa (1989) 3.
77. U. Helle and U. Onken, *Appl. Microbiol. Biotechnol.* **28** (1988) 553.
78. V.V. Balashova, I. Ya. Vedinina, G.E. Markosyan and G.A. Zavarzin, *Microbiology* **43** (1974) 491.
79. J.W. Brown, C.J. Daniels and J.N. Reeve, *Crit. Rev. Microbiol.* **16** (1989) 287.
80. D.E. Rawlings, I.M. Pretorious and D.R. Woods, in *Biohydrometallurgy, Proc. Int. Symp.*, eds. P.R. Norris and D.P. Kelly, Science and Technology Letters, Kew (1988) 161.
81. I.J. Fry and E. Garcia, in *Biohydrometallurgy '89*, eds. J. Salley, R.G.L. McCready, Canmet, Ottawa (1989) 171.
82. C. Inoue, K. Sugawara, T. Shiratori, T. Kusano and Y. Kitagawa, *Gene* **84** (1989) 47.
83. C. Inoue, K. Sugawara and T. Kusano, *Gene* **96** (1990) 115.
84. J.R. Yates and D.S. Holmes, *J. Bacteriol.* **169** (1987) 169.
85. J. Schrader and D.S. Holmes, *J. Bacteriol.* **170** (1988) 3915.
86. D.S. Holmes and R. Ul Haq, in *Biohydrometallurgy '89*, eds. J. Salley, R.G.L. McCready and P.L. Wichlacz, Canmet, Ottawa (1989) 115.
87. S. Yeats, P. McWilliam and W. Zillig, *EMBO J.* **1** (1982) 1035.
88. D.S. Holmes, J.H. Lobos, L.E. Bopp and G.C. Welch, *J. Bacteriol.* **157** (1984) 324.
89. M. Drolet, P. Zanga and P.C.K. Lau, *Molec. Microbiol.* **4** (1990) 1381.
90. D.E. Rawlings, C. Gawith, A. Petersen and D.R. Woods, in *Recent Progress in Biohydrometallurgy*, eds. G. Rossi and A.E. Torma, Associazione Mineraria Sarda, Iglesias (1983) 555.
91. P. Visca, P. Valenti and N. Orsi, in *Fundamental and Applied Biohydrometallurgy*, eds. R.W. Lawrence, R.M.R. Brannion and H.G. Ebner, Elsevier, Amsterdam (1986) 429.
92. M.W.H. Mao, P.R. Dugan, P.A.W. Martin and O.H. Tuovinen, *FEMS Microbiol. Lett.* **8** (1980) 121.
93. M.S. Davidson and A.O. Summers, *Appl. Environ. Microbiol.* **46** (1983) 565.

94. D.S. Holmes, J.R. Yates, J.H. Lobos and M.V. Doyle, *Biotechnol. Appl. Microbiol.* **8** (1986) 258.
95. F.F. Roberto, A.W. Glenn and T.E. Ward, in *Biohydrometallurgy '89*, eds. J. Salley, R.G.L. McCready and P.L. Wichlacz, Canmet, Ottawa (1989) 137.
96. P.A.W. Martin, P.R. Dugan and O.H. Tuovinen, *Eur. J. Appl. Microbiol. Biotechnol.* **18** (1983) 392.
97. R.A. Dorrington and D.E. Rawlings, *J. Bacteriol.* **171** (1989) 2735.
98. S. Horinouchi and B. Weisblum, *J. Bacteriol.* **150** (1982) 815.
99. T. McKenzie, T. Hoshino, T. Tanaka and N. Sueoka, *Plasmid* **15** (1986) 93.
100. D.E. Rawlings, R. Sewcharan and D.R. Woods, in *Fundamental and Applied Biohydrometallurgy*, eds. R.W. Lawrence, R.M.R. Brannion and H.G. Ebner, Elsevier, Amsterdam (1986) 419.
101. T.E. Ward, M.L. Rowland, D.F. Bruhn, C.S. Watkins and F.F. Roberto, in *Biohydrometallurgy '89*, eds. J. Salley, R.G.L. McCready and P.L. Wichlacz, Canmet, Ottawa (1989) 159.
102. G. Muyzer, A.C. de Bruyn, D.J.M. Schmedding, P. Bos, P. Westbroek and G.J. Kuenen, *Appl. Environ. Microbiol.* **53** (1987) 660.
103. J.R. Yates, J.H. Lobos and D.S. Holmes, *J. Ind. Microbiol.* **1** (1986) 129.
104. C.M. Detz and G. Barvinchak, *Mining Congress J.* **65** (1979) 75.
105. F. Kargi, *Biotechnol. Lett.* **12** (1987) 478.
106. L. Lama, B. Nicolaus, A. Tricone, P. Morzillo, M. de Rosa and A. Gambacorta, *Biotechnol. Lett.* **12** (1990) 431.

5 Alkaliphiles: ecology and biotechnological applications

W.D. GRANT and K. HORIKOSHI

5.1 Introduction

5.1.1 *Ecology and environments*

Organisms with optimal pH for growth in excess of pH 9, usually between 9 and 10, are properly defined as alkaliphiles. The term alkalophile is also widely used, although alkaliphile is probably more correct (the root is alkali derived from the Arabic *al-qualiy*). Obligate alkaliphiles are incapable of growth at neutrality and generally have an optimal pH for growth of around pH 9.5. The term alkalitolerant (or alkalitrophic) applies to those organisms capable of growth at high pH but with pH optima for growth in the acid or neutral region of the pH scale.

Obligate alkaliphiles are often capable of growth at pH values of 11, although growth at pH values greater than pH 11.5 has not been convincingly demonstrated because of often overlooked problems with pH control in high-pH media, in which the pH drops during prolonged incubation because of the absorption of atmospheric carbon dioxide [1].

Most of the organisms described as growing under very alkaline conditions are prokaryotes, but eukaryotic algae, particularly diatoms, are commonly described as a significant component of the phototroph population in certain alkaline lakes [2, 3]. Often, whether or not an organism is alkaliphilic or merely alkalitolerant is not rigorously determined.

Alkaliphilic (and alkalitrophic) prokaryotes are widely distributed and may be found in, for example, almost any soil environment, even in soils that are not considered alkaline when subjected to bulk pH measurements. This widespread occurrence of alkaliphiles in non-alkaline environments is some-thing of a puzzle. The only rational explanation would seem to be that in heterogeneous environments like soils localised areas occur where transient alkalinity is generated by biological processes such as ammonification.

Certain commerical activities create alkalinity in the environment. Thus cement manufacture and casting (calcium hydroxide), electroplating (sodium hydroxide) and the alkali process for removing potato skins (potassium hydroxide) all produce alkaline effluents, although in recent years more rigorous effluent control measures have been applied and highly alkaline effluents are less common. The Kraft process for paper and board manufacture, although also less common in recent times, also generates alkaline effluents because of the lignin removal stage, in which lignin is solubilised by a combination of sulphate and alkali (sodium hydroxide). All of these man-made environments harbour prokaryotes, although the range of types is limited.

Naturally occurring highly alkaline environments are of two main types. Both are produced by a combination of geological and climatic features.

5.1.1.1 *High-Ca^{2+} environments.* Highly alkaline ground waters persist in certain geological locations including California, Oman and Jordan [4]. This kind of groundwater is analogous to the effluent produced by cement manufacture where the environment is dominated by solid-phase calcium hydroxide. The *in-situ* weathering of calcium and magnesium silicates determines the eventual composition of the groundwater. In particular, the presence of the primary minerals olivine and pyroxene is of key importance. In near surface waters, carbon dioxide-charged waters decompose these minerals in the manner shown in Figure 5.1 in the process known as serpentinisation. The end result is the release of Ca^{2+} and OH^-, the immobilisation of Mg^{2+}, highly reducing conditions through the initial release of Fe^{2+} and the production of hydrogen by the oxidation of transient metal hydroxides.

In such an environment, carbonate quickly becomes depleted by precipitation as calcite (calcium carbonate), leading to a calcium hydroxide-dominated brine in which solid-phase calcium hydroxide in equilibrium with soluble phase material maintains a highly alkaline environment (pH 11.5). However, the solubility of calcium hydroxide is extremely low under most conditions (around 10 mM), and such waters have a low buffering capacity.

Table 5.1 indicates the ionic composition of calcium hydroxide-dominated groundwaters in Oman [4]. The chemistry is dilute (major ions present at 10 mM or less) and the carbon, nitrogen and phosphorus contents are in the range seen for oligotrophic waters.

$$MgFeSiO_4 + CO_2 + H_2O \rightarrow Mg^{2+} + HCO_3^- + H_4SiO_4 + Fe^{2+} + OH^-$$
Olivine

$$MgCaFeSiO_3 + CO_2 + H_2O \rightarrow Mg^{2+} + Ca^{2+} + HCO_3^- + H_4SiO_4 + Fe^{2+} + OH^-$$
Pyroxene

$$Mg^{2+} + H_4SiO_4 + H_2O \rightarrow Mg_3Si_2O_5(OH)_4 + H^+$$
Serpentine

$$MgFeSiO_4 + MgCaFeSiO_3 + H_2O \rightarrow Mg_3Si_2O_5(OH)_4 + Fe^{2+} + Ca^{2+} + OH^-$$

$$Fe(OH)_2 \rightarrow Fe_2O_3 + H_2 + H_2O$$

Figure 5.1   Geochemical evolution of Ca^{2+} springs (modified from ref. 4).

5.1.1.2 *Soda lakes and soda deserts.* Soda lakes and soda deserts represent the dominant naturally occurring highly alkaline environments. Table 5.2 lists a range of these together with geographical location. It can be seen that

Table 5.1 Hydrochemistry of selected Ca^{2+} springs in Oman (modified from ref. 4). Ion concentrations are in mg/l.

	Nizwa Jill	Karkin	Bahla
pH (field)	11.24	11.44	11.41
E_h (mv)	-173	-373	-372
Temperature (°C)	33	35.7	34.1
Na^+	218	258	189
K^+	9.2	11.2	8.4
Ca^{2+}	54.7	72.0	62.2
Mg^{2+}	<0.1	<0.1	<0.1
OH^- (field)	84	104	83
CO_3^{2-}	0	0	0
Cl^-	291	351	275
SO_4^{2-}	0.91	2.39	0.19

these environments are widely distributed, although relatively few areas have been explored from the microbiological point of view. These environments differ from the alkaline environments described earlier by being in areas significantly depleted in Ca^{2+} and Mg^{2+}. Soda lakes, as the name implies, are characterised by the presence of large amounts of sodium carbonate (as complexes of this salt such as $Na_2CO_3 \cdot 10H_2O$ or $Na_2CO_3 \cdot NaHCO_3 \cdot 2H_2O$). In addition, evaporative concentration usually plays a significant role in the genesis of these lakes and, as a result, other salts (particularly sodium chloride) also concentrate, giving rise to environments that are both alkaline and somewhat saline. Such lakes have great buffering capacity, and maintain their high pH (10–11.5) in seasons with widely differing rainfalls [2].

The genesis of alkalinity is undoubtedly a complex process. Papers by Eugster and colleagues [7, 8] should be consulted for detailed considerations of a variety of factors influencing a range of lakes. A considerably simplified version (Figure 5.2) indicates the key role of Ca^{2+} (and to a lesser extent Mg^{2+}) in the formation of the lakes. It is supposed that weathering and biological activity produces carbon dioxide-charged surface waters, and thus a HCO_3^-/CO_3^{2-} solution is produced which leaches surrounding minerals. Soda lakes are usually in tropical and subtropical areas dominated by alkaline trachyte lavas (high Na^+, low Ca^{2+}, low Mg^{2+}) in closed basins where evaporative concentration can take place. In such a situation, the concentration of CO_3^{2-} greatly exceeds that of any Ca^{2+} and Mg^{2+}, which are removed from solution as insoluble carbonates, leaving a Na^+- and CO_3^{2-}-dominated brine (usually with Cl^- as another major ion; left pathway, Figure 5.2). The alkalinity is a consequence of the shift in the sodium carbonate/sodium hydroxide/sodium bicarbonate/carbon dioxide equilibrium produced by concentration. Normal surface waters do not become alkaline because the concentrations of Ca^{2+} and Mg^{2+} exceed that of CO_3^{2-} and thus CO_3^- is removed by precipitation (centre and right pathways, Figure 5.2). Table 5.3 lists the compositions of typical soda lakes, and it should be noted that,

Table 5.2   Worldwide distribution of soda lakes and soda deserts (from refs. 5 and 6).

North America	
Canada	Manito
USA	Alkali Valley, Albert Lake, Lake Lenore, Soap Lake, Big Soda Lake, Owens Lake, Mono Lake, Searles Lake, Deep Springs, Rhodes Marsh, Harney Lake, Summer Lake, Surprise Valley, Pyramid Lake, Walker Lake.
Central America	
Mexico	Texcoco
South America	
Venezuela	Langunilla Valley
Chile	Antofagasta
Europe	
Hungary	Lake Feher
Yugoslavia	Pecena Slatina
Russia	Kulunda Steppe, Tanatar Lakes, Karakul, Araxes plain, Chita, Barnaul, Slavgerod
Asia	
Turkey	Van
India	Lake Looner, Lake Sambhar
China	Qinhgai Hu, Sui-Yuan, Heilungkiang, Kirin, Jehol, Chahar, Shansi, Shensi, Kansu
Africa	
Libya	Lake Fezzan
Egypt	Wadi Natrun
Ethiopia	Lake Aranguadi, Lake Kilotes, Lake Abiata, Lake Shala, Lake Chilu, Lake Hertale, Lake Metahara
Sudan	Dariba lakes
Kenya	Lake Bogoria, Lake Nakuru, Lake Elmentieta, Lake Magadi, Lake Simbi, Lake Sonachi
Tanzania	Lake Natron, Lake Embagi, Lake Magad, Lake Manyara, Lake Balangida, Basotu Crater Lakes, Lake Kusare, Lake Tulusia, El Kekhooito, Momela Lakes, Lake Lekandiro, Lake Reshitani, Lake Lgarya, Lake Ndutu, Lake Rukwa North
Uganda	Lake Katwe, Lake Mahega, Lake Kikorongo, Lake Nyamunuka, Lake Munyanyange, Lake Murumuli, Lake Bunyampaka
Chad	Lake Bodu, Lake Rombou, Lake Dijikare, Lake Momboio, Lake Yoan
Australia	Lake Corangamite, Red Rock Lake, Lake Werowrap. Lake Chidnup

unlike the Ca^{2+}-dominated groundwaters described earlier, molar rather than millimolar quantities of ions are present. In addition, the levels of nitrogen, phosphorus and carbon place these lakes as highly eutrophic [6, 9].

5.1.2   *Alkaliphile diversity*

In order to maximise the isolation of alkaliphilic rather than alkalitolerant micro-organisms it is preferable to culture in/on media at pH 10 or above.

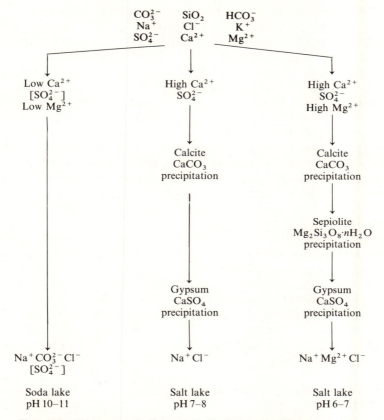

Figure 5.2   The genesis of saline and alkaline brines (modified from ref. 6).

Table 5.3   Hydrochemistry of soda lakes (from refs. 6 and 9). Ion concentrations are in mM.

	Wadi Natrun (Zugm)	Lake Magadi
pH (field)	11.0	10.5
Temperature ($^\circ$C)	40	38
Na^+	6177	4600
K^+	58	42
Ca^{2+}	0	< 0.01
Mg^{2+}	0	< 0.01
CO_3^{2-}	1120	640
Cl^-	4360	1950
SO_4^{2-}	235	—

High pH is widely achieved by the addition of sodium carbonate at between 1 and 5% (w/v). An inevitable consequence of high pH levels is that Ca^{2+} and Mg^{2+} are rendered essentially insoluble and it is pointless to add significant amounts of these ions. The equilibrium between NH_4^+ and ammonia at high pH results in the volatilisation of ammonia from media containing NH_4^+ salts, so nitrogen sources other than NH_4^+ should be used. Media containing sodium carbonate at the above concentrations have a pH of around 10.5 initially, equilibrating to pH 9.5 over a few days (the change is more rapid in agar media). It is, accordingly, not possible to maintain pH levels much above 9.5 for prolonged periods using this buffer system. Grant and Tindall [1] list a number of alternative buffer systems that perform better, but, in practice, long-term maintenance of high pH requires liquid media with regular, preferably automated, addition of alkali. A normal procedure would be to enrich at high pH and then subculture on to agar at somewhat lower pH.

Most of the alkaliphiles isolated to date have been aerobic organotrophs, largely cultured on rich media in which complex carbon sources such as yeast extract and/or peptone provide carbon, nitrogen and energy sources [1, 10]. It is necessary to sterilise alkali solutions (e.g. sodium carbonate) separate from the organic components. Agar media have to be mixed at 60°C and poured immediately to prevent alkaline hydrolysis of the agar.

Formulations at high pH are possible for a variety of other physiological groups including organotrophic anaerobes [11], anoxygenic phototrophs [9] and methanogens [12].

5.1.2.1 *Alkaliphile taxa from 'normal' soils and waters.* Enrichments or direct plating in/on peptone or yeast extract based media described by Horikoshi and Akiba [10] and Grant and Tindall [1] almost invariably yield endospore-forming rods that are currently classified in the genus *Bacillus*. The organisms are often pigmented, usually shades of yellow or red, and grow over the pH range 8–11.5 or so. Organisms of this type were first isolated by Vedder [13], who described them as a new species *B. alcalophilus*. To date, a very larger number of alkaliphilic bacilli have been isolated, notably by Horikoshi and co-workers, largely because of industrial interest in alkali-stable enzymes [2, 10, 14]. Most of these bear names that have not been validated and thus have no taxonomic standing. In 1982 Gordon and and Hyde [15] investigated a large number of these isolates and came to the conclusion that these organisms belonged to the *B. firmus–B. lentus* complex, although the group was significantly heterogeneous, in common with many *Bacillus* groups. Recently, Fritze *et al.* [16] have looked at the DNA base composition of 78 alkaliphilic bacilli and arranged these into three groups. Group 1 had a G + C content of 34–37.5 mol% and contained the type strain of *B. alcalophilus*. A second group had a G + C content of 38.2–40.8 mol%, and group 3 strains had a G + C content of 42.1–43.9 mol%. However, there was little DNA/DNA homology within the groups, nor were any of the strain significantly homologous with *Bacillus* type strains of similar G + C content (*B. megaterium, B. lentus, B. sphaericus*). These results indicate that the alkaliphilic bacilli comprise an extremely diverse group, probably representing a considerable number of taxa at the species level. Further work is needed to clarify the taxonomic position of these strains.

Enrichments carried out in similar media where the pH is poised by contact with solid-phase calcium hydroxide or crushed concrete also yield spore-forming bacilli in the main, although these have not yet been characterised in detail (D. Widdowson and W.D. Grant, unpublished results). To date, all these isolates also grow on sodium carbonate-based media.

Occasionally, non-spore-forming isolates are obtained from soils and water; and there are reports of pseudomonads [16], paracocci [17], micrococci [18], aeromonads [19], corynebacteria [20] and streptomycetes [21, 22]. It will be noted that Gram-negative isolates are rare, unlike the situation for soda lakes (see later).

Isolates from man-made alkaline environments include the Gram-positive *Exiguobacterium aurantiacum* from potato-processing waste [23] and a Gram-negative organism described as an *Ancylobacter* sp. from Kraft paper process effluents [24].

5.1.2.2 *Alkaliphile taxa from Ca^{2+} rich groundwaters?*   Several Ca^{2+} ground-water springs in Oman have been subjected to preliminary microbiological analysis [4]. This type of groundwater is of interest because the hydrochemical characteristics resemble the inferred conditions in cement pore waters and might be a model system in which to examine microbial activity that might be deleterious to concrete structures. Analyses carried out by Bath *et al.* [4] indicate a low population in the Oman springs, from 10^1 to 10^4 c.f.u./ml. Microbial genera detected were similar to those encountered in 'normal' environments. However, it is not clear if any of the organisms described are actually indigenous to the groundwaters, or merely surface contaminants. No attempt was made to culture organisms under conditions comparable to that of the chemistry of the springs, so it is not clear if these waters do support a population of oligotrophic anaerobic alkaliphiles.

The alkaline spring in California investigated by Souza and Deal [20] that yielded a coryneform organism is likely to have a similar genesis and chemistry to the Oman springs, but again it is not clear if the organism is adapted to grow in the spring waters.

5.1.2.3 *Alkaliphile taxa from soda lakes.*   Soda lakes exhibit very high productivity rates (often $> 10 \, g \, C \, m^{-2} \, d^{-1}$) [2, 25, 26]. Such rates are not encountered elsewhere, except in polluted inland waterways. Accordingly, these tropical and semitropical lakes support large-standing crops of a diverse range of micro-organisms that are undoubtedly indigenous to these lakes. There is a considerable body of limnological work on many of these lakes, but detailed microbiological analyses are rare, being largely confined to the

Kenyan Rift Valley [2, 16] and the Wadi Naturn [9]. Less detailed information is to be found relating to Russian [27] and Tibetan [28] soda lakes.

The primary productivity in these lakes is, unusually, almost entirely due to prokaryotes. Dense populations of cyanobacteria dominate the more dilute lakes, and often the blooms may consist of only one species, usually a *Spirulina* sp. Other cyanobacteria that participate include *Anabaenopsis* (*Cyanospira*) and *Chroococcus* spp., occasionally with some contribution from eukaryotic diatoms [2]. There is also an unquantified contribution to primary productivity by anoxygenic phototrophic bacteria of the genus *Ectothiorhodospira* (usually *E. mobilis* and *E. vacuolata*) [6, 11]. Phototrophs from these lakes may be cultured in conventional media made alkaline with sodium carbonate, but there are profound technical difficulties in obtaining axenic cultures of cyanobacteria.

Rather little attention has been paid to the organotrophic bacteria in soda lakes, although these are readily cultured. The peptone and yeast extract media referred to earlier approximate quite well to the conditions pertaining in a range of dilute East African soda lakes, provided an appropriate level of sodium chloride is added. Total counts indicate bacterial populations of 10^7–10^8/ml [2] and viable counts of aerobic organotrophs yield 10^5–10^6/ml. Counts made on site yield a more diverse population, whereas stored samples show an increase in viable count, but a less diverse population (W.E. Mwatha and W.D. Grant, unpublished results). The bacterial population in these lakes remains at a stable level despite significant changes in conductivity and periodic developments of cyanobacterial blooms [2].

Bacterial analyses have been largely confined to aerobic organotrophs, although a start has been made on anaerobes [11]. Unlike 'normal' soil, soda lake samples yield largely Gram-negative isolates (N. Collins, B.E. Jones and W.D. Grant, unpublished results), as determined by quinone analysis [28] and sensitivity to potassium hydroxide [30]. Gram stain procedures are not helpful, since organisms from soda lakes (and other extreme environments) invariably stain negative regardless of cell wall structure. The taxonomic position of these isolates remains to be established.

Slopes of these isolates (and indeed alkaliphilic bacilli from soil) may be kept in the fridge for many months without losing viability. Long-term storage can be achieved on ceramic beads at 70°C [31] suspended in growth medium supplemented with 15% (w/v) glycerol. Freeze-drying has not been systematically explored.

Highly saline alkaline environments such as Lake Magadi (Kenya), Owens Lake (California) and the Wadi Natrun (Egypt) harbour a different population of prokaryotes. These very saline lakes are coloured red by large numbers of haloalkaliphilic archaeobacteria. Two genera are currently recognised, *Natronobacterium* and *Natronococcus*. These organisms are clearly strictly halobacteria, but represent distinct lines within the halobacterial group [32]. Organisms of this type have been isolated from Lake Magadi [33], Wadi

Natrun [34], Owens Lake [35] Tibetan lakes (H. Jannasch and W.D. Grant, unpublished results) and Russian [36] and Chinese sites [28]. The organisms are readily enriched in the medium described originally by Tindall *et al.* [37], but variations on the medium have also been successfully used [35].

Alkaliphilic and halotolerant eubacteria are also to be found in these environments [2, 38], but tend to be overgrown by archaeobacteria in the media described. So far, no halophilic and alkaliphilic isolates are known, but a range of halotolerant types can be isolated, particularly by increasing the proporation of sodium carbonate to sodium chloride in the growth medium (R. Gemmell and W.D. Grant, unpublished results). Many of the organisms almost certainly represent new taxa. The only examples ascribed to an extant taxon at the moment are bacilli from the Wadi Natrun [16].

Archaeobacterial blooms in these highly saline lakes represent secondary productivity. Primary productivity is almost certainly due to halotolerant *Ectothiorhodospira* spp. (*E. halochloris*, *E. halophila*) which can be readily isolated [6, 9].

Soda lakes may also show methanogenesis. A number of slightly alkaliphilic and halophilic methanogens have been isolated from these environments [39, 40], although none of these is as alkaliphilic or halophilic as natronobacteria. A new genus, *Methanohalophilus*, has been created to accommodate some of these isolates [41].

5.1.3 *Alkaliphile physiology*

5.1.3.1 *Bioenergetics.* Alkaliphilic bacteria clearly must have functional membranes where the outside surface is in an extremely alkaline solution, apparently posing problems for ion movements, particularly proton movements. Alkaliphiles thus provide a unique system in which to test the dogmas surrounding the chemiosmotic theory of energy generation.

The theory supposes that prokaryotes generate an electrochemical potential ($\Delta\mu_H^+$) across the cell membrane by the active extrusion of H^+. Both the H^+ gradient itself (Δ pH, outside acid) and the transmembrane electrical potential ($\Delta\psi$) generated by charge partition (inside negative) contribute to $\Delta\mu_H^+$ in the following relationship $\Delta\mu_H^+ = \Delta\psi - Z \Delta pH$ where Z is a constant.

There is little reason to doubt that alkaliphiles regulate their internal pH within the range pH 7–9 [42]. A simple calculation indicates that an alkaliphile growing in a growth medium at pH 11.5 has a 'reversed' ΔpH of 2.5 units, which corresponds to a force of around 150 mV. A major elevation of $\Delta\psi$ might compensate for this problem, but it now seems clear that alkaliphile $\Delta\psi$ values do not offset increasingly adverse ΔpH values [42, 43] (although measurements are largely confined to a limited range of alkaliphilic bacilli). This problem for the chemiosmotic theory has been addressed several times over the years, notably by Krulwich and colleagues [42, 43]. There seem to be several possibilities.

(1) For a large number of solutes, alkaliphiles bypass the problem of low $\Delta\mu_H^+$ values by utilising Na^+ as a coupling ion for active transport. Alkaliphiles require Na^+ in the growth medium, and it is clear that cell pH homeostasis depends on the presence of an H^+/Na^+ antiporter. Motility is also dependent on the use of Na^+ as a coupling ion.

(2) Some mildly alkaliphilic and halophilic vibrios appear to possess an Na^+-translocating ATPase that functions as an ATP synthase at high pH. Skulachev and colleagues [44] have extensively characterised this system.

(3) Alkaliphilic bacilli, on the other hand, appear to lack Na^+-ATPases, appearing to possess normal $F_1 F_0$ ATPases, and may have an electron transport chain that can deliver H^+ to the ATPase without equilibration with bulk phase H^+, i.e. localised coupling [45].

5.1.3.2 *Cell lipids.* Bioenergetics alone do not explain the basis of obligate alkaliphily. Krulwich and colleagues [42] speculate that the membrane integrity of obligate alkaliphiles may be compromised at low pH. Studies of the complex lipids of obligately alkaliphilic bacilli and alkalitolerant isolates indicate that the main lipids are phosphatidyl glycerol, phosphatidyl ethanolamine and cardiolipin with small amounts of phosphatidic acid together with substantial amounts of squalenes and C_{40} and C_{30} isoprenoids. The neutral lipids may have a cholesterol-like function in membrane rigidity operational only at high pH. Obligate alkaliphiles also seem to have more non-esterified fatty acids [42]. There is some evidence that alkaliphile membranes become leaky to protons at low pH [42]. The evidence is therefore that membrane structure and function is likely to be a key factor in alkaliphily.

5.1.3.3 *Cell walls of alkaliphilic* Bacillus *strains.* Cell walls are also directly exposed to highly alkaline environments. Recently it has been observed that protoplasts of alkaliphilic *Bacillus* strains lose their stability against alkaline environments. This fact suggests that the cell wall may also play some role in protecting the cell from alkaline environments.

Components of the cell walls of several alkaliphilic *Bacillus* bacteria have been investigated in comparison with those of neutrophilic *Bacillus subtilis* [46–48]. The walls were shown to be composed of peptidoglycan and acidic compounds. The peptidoglycan was essentially similar in composition to that of *B. subtilis*. After hydrolysis, the acidic compounds detected were galacturonic acid, glucuronic acid, glutamic acid, aspartic acid and phosphoric acid. The strains tested could be divided into three groups as follows: group 1 bacilli had high amounts of glucuronic acid and hexosamine; no growth was observed at pH 7.0; Na^+ was essential for their growth. Group 2 bacilli had large amounts of glutamic acid, aspartic acid, galacturonic acid and glucuronic acid. For this group, growth at higher pH values increased the content of acidic compounds; growth was observed at pH 7.0; Na^+ was essential for their growth. Group 3 bacilli contained phosphoric acid and no remarkable differences were detected in comparison with *B. subtilis*; growth was observed in the presence of Na^+ or K^+ at pH 7.0 and 10.2.

All peptidoglycans appeared to be similar in composition to that of *Bacillus subtilis*. However, the composition was characterised by an excess of hexosamines and amino acids in the cell walls. Major constituents detected commonly in hydrolysates of peptidoglycans were glucosamine, muramic acid, D- and L-alanine, D-glutamic acid, *meso*-diaminopimelic acid and acetic acid [49]. The composition of peptidoglycan was not changed whether the strain was cultured at pH 7 or 10. The peptides are likely to be cross-linked directly between *meso*-diaminopimelic acid and D-alanine because of the lack of amino acids known to be involved in an interpeptide bridge.

It was concluded that all of the peptidoglycans of the alkaliphilic *Bacillus* strains so far examined were of the A1-γ type of peptidogylcan, which is found in the vast majority of strains of the genus *Bacillus*. The variation in the amide content was similar to the variation which is known in neutrophilic *Bacillus* species.

The alkaliphilic strains of *Bacillus* in group 1 also contain teichuronic acid in their cell walls. The walls of group 3 strains contain teichoic acid, whereas the cell walls of group 2 organisms contain large amounts of acidic amino acids and uronic acids [49]. The same acidic compounds are found in much smaller quantities in the walls prepared from bacteria grown at a neutral pH. This may indicate that the acidic components in the outermost layers have a function in supporting growth at alkaline pH. In support of this view, one of the alkaliphilic *Bacillus* strains, *Bacillus* sp. strain C-125, which is a producer of xylanase, was classified in group 2 and grows well at neutral pH values. The cells of *Bacillus* C-125 share the A1-γ type of peptidoglycan and the peptidoglycan is enclosed by at least two acidic polymers with highly negative charges [50–52]. It is assumed that the negative charges on the acidic non-peptidoglycan components give the cell surface its ability to adsorb sodium and hydronium ions and repulse hydroxide ions, and as a result may enable the cells to grow in alkaline environments.

5.2 Alkaliphiles and industry

5.2.1 *Enzymes*

5.2.1.1 *Detergent enzymes.* Alkaliphilic micro-organisms have already made a considerable impact in the application of biotechnology for the manufacture of consumer products [2, 14, 53]. With one or two notable exceptions, the main industrial applications for alkaliphilic enzymes are in the detergent industry. Biological detergents contain enzymes that have been obtained (usually) from alkaliphilic or alkalitolerant bacteria. Detergent enzymes account for 25% of worldwide enzyme production, and the world market is current worth > $200 million [2].

For an enzyme to be an effective addition to a laundry detergent it has to be active at high pH (most detergents are alkaline) and compatible with a range of additives such as builders, bleaches, brighteners, perfumes and so on [2] Consumer demand in terms of washing temperature also affects the choice of an enzyme for a particular formulation. The recent penetration of thixotrophic liquid detergents (which are more readily miscible at lower temperatures) is another factor to be taken into account in choosing an enzyme. These developments present new problems that can often best be met by the development of a new enzyme.

The most widely used enzymes in detergents are proteases and amylases. Proteases have the largest market sector. Alkaline proteases sensitive to di-isopropylphosphofluoridate (DFP, which is a serine-directed reagent) are classified as serine endoproteases, and these are widely used in detergent formulations. Examples that have been on the market for some time include Alcalase derived from alkalitolerant *B. licheniformis*, originally marketed in the presoak detergent Biotex [2]. More recently introduced proteases include Esperase, Savinase, Maxatase and Maxacal, derived from alkaliphilic bacilli. Proteases hydrolyse stains such as blood, grass, milk, etc. and ensure that coagulated proteins are not deposited back on the fabric giving a grey appearance.

Amylases remove starch-based stains, often working together with protease. Products include Termamyl and Maxamyl, which are α-amylases effective at high pH and high temperature, again derived from alkaliphilic or alkalitolerant bacilli [2, 14].

Use has also been made of proteases in leather tanning. Traditional tanning depends on bacterial growth during the dehairing and bating (softening) of hides [2, 14]. The application of enzymes results in a quicker and more reliable process. Again, toleration of alkaline conditions is required since salted, cured hides are soaked in alkali to soften and swell the skins. Enzymes that can be used include Alcalase, Milezyme, Rapidermase and Batinase. Effluent problems can also be reduced, since some of the more noxious aspects of the traditional process (e.g. dung addition) can be avoided.

As the demand for proteases has increased to compensate for lower washing temperature, so has the demand for other detergent enzymes such as cellulases and lipases. Alkaliphilic cellulases are believed to modify the non-crystalline part of the cellulose matrix in linens and cottons (the area most susceptible to soiling). The recently introduction cellulose-containing detergent Attack has captured a significant proportion of the Japanese market [14]. The enzyme is derived from an alkaliphilic *Bacillus*. Cellulases also have a role in the 'stone-washing' and softening of denims.

Preliminary studies have been carried out with detergent lipases (which solubilise fats at low washing temperatures where they would not melt), but these have yet to be widely accepted [2].

There is a considerable body of work concerned with the cloning and expression of a variety of potential detergent enzymes [14, 53] with a view to the improvement of yields and possible circumvention of existing patents.

5.2.1.2 *GCTases and cyclodextrin production.*   The cyclodextrins, Schardinger dextrins, are a group of homologous oligosaccharides, obtained from starch by the action of cyclomaltodextrin glucanotransferases (CGTase). These curious compounds are fascinating to carbohydrate chemists because they have properties markedly different from other dextrins. Some of these unique properties are as follows:

(1) Cyclodextrins (CD) are homogeneous cyclic molecules composed of six or more α-1,4-linked glucose units as in amylose.
(2) As a consequence of the cyclic arrangement they have neither a reducing end-group nor a non-reducing end-group and are not decomposed by hot aqueous alkali.
(3) They are relatively resistant to acid hydrolysis and the common starch-splitting amylases.
(4) They crystallise very well from water and from aqueous alcohol solutions.
(5) They form an abundance of crystalline complexes called inclusion compounds with organic substances.
(6) They form a variety of inorganic complexes with neutral salts, halogens and bases.

Mass production of these unique compounds on an industrial scale has been attempted several times in the past. In 1969, Corn Products International, USA, produced β-CD (seven glucose units) using *Bacillus macerans* CGTase. The Teijin Company in Japan also produced β-CD by using the macerans enzyme in a pilot plant. There have been serious problems in both production processes.

(1) CGTase from *B. macerans* is not suitable for industrial use, because the enzyme is not sufficiently thermostable.
(2) The yield of CD from starch is not high, usually 20–30% on an industrial scale.
(3) Toxic organic solvents such as trichloroethylene, bromobenzene, toluene, etc. were used to precipitate CD because of the low conversion rate. Therefore, the development of large-volume applications has been quite limited.

However, several CGTases were isolated from alkaliphilic *Bacillus* strains [54, 55]. One of these, a CGTase produced by alkaliphilic *Bacillus* sp. no. 38–2, overcame all these weak points and resulted in the mass production of crystalline α- (six glucose units), β-, γ-CD (eight-glucose units) and CD mixtures at low cost without using organic solvents [56]. This relatively simple process reduced the cost of β-CD from £500 to £15/kg, and that of α-CD to within £150/kg. This success has paved the way for their use in large quantities in foodstuffs, chemicals and pharmaceuticals.

5.2.1.3 *Mannan-degrading enzymes.* Three extracellular β-mannanases were produced by alkaliphilic *Bacillus* sp. AM-001 isolated from a soil sample [57].

These alkaline extracellular β-mannanases were most active at pH 8.5–9.0, at pH 8.5 hydrolysing β-1,4-manno-oligosaccharides larger than mannotriose, the major components in the digest being di-, tri- and tetrasaccharides [58, 59]. These enzymes are of interest because of their activity towards mannans used as food additives or products.

The crude enzyme hydrolysed guar gum as well as konjak mannan (a Japanese food product). The hydrolysate of konjak mannan contained oligosaccharides having a degree of polymerisation of 2–7 and exhibiting faint sweetness. The guar gum hydrolysate product was a water-soluble polysaccharide which had an average molecular weight of 5000 and did not show sweetness. Guar gum hydrolysate will be available as a dietary fibre in the near future.

5.2.1.4 *Pectinases.* Pectinolytic enzymes i.e. pectinases, which degrade pectic polysaccharides such as pectin and pectic acid, are distributed in micro-organisms and higher plants but are not found in higher animals. Pectinases produced by micro-organisms are widely used in the fruit and vegetable processing industries.

The first paper on alkaline endopolygalacturonase (pectinase) produced by alkaliphilic *Bacillus* sp. no. P-4-N was published in 1972 [62]. *Bacillus* sp. no. P-4-N, could grow in the pH range from 7.0 to 11.0, but the most active growth was observed at pH 10 in a medium containing 1% (w/v) sodium carbonate. The optimum pH of the enzyme towards pectic acid was 10.0, a unique property which is different from other polygalacturonases.

Another *Bacillus* species isolated from garden soil was also shown to produce an endopolygalacturonate lyase [61]. The optimum pH of the enzyme was 10.0 using acid-soluble pectic acid as substrate.

The first application of alkaline pectinase-producing bacteria in the retting of Mitsumata bast (*Edgeworthia papyrifera*) for the production of Japanese paper was reported by Yoshihara and Kobayashi [62]. They isolated about 800 alkaliphilic bacteria from soil, sewage and manure samples obtained from Japan and Thailand. *Bacillus* sp. GIR-277 had strong macerating activity toward Mitsumata bast. This bacterium produced pectate lyase which had an optimum pH for enzyme action at 9.5.

Japanese paper was produced by suspending 4 g of Mitsumata bast in 100 ml of a culture medium containing 0.05% (w/v) yeast extract, 0.05% (w/v) casamino acids, 0.2% (w/v) ammonium chloride, 0.1% (w/v) dipotassium hydrogen phosphate, 0.05% (w/v) magnesium sulphate and 1.5% (w/v) sodium carbonate. *Bacillus* sp. GIR-277 was inoculated into the medium. After 5 days of cultivation at 34°C with shaking, the retted basts were harvested and Japanese paper was prepared by the method described in

Japanese Industrial Standard P8209. The strength of the unbeaten pulp resulting from bacterial retting was higher than that obtained by the conventional soda ash-cooking method. The paper sheets produced were uniform and soft to touch compared with those prepared by the conventional process.

Waste water from the citrus-processing industry contains pectinaceous materials that are only slightly degraded by micro-organisms during activated-sludge treatment, and Tanabe *et al.* [63, 64] have tried to develop a new waste treatment by using alkaliphilic micro-organisms. They isolated an alkaliphilic *Bacillus* sp. GIR-621 from soil in Thailand which produced an extracellular endopectate lyase in alkaline media at pH 10.0. The strain GIR-621 was applied to the pretreatment of waste water from an orange-canning factory which contained pectinaceous material. Treatment with *Bacillus* sp. strain GIR-621-7 proved to be useful as a pretreatment of the waste water to remove pectic substances, removing 93% of uronic acid within 20 hours.

5.2.1.5 *Xylanases*. Xylan biomass is abundant in agricultural residues such as rice straw, barley straw and corn cobs. Xylanases have been shown to be widely distributed in bacteria and fungi. Alkaline xylanases are of interest from the industrial point of view, since xylan is soluble in alkaline solution.

Bacillus strains which produced xylanase have been isolated: *Bacillus* sp. no. C-59-2 [62] and C-11 [66]. Both *Bacillus* strains could grow well and produce xylanase in alkaline media at pH 10. The purified xylanase of *Bacillus* sp. no. C-59-2 exhibited a broad pH optimum for enzyme action (6.0–9.0). The maximum degree of hydrolysis of xylan recorded was ∼ 40% either at pH 6.0 or at pH 9.0 [65]. The optimum pH for xylanase activity from *Bacillus* sp. no. C-11 was 7.0, and approximately 37% of the activity remained even at pH 10.0 [66].

Four strains (W1, W2, W3 and W4) of alkaliphilic and thermophilic bacilli which produced xylanase have also been isolated from soils [67]. The enzymes were stable up to 60°C and the addition of 5 mM calcium chloride had no effect on their thermal stability [68]. All of these xylanases hydrolysed xylan to yield xylose and xylobiose. The degree of hydrolysis of xylan was about 70% after 24 hours' incubation.

There is interest in micro-organisms that can utilise xylan-containing industrial wastes as a carbon source, and alkaliphilic *Bacillus* sp. no. C-11 has been shown to utilise rayon waste under alkaline conditions [66].

Recently, another industrial application has been established by a Japanese company. A beverage containing water-soluble dietary fibre prepared from corn fibre hydrolysed with the xylanase produced by *Bacillus* sp. no. C-59-2 has been developed. The market size is not large, but this is a novel application of alkaline xylanases.

5.2.2   Spirulina

Filamentous cyanobacteria that grow in the form of tight or nearly tight right- or left-handed helices comprise the dominant population in many soda lakes. The organism (or organisms) comprise a group known interchangeably as *Spirulina* or *Arthrospira* [69] and there is still confusion over the correct name and the number of species involved. Thus the East African lakes are said to harbour *S. platensis*, whereas Mexican lakes harbour *S. maxima*, although they may both be the same organism [70, 71].

Vast numbers of flamingos feed almost exclusively on *Spirulina* [72]. Even more interesting is the well-documented utilisation of *Spirulina* as a human food source. Dried *Spirulina* has been consumed by the native population in the Kanen region near Lake Chad in the form of a supplement added to millet meal (Dihé). Historical research has indicated similar use of *Spirulina* harvested from Lake Texcoco by the Aztecs (tecuilaitl). The reviews by Ciferri [70] and Ciferi and Tiboni [71] should be consulted for details of history, biochemistry and industrial potential.

Extensive analyses have indicated an uncommonly high protein content (70%), a low nucleic acid content and a good amino acid spectrum, ranking amongst the best in the plant world [73, 74]. *Spirulina* has been extensively used as a diet supplement for laboratory and farm animals and for the aquaculture of fish, shrimps and prawns. In all cases the results have been favourable, and, in addition, the high concentration of β-carotene and other carotenoids has resulted in an advantageous enhancement in pigmentation, which has also been utilised in the rearing of ornamental fish [73, 75]. Recent studies include poultry [76] and pigmentation in Nile perch [77].

In many countries *Spirulina* has attained the status of a health food, and several small-scale operations, notably at Lake Texcoco, depend on its use as a high-cost health food, ranging from use by Olympic athletes [71, 74] to use as a space diet supplement [78]. Naturally concentrated brines provide essential nutrients, although nitrate is added to increase productivity [73]. Biomass is recovered by filtration and spray- or drum-dried. Other operations consist of artificial plastic-lined ponds or troughs (Japan, Thailand, Israel) [73], sometimes covered by plastic hoods. The average yield is $2–3\,\mathrm{kg\,m^{-2}}$ $\mathrm{year^{-1}}$ [73]. There is a considerable body of research currently being undertaken concerned with maximising yield and optimising biomass composition using continuous culture procedures [73, 79]. Various other culture methods including sea water [80] and the reclamation of waste waters [81] have been used, and there is encouragement to utilise Third-World technologies to produce valuable biomass in saline and arid areas [82, 83].

There is now increasing interest in *Spirulina* as a source of fine chemicals. Besides a high content of water-soluble vitamins [74, 84], *Spirulina* is noteworthy for the high content of carotenoids and xanthophylls, notably β-carotene [74, 84]. In recent years the use of *Spirulina* for the production of essential fatty acids such as γ-linolenic acid has been extensively explored

[84–86]. The organism also synthesises polyhydroxybutyrate (PHB) [87], a potentially useful β-glucan [88,89], and there are recent reports of an antitumour antibiotic [90,91]. Genebanks have been prepared and the technologies for spheroplast formation explored [71,92].

It is clear, therefore, that *Spirulina* has considerable potential in a diverse range of applications. This cyanobacterium grows in an alkaline medium that excludes potentially harmful organisms, is gas vacuolate, facilitating cell harvesting, and since it has been used as a food for many centuries there is no toxicity problem. Furthermore, the photosynthetic yield surpasses anything yet achieved in the terrestrial environment [74,79].

5.2.3 Secretion vectors

One of the most important processes in the fermentation industry is the production of extracellular proteins. *Escherichia coli* is widely used in genetic engineering experiments because it has been extensively studied and much is known about its genetics and biochemistry. Unfortunately, with the exception of a few proteins such as colicins [93], cloacin [94] and haemolysin [95], *E. coli* does not secrete gene products from the cell. If *E. coli* could be modified to secrete recombinant DNA products it would be of considerable interest from the industrial point of view. In the course of an analysis of an alkaliphilic penicillinase, it was noted that most of the product of a penicillinase gene derived from an alkaliphilic *Bacillus* sp. 170 cloned into *E. coli* was released into the culture medium by *E. coli* harbouring a penicillinase gene-carrying plasmid vector based on pMB9 [96]. *E. coli* periplasmic proteins were also excreted. This proved to be due to the insertional activation of a dormant *kil* gene in pMB9 by the DNA of the alkaliphilic *Bacillus*. The *kil* gene [97] is responsible for colicin El release by cell lysis but is not expressed in pMB9 because it lacks a promoter. The weakly activated *kil* gene [98] produced by insertion of a promoter-like DNA sequence from the alkaliphilic *Bacillus* forms the basis of secretion vectors that have been used for the extracellular production of xylanases [49], cellulases [99], human growth hormone [100] and human Fc protein [101]. Using appropriate vectors some 60–70% of the total product is secreted in the active broth C9; the yield of human growth hormone was about 60–70 mg/ml. Western blots have indicated that the secreted proteins are immunologically identical to those of the authentic products. Accordingly, the genetic analysis of one alkaliphilic enzyme has produced an unexpected commercial success in a much wider sense.

5.2.4 Future trends

The detergent industry is a rapidly changing and growing industry with considerable potential for innovation and improvement. One way forward would be to engineer improved enzymes. Site-directed alterations are already part of the enzyme engineer's repertoire [102, 103]. However, there is still room for the classical approach. The choice of environments as source material for alkaliphiles has been remarkably conservative. Almost all enzymes are derived from soil-dwelling bacilli. Other environments have much greater potential, notably the soda lakes extensively discussed in this article.

One might also speculate over possibilities in waste treatment, e.g. the Kraft process, food processing, leather processing, etc. where there is the possibility both of rendering the waste less offensive and obtaining valuable biomass or products.

The catalytic properties of alkaliphiles for biotransformations are also receiving attention. There is a growing demand for pharmaceutical and agricultural chemicals to be produced in an optically pure form. There is already an extensive body of work in chiral transformation—lipases are particularly useful in this area [104]. It is possible to foresee that a highly alkaline milieu could well exert considerable selective pressure on the maintenance of one chiral form, whether it be substrate or product, and there might also be advantages in improved solubilities of certain classes of compound. One might also expect that enzymes suited to the harsh conditions of high pH (and high salt) might be intrinsically more stable in the presence of solvents.

Lastly, one should not forget that 'academic' research may have unforeseen industrial benefits—the discovery of the secretion vector described in this article bears testimony to the power of serendipity.

References

1. W.D. Grant and B.J. Tindall, in *Microbial Growth and Survival in Extremes of Environment*, eds. G.W. Gould and J.C.L. Corry, Academic Press, New York (1980) 27.
2. W.D. Grant, W.E. Mwatha and B.E. Jones, *FEMS Microbiol. Rev.* **75** (1990) 255.
3. R.E. Hecky and P. Kilham, *Limnol Oceanogr.* **18** (1973) 53.
4. A.H. Bath, N. Christofi, C. Neal, J.C. Philip, I.G. McKinley and U. Berner, *Rep. Fluid Processes Res. Group. Br. Geol. Surv.* FLPU 82–2 (1987).
5. B. J. Tindall, in *Halophilic Bacteria*, Vol. 1, ed. F. Rodriguez-Valera. CRC Press, Boca Raton, FL (1988) 31.
6. W.D. Grant and B.J. Tindall, in *Microbes in Extreme Environments*, eds. R.A. Herbert and G.A. Codd. Academic Press, New York (1986) 22.
7. H.P. Eugster and L.A. Hardie, in *Lakes: Chemistry, Geology and Physics*, ed. A. Lermann, Springer, Berlin (1978) 237.
8. L.A. Hardie and H.P. Eugster, *Mineral Soc. Am. Spec. Pap.* **3**, 273.
9. J.F. Imhoff, H.G. Sahl, G.S.H. Soliman and H.G. Truper, *Geomicrobiology* **1** (1979) 219.
10. K. Horikoshi and T. Akiba, *Alkalophilic Microorganism*, Springer, Berlin (1982).
11. H. Shiba, in *Superbugs: Microorganisms in Extreme Environments*, eds. K. Horikoshi and W.D. Grant. Japan Scientific Press/Springer, Tokyo/Berlin (1991) 191.
12. I.M. Mathrani, D.R. Boone, R.A. Mah, G.E. Fox and R.P. Lau, *Int. J. Syst. Bacteriol.* **38** (1988) 139.
13. A. Vedder, *J. Anton van Leevenhoek, Microbiol. Serol.* **1** (1934) 143.
14. W.D. Grant and K. Horikoshi, in *Microbiology of Extreme Environments and its Potential for Biotechnology*, eds. M.S. DaCosta, J.C. Duarte and R.A.D. Williams, Elsevier, Amsterdam (1989) 346.
15. R.E. Gordon and J.L. Hyde, *J. Gen. Microbiol.* **128** (1982) 1109.

16. D. Fritze, J. Flossdorf and D. Claus, *Int. J. Syst. Bacteriol.* **40** (1990) 92.
17. T. Urakami, J. Tamaoka, K.I. Suzuki and K. Komagata, *Int. J. Syst. Bacteriol.* **39** (1989) 116.
18. T. Kimura and K. Horikoshi, *FEMS Microbiol. Lett.* **71** (1990) 35.
19. A. Ohkoshi, T. Kudo, T. Mase and K. Horikoshi, *Agric. Biol. Chem.* **49** (1985) 3037.
20. K.A. Souza and P.H. Deal. *J. Gen. Microbiol.* **101** (1977) 103.
21. V. Chauthaiwale, S. Phadatare, V. Deshpande and M.C. Srinivasan. *Biotechnol. Lett.* **12** (1990) 225.
22. H. Tsujibo, T. Sakamoto, H. Nishino, T. Hasegawa and Y. Inamori, *J. Appl. Bacteriol.* **69** (1990) 398.
23. M.D Collins, B.M. Lund, J.A.E. Farrow and K.H. Schliefer, *J. Gen. Microbiol.* **129** (1983) 2037.
24. S.E. Strand, J. Dykes and V. Chiang, *Appl. Env. Microbiol.* **47** (1984) 268.
25. J.F. Talling, R.B. Wood, M.V. Prosser and R.M. Baxter, *Freshwater Biol.* **3** (1973) 53.
26. J.M. Melack and P. Kilham. *Limnol. Oceanogr.* **19** (1974) 743.
27. B.L. Isachenko, *Selected Works*, Vol. 2, Academiya Nauk, Leningrad (1951) 147.
28. D. Wang and Q. Tang, in *Recent Advances in Microbial Ecology*, eds. T. Hattori, Y. Ishida, Y. Markyama, R.Y. Morita and A. Uchida. Japan Scientific Press, Tokyo (1989) 68.
29. M.D. Collins, in *Chemical Methods in Bacterial Systematics*, eds. M. Goodfellow and D.E. Minnikin. Academic Press, New York (1985) 267.
30. S. Halebian, B. Harris, S.M. Finegold and R.D. Wolfe, *J. Clin. Microbiol.* **13** (1981) 444.
31. R.K.A Feltham, A.K. Power, P.A. Pell and P.H.A. Sneath, *J. Appl. Bacteriol.* **44** (1976) 313.
32. W.D. Grant and H. Larsen, in *Bergey's Manual of Systematic Bacteriology*, Vol. 3, eds. J.T. Staley, M.P. Bryant, N. Pfennig and J.G. Holt. Williams & Wilkins, Baltimore (1989) 2216.
33. B.J. Tindall, H.N.M. Ross and W.D. Grant, *Syst. Appl. Microbiol.* **5** (1984) 41.
34. G.S.H. Soliman and H.G. Trüper, *Zentralbl. Bakteriol. Microbiol. Hyg. I. Abt. Orig.* **C3** (1982) 318.
35. S. Morth and B.J. Tindall, *Syst. Appl. Microbiol.* **6** (1985) 247.
36. S. Zvyagintseva and A.L. Tarasov, *Microbiology* **56** (1988) 664.
37. B.J. Tindall, A.A. Mills and W.D. Grant, *J. Gen. Microbiol.* **116** (1980) 257.
38. W.D. Grant and W.E. Mwatha, in *Recent Advances in Microbial Ecology*, eds. T. Hattori, Y. Ishida, Y. Maruyama, R.Y. Morita and A. Uchida, Japan Scientific Press, Tokyo (1989) 64.
39. T.N. Zhilina, *Syst. Appl. Microbiol.* **7** (1986) 216.
40. J.R. Paterek and P.H. Smith, *Appl. Env. Microbiol.* **50** (1985) 877.
41. J.R. Paterek and P.H. Smith, *Int. J. Syst. Bacteriol* **38** (1988) 122.
42. T.A. Krulwich, D.B. Hicks, D. Seto-Young and A.A. Guffanti, *CRC Crit. Rev. Microbiol.* **16** (1998) 15.
43. T.A. Krulwich and A.A. Guffanti, *Ann. Rev. Microbiol.* **43** (1989) 435.
44. P. Pibrov, R.L. Lazarova, V.P. Skulachev and M.L. Verkhjovskaga, *Biochim. Biophys. Acta* **850** (1986) 458.
45. A.A. Guffanti and T.A. Krulwich, *J. Biol. Chem.* **203** (1988) 14748.
46. R. Aono and K. Horikoshi, *J. Gen. Microbiol.* **129** (1983) 1083.
47. R. Aono, K. Horikoshi and S. Goto, *J. Bacteriol.* **157** (1984) 688.
48. Y. Ikura and K. Horikoshi, *Agric. Biol. Chem.* **47** (1983) 681.
49. R. Aono, *J. Gen. Microbiol.* **131** (1985) 105.
50. R. Aono and M. Uramoto, *Biochem. J.* **233** (1986) 291.
51. R. Aono, *Biochem. J.* **245** (1987) 467.
52. R. Aono, *J. Gen. Microbiol.* **135** (1989) 265.
53. K. Horikoshi and W.D. Grant, *Superbugs: Microorganisms for Extreme Environments*. Japan Scientific Press/Springer, Tokyo/Berlin (1991).
54. N. Nakamura and K. Horikoshi, *Agric. Biol. Chem.* **40** (1976) 753.
55. N. Nakamura and K. Horikoshi, *Agric. Biol. Chem.* **40** (1976) 935.
56. M. Matsuzawa, M. Kaedno, N. Nakamura and K. Horikoshi, *Die Staerke* **27** (1975) 410.
57. T. Akino, N. Nakamura and K. Horikoshi, *Appl. Microbiol. Biotechnol.* **26** (1987) 323.
58. T. Akino, N. Nakamura and K. Horikoshi, *Agric. Biol. Chem.* **52** (1988) 773.
59. T. Akino, N. Nakamura and K. Horikoshi, *Agric. Biol. Chem.* **52** (1988) 1459.
60. K. Horikoshi, *Agric. Biol. Chem.* **36** (1972) 285.
61. C.T. Kelly and W.M. Fogarty. *Can. J. Microbiol.* **24** (1978) 1164.
62. K. Yoshihara and Y. Kobayashi, *Agric. Biol. Chem.* **46** (1982) 109.

63. H. Tanabe, K. Yoshihara, K. Tamura, Y. Kobayashi, I. Akamatsu, N. Niyomwan and P. Footrakul, *J. Ferment. Technol.* **65** (1987) 243.
64. H. Tanabe, Y. Kobayashi and I. Akamatsu, *Agric. Biol. Chem.* **52** (1988) 1855.
65. K. Horikoshi and Y. Atsukawa, *Agric. Biol. Chem.* **37** (1973) 2097.
66. Y. Ikura and K. Horikoshi, *Agric. Biol. Chem.* **41** (1977) 1373.
67. W. Okazaki, T. Akiba, K. Horikoshi and R. Akahoshi, *Appl. Microbiol. Biotechnol.* **19** (1984) 335.
68. W. Okazaki, T. Akiba, K. Horikoshi and R. Akahoshi, *Agric. Biol. Chem.* **49** (1985) 2033.
69. R.W. Castenholz, in *Bergey's Manual of Determinative Bacteriology*, Vol. 3, eds. J.T. Staley, M.P. Bryant, N. Pfennig and J.G. Holt. Williams & Williams, Baltimore (1989) 1771.
70. O. Ciferri, *Microbiol. Rev.* **47** (1983) 551.
71. O. Ciferri and O. Tiboni, *Ann. Rev. Microbiol.* **39** (1985) 503.
72. E. Vareschi, *Oecologia* **32** (1978) 11.
73. A. Richmond, in *Micro-Algal Biotechnology* eds. M.A. Borowitzka and L.S. Borowitzka. Cambridge University Press (1988) 85.
74. L.E. Shubert, in *Progress in Phycological Research*, Vol. 6, eds. F.E. Round and D.J. Chapman, Biopress, Bristol (1988) 237.
75. W. Miki, K. Yamaguchi and S. Konosu, *Bull. Jpn. Soc. Sci. Fish* **52** (1986) 1225.
76. E. Ross and W. Dominy, *Poultry Sci.* **69** (1990) 794.
77. M. Boonyaratpalin and N. Unprasert, *Aquaculture* **79** (1989) 375.
78. Z. Nakhost and M. Karel, *Sci. Aliments* **9** (1989) 491.
79. A. Vonshak and A. Richmond, *Biomass* **15** (1988) 233.
80. M.R. Tredici, T. Papuzzo and L. Tomaselli, *Appl. Microbiol. Biotechnol.* **24** (1986) 47.
81. F. Ayala and T. Vargas, *Hydiobiologia* **151–152** (1987) 91.
82. B. Goldstein, in *Arid Lands: Today and Tomorrow*, ed. E.E. Whitehead. Westview Press, Coloraco (1985) 755.
83. R.D. Fox, in *Algal Biotechnology*, ed. T. Statler, Elsevier, Amsterdam (1988) 355.
84. M.A. Borowitzka, in *Micro-Algal Biotechnology*, eds. M.A. Borowitzka and L.J. Borowitzka. Cambridge University Press (1988) 162.
85. P.G. Roughan, *J. Sci. Food Agric.* **47** (1989) 85.
86. M. Hirano, H. Mori, Y. Miura, N. Nakamura and T. Matsunaga, *Appl. Biochem. Biotechnol.* **24–25** (1990) 183.
87. M. Vincenzini, C. Sili, R. De Philippis and R. Materassi, *J. Bacteriol.* **172** (1990) 2791.
88. K.M. Sekharam, L.V. Venkataram and P.V. Salimath, *Food Chem.* **31** (1989) 85.
89. M. Kubota, T. Koyano, K. Shinohara and K. Nishinari, *J. Jpn. Food Sci. Technol.* **36** (1989) 569.
90. G. Shklar and J. Schwarz, *Eur. J. Cancer Clin. Oncol.* **24** (1988) 839.
91. C.J. Mo and W.X. Wang, *Chin. J. Antibiot.* **14** (1989) 135.
92. L. Lanfaloni, R. Grigfantini, A. Petris and C.O. Gualerzi, *FEMS Microbiol. Lett.* **59** (1989), 141.
93. S. Zhang, A. Faro and G. Zubay, *J. Bacteriol.* **163** (1985) 174.
94. F.K. De Graaf and B. Oudega, *Current Topics in Microbiology and Immunology* **125** (1985) 183.
95. N. Mackman, J.M. Nicaud, L. Gray and I.B. Holland, *Current Topics in Microbiology and Immunology* **125** (1985) 159.
96. T. Kudo, C. Kato and K. Horikoshi, *J. Bacteriol.* **156** (1983) 949.
97. P.T. Chan, H. Ohmori, J. Tomizawa and J. Lebowitz, *J. Biol. Chem.* **260** (1985) 8925.
98. T. Kobayashi, C. Kato, T. Kudo and K. Horikoshi, *J. Bacteriol.* **166** (1986) 8925.
99. C. Kato, T. Kobayashi, T. Kudo and K. Horikoshi, *FEMS Microbiol. Lett.* **36** (1986) 31.
100. C. Kato, T. Kobayashi, T. Kudo, T. Furusato, Y. Murakami, T. Tanaka, H. Baba, T. Oishi, E. Otsuka, M. Ikehara, T. Yanagida, H. Kato, S. Moriyama and K. Horikoshi, *Gene* **54** (1987) 197.
101. K. Kitai, T. Kudo, S. Nakamura, T. Masegi, Y. Ichikaga and K. Horikoshi, *Appl. Microbiol. Biotechnol.* **28** (1988) 52.
102. J.H. van Ee, L.S.J.M. Mulleners, W.C. Rijswijk and R.A. Cuperns, *Comm. J. Con. Esp. Deterg.* **19** (1988) 257.
103. M.W. Pantoliano, R.C. Ladner, P.N. Bryan, M.L. Pollence, J.F. Wood and T.L. Poulds, *Biochemistry* **26** (1987) 2077.
104. J. Harwood, *Trends Biol. Sci.* **14** (1989) 125.

6 Physiology and biotechnological potential of deep-sea bacteria

D. PRIEUR

6.1 Introduction

The expedition of the *Travailleur* and *Talisman* (1882–1883) was the first oceanographic expedition to demonstrate the occurrence of bacteria living in oceanic waters and sediments at depths of 5000 m [1]. However, even though these initial studies on marine microbiology dealt with deep-sea microbiology, the study of deep-sea bacteria still represents a small part of the research performed on micro-organisms living in sea water.

If the definition of the deep sea is accepted as waters > 1000 m deep [2], it has been calculated that the deep sea represents 88% of the earth's area covered by sea water and 75% of the total volume of the oceans. Consequently, this particular marine environment, which represents 62% of the biosphere, is not space-limited but is a major component of the planet's biotopes, and is probably the most extended extreme environment on earth.

The deep sea is influenced by three main parameters which, together, make it particularly extreme: low temperature, low nutrient concentrations and high hydrostatic pressure.

Low temperatures (< 5°C) are the dominant features of the biosphere, taking into account that polar regions and the oceans represent 14 and 71% of the earth's surface respectively [3]. Expressed in terms of volume, 90% of the oceanic waters have a temperature < 5°C, and at the average depth of the oceans (3800 m), the mean temperature is 2°C.

Solar radiation does not penetrate into the deep layers of the oceans, and the consequence is the lack of photosynthetic primary production in the deep sea. From a trophic viewpoint, deep-sea communities depend on the material sinking from the euphotic zone. Approximately 95% of the organic matter produced in surface waters is degraded in the upper layers of the oceans [2] and the remaining material is almost completely degraded before reaching a depth of 1200 m [2]. Concentrations of particulate organic matter and dissolved organic matter below depths of 300 m have been estimated to be in the range 3–10 μg C/l and 0.35–0.70 mg C/l respectively [4]. Because of this low input of nutrients, the deep-sea ecosystems are oligotrophic. However, some large particles (from faecal pellets to marine mammal carcasses) sink fast enough to reach the sea floor before being completely degraded [2], and so constitute the food of scavenger communities. The percentage contribution of large particles to the total carbon input has yet to be evaluated.

Hydrostatic pressure increases by approximately one atmosphere per 10 m depth, that is 380 atmospheres at the average depth of the oceans, and 1100 atmospheres in the deepest trenches. Of the three parameters governing the deep sea, hydrostatic pressure is a parameter unique to this environment.

The discovery of deep-sea hydrothermal vents in 1977 has significantly modified the established views on deep-sea biology [5]. Principally, the hydrothermal fluids contain inorganic energy sources that chemoautotrophic bacteria are able to utilise and hence provide a microbial-based food web [6]. In some cases, densities of giant tube worms *Riftia pachyptila* of up to 300 individuals per m^2 have been recorded, which is in sharp contrast with the oligotrophic system described previously [7]. While high hydrostatic pressure remains a main feature (although little studied in this case) of these newly discovered systems, the warm temperatures (10–25 °C) of these environments are less extreme. However, because of the hydrostatic pressure, sea water remains liquid at temperatures up to 350°C, and emissions of hydrothermal fluids at such temperatures have been recorded [8]. In this situation, temperature is extremely high rather than extremely low. In addition, warm and hot hydrothermal fluids contain toxic mineral components (hydrogen sulphide and heavy metals) in high concentrations [9], which contribute to the extreme conditions of these newly discovered ecosystems.

6.2   Deep-sea bacteria

Deep-sea microbiology as a subject developed in the 1950s, more than a half century after the pioneering works of Certes and Fischer [1, 10]. Zobell and Morita were the initiators of these studies, and published the first extensive reviews on deep-sea microbiology [11–14].

After the isolation of viable bacteria in water and sediment samples from the deep sea had been achieved, and preliminary studies on their taxonomy had been undertaken [15], the study of microbial heterotrophic activity constituted the primary approach of deep-sea microbiologists.

Marine bacteria isolated at or near the surface were used to degrade several organic substrates in the deep sea. These experiments indicated that the microbial degradation was 'considerably retarded', suggesting the existence of a deep-sea microbial flora, with presumably different characteristics [16]. *In-situ* experiments using the submersible *Alvin* [17] were then designed to study the response of microbial communities to organic enrichments of deep-sea waters and sediments, within a 2-year incubation time. The conclusion was that deep-sea micro-organisms have extremely low metabolic rates. These results were confirmed by several authors, performing bacterial activity measurements by shipboard or laboratory incubation experiments using simultaneously decompressed and undecompressed samples [17–21]. These studies have been extensively reviewed and should be consulted by the reader [2, 22, 23]. The data obtained indicated that respiration of deep-sea bacteria was less affected by deep-sea conditions [18, 24] than substrate incorporation, suggesting that bacterial metabolism in the deep sea was primarily concerned with maintenance. From these observations, it was concluded that probably no free-living bacteria adapted to deep-sea conditions existed [25]. Despite the low metabolic activity of the microflora demonstrated by these studies [17], further attempts were made to isolate truly barophilic marine bacteria from the digestive tracts of deep-sea invertebrates. These proved successful when Schwarz *et al.* [26] studied bacteria from the gut of a deep-sea amphipod that showed under simulated *in-situ* conditions growth rates and substrate uptake rates equal to or greater than those recorded under atmospheric conditions. These results have been subsequently confirmed by several other investigators [27–30] who have also reported barophilic responses to organic enrichments of mixed microbial flora from the intestinal tract of amphipods, fish and holothurians.

Although the differences recorded in the barophilic responses to organic enrichment of deep-seawater and sediment microbial flora may vary according to the nature of the substrates [21], it was concluded that the failure to discover barophilic activity in deep-sea bacteria could be due to the domination of barophilic bacteria by the more numerous non-barophilic bacteria accompanying particles sinking from the surface [21]. Obviously, sinking particles constitute a link between the surface layers (euphotic zone) and the deep-sea bottom. The study of the microbial flora of sinking particles has provided interesting results. Short-time incubation experiments (5 days) of deep-sea surface sediment and deep sediment trap material, in the presence of [^{14}C] glutamic acid indicated, in both cases, that high pressure enhanced microbial utilisation of the organic substrate [31]. However, no significant bacterial growth was detected. The authors concluded that 'barophilic bacteria play a role in the turnover of naturally low levels of glutamic acid, and the potential for intense microbial activity upon nutrient enrichment was more likely to occur in association with recently settled particulates, especially faecal pellets'. A more extensive study of particle-trapped microflora [32] indicated that surface-borne bacteria, sinking with particles, lost their activity when reaching cold deep waters and showed an 'increasingly significant fraction of deep-sea bacteria, in faecal pellet samples trapped at increasing depth'. The growth rates calculated for the deep-sea bacterial flora isolated from different kinds of organic matter illustrate their adaptation, or lack of adaptation, to the deep-sea conditions. While the surface sediment microflora have doubling times of weeks or months, bacteria from trapped material show doubling times ranging from days in unenriched samples to hours in enriched samples [32]. More recent studies using free sampling vehicles have led to a general conclusion that deep-sea sediments contain barophilic and barotolerant organisms in different proportions [33]. The low molecular weight organic matter contained in the sedimenting particles which reach the bottom can be degraded locally, which means that bacterial acitivity is intermittent. Phytodetritus containing active cyanobacteria (indicating their fast rate of sinking) have been found at 4500 m depth [34], and it has been demonstrated that this detritus can be rapidly utilised under *in-situ* conditons by well-adapted microbial communities. The low concentrations of bacterial cells in incubation vessels can be explained by the grazing of a strictly barophilic flagellate population.

The availability of organic material reaching the bottom is irregular in time and space so most deep-sea bacteria are exposed to starvation. Responses of marine bacteria to starvation have been studied by several groups of workers [35], but the studies carried out by Morita since 1976 have also considered the effect of hydrostatic pressure, and are more relevant to deep-sea microbiology. The bacterium most intensively studied is a psychrophilic *Vibrio* sp., named ANT300, isolated from the Antarctic convergence, a water mass that sinks to depths of 1500 m, becoming the Antarctic intermediate water [36], which is cold, nutrient-poor and subject to hydrostatic pressures of between 150 and 200 atmospheres. When cells of *Vibrio* ANT300 are starved (i.e. inoculated into a mineral salts medium devoid of organic matter), they increased in number during the first week (from 100 to 800%) and their size is reduced considerably: rod-shaped cells of $1 \times 4 \mu m$ become spheres of $0.4 \mu m$ in diameter [37, 38]. The respiration rate of the starved cells decreased to 20% during the first 2 days, and further decreased to 0.0071% of total carbon respired per hour after 7 days [38]. The first week of starvation was also marked by a sharp decrease in protein, DNA and RNA content of the cells. Subsequently protein and DNA increased slightly and stabilised at levels lower than prestarvation values, whilst RNA increased linearly, but no clear explanation for this phenomenon has yet been given [39].

Similar starvation patterns have been reported for several marine bacteria, mostly originating from the deep sea [40]. The viability of bacterial cells subjected to starvation periods >1 year has been demonstrated [38]. Recovery from starvation state has been observed after the addition of nutrients, and protein, DNA and RNA increased to the maximal values prior to cell division [41]. It has also been observed that the lag phase during recovery is 'directly proportional to the length of the starvation period', the dormancy of the cells appearing deeper with increasing starvation time. The term dormancy is more appropriate to describe this physiological state since it has been demonstrated for *Vibrio* ANT300 [42] and for other marine bacteria [43] that specific starvation proteins are synthesised during the starvation period.

The experiments reported above were performed on cells previously grown in batch cultures in nutrient-rich media. More recently, DNA, RNA and protein concentrations of *Vibrio* ANT300 cells were monitored during starvation experiments on cells grown at different growth rates [44]. The most surprising data obtained from these experiments was the DNA concentration, expressed per total cell, which decreased to 5–10% of the initial concentration. Transmission electron microscopy (TEM) observation of DNA of individual starved and unstarved cells clearly indicated a reduction in size of the starved cell DNA. Moyer and Morita [44] postulated that 'the decrease in cellular DNA may be due to the cells reducing DNA content to a single genome, or at least to the DNA required to maintain function throughout the metabolic arrest state. This remaining DNA must also allow for growth initiation and reproduction when conditions permit'. However, it is not clear whether or not this decrease corresponds to a decrease in genome size or in the number of genome copies.

Surface-water bacteria, like *Vibrio* ANT300, are progressively exposed to increasing hydrostatic pressures when the water masses they inhabit sink into the deepest ocean. Cultures of *Vibrio* ANT300 lost 75% of their viability when exposed to a pressure of 250 atmospheres under nutrient-rich conditions and starvation conditions, within 3 and 2 days respectively [36]. However, when cells of this bacterium were starved for 1 week before being pressurised to 250 atmospheres, they remained viable. These data indicate that barotolerance is affected by nutrient status, and is particularly enhanced by starvation. However, whilst *Vibrio* ANT300 seems clearly adapted to its changing environment, it represents an individual case.

The most important effort in searching for barophilic bacteria has been carried out by Yayanos and his group since 1978. The term 'barophilic' [45] is defined as 'a requirement for increased pressure for growth, or by increased growth at pressures higher than atmospheric pressure' [22]. From dead and decomposing amphipods previously caught in a pressure-retaining trap at 5700 m depth [46], a barophilic bacterium CNPT3 was isolated at 2°C and 570 atmospheres, using silica gel [47, 48]. This *Spirillum* spp. grew optimally at 500 atmospheres, at temperatures between 2 and 4°C, with a doubling time between 4 and 13 hours. In contrast, at the same temperatures and atmospheric pressure, the doubling time was 3–4 days. The first isolation of an obligately barophilic bacterium was published in 1981 [49]. This bacterium, named MT41, was cultivated from the decomposing amphipod *Hirondella gigas* from a depth of 10 476 m. No growth was observed under pressure below 380 atmospheres. The optimum doubling time (25 hours) was found at 2°C and 690 atmospheres. Although this bacterium had only been exposed for short periods (30–60 minutes) to atmospheric pressure during transfer inoculations, it appeared to lose its ability to form colonies under *in-situ* conditions after being exposed for several hours to atmospheric pressure.

The study of several bacterial strains isolated from samples collected at depths ranging from 1957 to 10 476 m led Yayanos *et al.* [50] to the following conclusions. Barophily is a general characteristic of bacteria from cold deep seas (1957–10 476 m). Hydrostatic pressure allowing the maximum growth rate at 2°C is always below the pressure corresponding to the capture depth. The pressure corresponding to the highest growth rate may be an indicator of the original depth of a particular strain. Bacteria isolated from 5600–5900 m have the broadest pressure tolerance range.

The pressure sensitivity experiments reported above [50] were performed at 2°C, the mean temperature of the deep ocean. Therefore, deep-sea bacteria are psychrophilic, i.e. 'organisms having an optimal temperature for growth at about 15°C or lower, a maximal temperature for growth at about 20°C, and a minimal temperature for growth at 0°C or below' [3]. This definition is still too wide for bacteria such as CNPT3 [49, 51], which are inactivated by exposures to temperatures between 10 and 32°C at atmospheric pressure. Because psychrophilic bacteria are generally very sensitive to elevated temperatures, the quality of deep water samples has been discussed [51]. In addition to the difficulties encountered as a result of the coexistence in unknown proportions of surface barotolerant and deep barophilic bacteria in a particular sample, the use of poorly insulated deep-water samples may also reduce or eliminate psychro-barophilic organisms [51]. The pressure and temperature limits of growth for the bacterium CNPT3 have been determined, using a 'high pressure temperature gradient instrument' [52]. The psychrophily of the bacterium studied appeared to be dependent on the pressure. The organism was more thermotolerant (-1.5–17°C) at 600 atmospheres (a pressure close to its origin pressure), and more barotolerant (1–1000 atmospheres) at 7°C. At about 8°C, the organism became obligately barophilic. The conclusion from these experiments is that barophily, or psychrophily, must be defined for a particular organism at standardised temperatures and pressures.

The generally accepted conclusions—that bacteria from the cold deep sea are barophiles, with a threshold at about 2000 m [53], while the upper layers are dominated by barotolerant species [54], which are also psychrophilic and stenothermal—are in agreement with characteristics of the deep sea: low temperature and high pressure. However, the majority of barophilic bacteria that have been isolated from deep-sea invertebrates appear to exist mainly (if not only) in nutrient-rich environments such as invertebrate digestive tracts or faecal pellets [54]. These conditions are space–time limited, while low nutrient concentrations are more characteristic of the deep sea. Complete mineralisation of organic matter requires the participation of oligoheterotrophs and chemoautotrophs. The description of barophilic activity in such organisms is required if the concept that 'barophily is an ubiquitous and essential character of the true bacteria inhabitants of the deep cold ocean' is to be accepted [55].

Before true barophilic bacterial activities were demonstrated and true barophiles and obligate barophiles were isolated in pure culture, the effects of hydrostatic pressure on microbial life processes were investigated, often using terrestrial strains, including *Escherichia coli*. The choice of such an organism was supported by studies showing that not only *E. coli* but also *Streptococcus faecalis* survived better at 250 or 500 atmospheres than at 1 atmosphere, at 4°C [56]. These studies dealing with the effects of hydrostatic pressure on protein, DNA and RNA synthesis in enteric bacteria have been well summarised in a previous review [23]. Protein synthesis in *E. coli* or *Vibrio marinus* appeared to be the most barosensitive process [57–59]. Further studies on cell-free extracts from these bacteria together with those prepared from *Pseudomonas fluorescens* and also a deep-sea bacterium (*Pseudomonas bathycetes*) demonstrated the role of ribosomes (barotolerant for *P. fluorescens*, barosensitive for *E. coli*) in barotolerance [60], and particularly the role of the 30S ribosomal subunit [61]. In the case of *P. bathycetes*, ribosome-associated barotolerance was dependent on Na^+ and Mg^{2+} concentrations [62].

The availability of barophilic bacteria grown in pure cultures has made it possible to understand better the physiological response of micro-organisms to hydrostatic pressure. One approach has been to study the fatty acid composition of the cell membrane of barophilic bacteria as a function of hydrostatic pressure. The bacterium strain CNPT3, previously described, was grown at 2°C at pressures ranging from 1 to 690 atmospheres and its membrane fatty acid composition analysed [63]. The results showed that the fatty acid composition changed as a function of pressure. Higher proportions of $C_{16:1}$ and $C_{18:1}$ fatty acids were found at higher pressures, while $C_{14:1}$, $C_{16:0}$ and $C_{14:0}$ decreased. From 1 to 680 atmospheres, the ratio of total unsaturated fatty acids to total saturated fatty acids increased from 1.9 to 3.0 A similar increase of unsaturated fatty acids has been previously reported for *Vibrio marinus* when temperatures for growth decreased from 25 to 15°C [64]. By comparison with studies on other organisms which showed that changes in the lipid composition were correlated with the maintenance of membrane fluidity and function, a similar effect of hydrostatic pressure on the membrane of deep-sea bacteria has been proposed [63]. In further studies, 11 marine strains, isolated from depths ranging from 1200 to 10 476 m, were analysed in the same way [65]. All the strains studied revealed fatty acids commonly found in marine vibrios, especially long-chain (up to C_{22}) polyunsaturated fatty acids. Three of these isolates, grown at 2°C under increasing pressures, contained more polyunsaturated fatty acids at their optimal growth pressure. Similar results have been reported for the psychrophilic bacterium *V. marinus*, when exposed to decreasing temperatures at 1 atmosphere. These results were unexpected because polyunsaturated fatty acids were previously thought to be absent (with the exception of a marine strain of a *Flexibacter* sp.) in prokaryotic organisms. In addition to their role in the maintenance of an optimal membrane fluidity, it has been demonstrated [65] that those lipids which are essential dietary components of invertebrates but mainly synthesised by photosynthetic organisms could be provided to deep-sea fauna through the consumption of barophilic bacteria.

Delong and Yayanos [66] have also studied the properties of an enzymatic system (PTS: sugar phosphotransferase system) involved in sugar transport in two barophilic strains, PE36 and CNPT3, and *Vibrio marinus*. In CNPT3, and particularly *V. marinus*, an increase of hydrostatic pressure inhibited the uptake of methyl-glucoside (a sugar analogue), while uptake was stimulated for isolate PE36. However, conflicting results were obtained for *in-vitro* methyl α-glucoside phosphorylation which precedes the sugar uptake. Pressure inhibited phosphorylation in *V. marinus* and strain PE36 but enhanced this activity for the barophile CNPT3. The authors proposed that 'the difference in pressure effects on uptake were due to differences in the pressure response of the membrane-bound activity'. Since the membrane fatty acid composition of *V. marinus* and PE36 are almost identical [65], the membrane lipids are probably not responsible for the differences observed. The role of an integral membrane protein, enzyme II^{Gcl}, which transports the sugar across the membrane has been proposed [66]. The properties of this protein may be less affected by pressure in the case of PE36 than *V. marinus* However, although it remained functional at higher pressures than that of *V. marinus*, the glucose uptake system of CNPT3 is sensitive to pressure, suggesting that in this bacterium glucose transport is not a major factor involved in the adaptation to barophily.

Several investigators have examined the effect of hydrostatic pressure on extracellular enzymatic activities, and especially chitinase, an exoenzyme involved in chitin degradation, a major component of invertebrate skeletons. Helmke and Weyland [67] showed that chitinase synthesis was inhibited by elevated pressures (400 atmospheres), particularly in psychrotrophic bacteria. However, the chitinase activity proposed by the different bacteria investigated was barotolerant and rather similar. More recently, the effect of pressure on enzymatic activity of barophilic bacteria has been studied using a diagnostic test kit [68]. The commercially available API ZYME assay kit has been modified for use at *in-situ* pressures and temperatures in pressurised stainless-steel vessels. This diagnostic kit allows the detection and approximate quantification of 19 different enzymatic activities. Ten facultative barophiles have been studied by this method, and nine had their phenotype affected in at least one enzyme reaction when tested under pressure. Enzymes showing strong activity were not affected by pressure, but others, for example esterase–lipase, leucine aminopeptidase, β-galactosidase and *N*-acetyl-β-glucosaminidase, showed higher activities in particular strains when tested under pressure. Clearly, this technique is an interesting and easy method to determine the biochemical phenotype of deep-sea bacteria, under *in-situ* conditions.

The recent molecular approach to the study of barophily may explain some hitherto unexplained differences. Strain SS9 is a Gram-negative bacterium isolated from a depth of 2500 m in the Sulu Sea. This bacterium grows optimally at 280 atmospheres but will also grow at 1 atmosphere and temperatures up to 23°C [69]. Polyacrylamide gel electrophoresis of total proteins synthesised under both atmospheric and *in-situ* pressure conditions have been compared. Protein patterns were similar, except for a particular band corresponding to a molecular mass of 37 kDa. This protein has been found to be located in the outer membrane, and named *ompH* (outer membrane protein, high pressure). The *ompH* gene has been cloned from an expression library in *E. coli*. Further experiments have indicated that transcription of the *ompH* gene is regulated by pressure, although the possibility that pressure may influence the stability of the *ompH* mRNA cannot be excluded. Preliminary comparisons of *ompH* amino acid sequences indicate that this protein is related to the outer membrane porin protein of *Neisseria gonorrhoeae* [69]. It has been postulated that 'porin channels which function at low pressure could be perturbated by high pressure which would necessitate production of a new porin molecule such as *ompH*', and this has 'suggested the existence of a genetic programme which couples sensing of pressure to production of appropriate activities'. These findings may account for the differences observed between a barophilic response of whole cells, and a non-barophilic response of free cell extracts containing exoenzymes.

A few deep-sea barophiles obtained in pure cultures have also been studied from a phylogenetic point of view. Two Gram-negative bacteria isolated from digestive tracts of animals living at 5900 and 4300 m have been assigned to the genus *Vibrio*, based on their phenotypic characteristics and G + C molar ratios [70]. The relationships of these isolates to the *Vibrio–Photobacterium* group, represented by *Vibrio harveyi* and *Photobacterium phosphoreum* have been confirmed by the determination of 5S rRNA sequences [70]. The number of sequence differences between the two barophiles and the two reference strains appeared greater than differences between the reference strains themselves, and the possibility that the two barophiles studied belong to a new genus has been proposed. Further phylogenetic studies [71] have assigned these strains to a new genus, *Shewanella*, with the type strain being designated *Shewanella benthica*. More recently, the 5S rRNA sequence of an obligate barophile isolated from particle-rich sea water at a depth of 7410 m has been determined [72]. This obligately barophilic bacterium appeared similar to *Vibrio psychroerythrus*, both strains being distinct from other vibrios. It has been proposed that these isolates should be incorporated in a new genus, *Colwellia*, the barophile being assigned to the species *hadaliensis*, while the renaming of *V. psychroerythrus* as *Colwellia psychroerythrus* has also been recommended [72].

In recent years, several deep-sea bacteria have also been screened for the presence of plasmids [73]. The bacterial strains studied were isolated from deep-sea amphipods, caught at a depth of 4300 m in the Bay of Biscay. Eleven of the 16 isolates studied carried one or more plasmids, ranging from 2.9 to 63 MDa. The occurrence of plasmids with the same molecular weight in strains showing different phenotypic characters demonstrated, after agarose gel electrophoresis, DNA hybridisation and restriction analysis, that plasmid transmission may occur in particular deep-sea habitats. However, the significance of the plasmids in these bacteria has yet to be established.

6.3   Hydrothermal vents

The discovery of deep-sea hydrothermal vents (2000–3000 m deep) and their associated animal communities along the East Pacific Rise (EPR), is one of the major scientific events of the last 15 years. In February 1977, the submersible *Alvin* discovered surprisingly dense animal communities, mainly composed of new and large species at a depth of 2500 m near the Galapagos Islands [5]. This discovery was unexpected because the cruises had been organised to investigate the occurrence of hydrothermal activity, as suggested by previous reports of physical and chemical seawater anomalies, and no new biological data were anticipated.

Several other vent sites were found in the eastern Pacific in the following years by American, French and Canadian expeditions. All the sites have in common animal communities with a high biomass which depend on hydrothermally active springs for their existence. The fauna is dominated by new species with low diversity and a relatively stable species composition [74]. To explain the high biomass associated with the vents, it was initially proposed that venting of hot hydrothermal fluids could produce advective bottom currents that could concentrate particulate matter sinking from the euphotic zone [75]. Although this hypothesis cannot be completely discounted, numerous studies have demonstrated the key role of chemosynthetic bacteria in the hydrothermal vent food web [76–78]. The bacteria present are morphologically diverse and may be either free-living, attached to surfaces or growing as symbionts [79–82]. These include sulphur-oxidising bacteria which use hydrogen sulphide emitted in the hydrothermal fluids as an energy source.

In addition, many other modes of bacterial metabolism (autotrophic, mixotrophic, heterotrophic, aerobic or anaerobic) have been reported at the different sites [83]. In particular, a number of studies have indicated that some of the methane or hydrogen sulphide present in the vent fluids may be of biological origin [9], and research works were initiated to determine whether ultrathermophilic micro-organisms were associated with the hot vent fluids. Most of the microbiological data presently available concern hydrothermal vents of the eastern Pacific. The newly discovered sites in the mid-Atlantic and the western Pacific will be also described, although fewer microbiological studies have been performed at these sites. Cold seeps and subduction zones which have been recently explored also reveal bacterial chemosynthesis-dependent animal communities and they will be discussed in this section.

6.3.1   *Distribution of vent fields and their main features*

The Galapagos site ($00° 18'$ N; $86° 13'$ W) was discovered in 1977 by the submersible *Alvin* at a depth 2480 m. Hydrothermal fluids at this site have a maximum temperature of $23°C$ and vent out through cracks in the seafloor basalt. The maximum temperature of zones colonised by invertebrates ('Garden of Eden' site) is 17°C. The dominant species of primary consumers are the vestimentiferan tubeworms *Riftia pachyptila* [84] and the large bivalves *Calyptogena magnifica* [85] and *Bathymodiolus thermophilus* [86]. Most of the vent microbiology studies have been performed at this site [76, 77, 79, 83, 87].

Two years later, at 21°N on the East Pacific Rise (20° 50′ N; 109° 06′ W), sulphide chimneys (black smokers) venting hot fluids with temperatures up to 350°C were also discovered by the submersible *Alvin* [8] at a depth of 2616 m. Lower temperature vents (23°C) also exist and are colonised by the same animal communities as those found at the Galapagos site. In addition, walls of the active sulphide chimneys are colonised by the polychaete *Alvinella pompejana* [88]. Thermophilic bacterial communities were reported in fluids sampled from these smokers [9].

At approximately 13°N, active sites (12° 48′ N; 103° 56′ W) were found in 1982 and 1984 at an average depth of 2630 m by French scientists using the submersible *Cyana* [89]. Except for the absence of *Calyptogena magnifica*, only represented by dead shells, the 13°N site has geological and biological features similar to those described for the 21°N site [74].

The Guaymas site (27° 02′ N; 111° 22′ W; 2020 m depth) in the Gulf of California, explored since 1982 by the *Alvin* is rather different. Seafloor basalts are covered by a thick layer (several hundred metres) of sediments. Hot fluids (315°C) vent out through those sediments or from pagoda structures [8]. The vestimentiferan tubeworm (*Riftia pachyptila*) also inhabits the Guaymas site, while bivalves are represented by *Calyptogena pacifica* [90]. In many places, the sediments are covered by extremely large bacterial mats, rich in *Beggiatoa*-like cells [90].

In 1983, Canadian scientists discovered new sites, with different associated fauna, at the Juan de Fuca Ridge (45° 57′ N to 46° 53′ N; 129° 17′ W to 130° 01′ W; 1544 to 2370 m depth). Vestmentifera are represented by *Ridgeia piscesae*, and polychaetes by species of the genus *Paralvinella*, while bivalves are almost absent [91].

Although temperature anomalies have been recorded on the mid-Atlantic ridge since 1975 [92], active hydrothermal vent sites were only reported in 1985 [93]. Two sites, TAG (26° 08′ N; 44° 49′ W; 3630 m) and Snake Pit (23° 22′ N; 44° 57′ W; 3500 m) have been explored and have revealed a new kind of associated fauna, dominated by caridean shrimps of the genus *Rimicaris*.

The previous hydrothermal vents were associated with either fast- (east Pacific) or slow- (Atlantic) expanding ridge. Other tectonic situations have also revealed hydrothermal activities, particularly back-arc basins. Active vent sites were discovered in the Mariana back-arc basin (18° 11′ N; 144° 43′ W; 3650 m depth) in 1987 [94], in the north Fiji Basin [95, 96] in 1988 (173° 56′ E; 16° 59′ S at 2000 m) and in the Lau Basin (176° 43′ W; 21° 25′–22° 40′ S) in 1989 [97]. The macrofauna associated with these west and south-west Pacific hydrothermal vents are composed of gastropods and bivalves, but very few vestimentiferans.

Hydrothermal activities have also been found associated with submarine volcanoes [98] and dormant seamounts [99]. In these cases, no associated macrofauna was found, even at Loihi Seamount, an active submarine mid-plate, hot-spot volcano, near Hawaii. However, bacterial mats have been observed and have been sampled [98].

A few years following the discovery of deep-sea hydrothermal vents, invertebrate communities resembling those of deep-sea vents were discovered in several cold-seep areas. Dense biological communities, including mussels and vesicomyid bivalves, have been found [100] on a passive continental margin at the base of the Florida escarpment in the Gulf of Mexico (26° 02′ N; 84° 55′ W at a depth of 3266 m). Hydrocarbon seep vent-type taxa, including different species of bivalves (Lucinidae and Vesicomyidae), have been discovered [101] on the Louisiana slope in the Gulf of Mexico (27° 41′ N; 91° 32′ W at 600–700 m depth). Vent-type animals were also found [102] in the Oregon subduction zone (44° 41′ N; 125° 18′ W at a depth of 2036 m), including bivalves of the genera *Calyptogena* and *Solemya*. In the Japan subduction zones of the western Pacific [103, 104], beds of *Calyptogena* at depths between 3850 and 6000 m have been reported (Nankai Trough: 33° 37′ N; 137° 32′ E; 3830 m; Japan Trench at Kashima Seamount: 35° 54′ N; 142° 31 E; 5640 m; Japan Trench: 40° 06′ N; 144° 10′ E; 5900 m). Clear similarities exist between those newly discovered communities and the EPR hydrothermal vent communities, including the chemical composition of seeping fluids and the stable isotope composition of the invertebrates, suggesting that the trophic resources of these bivalves were probably chemosynthetic.

6.3.2   *Chemical features of hydrothermal fluids and expected metabolisms*

Hydrothermal physiochemical processes have been very well documented and reviewed by several authors [105–107]. They can be summarized as follows. Sea water penetrates the oceanic crust down to several kilometers beneath the sea bottom and interacts with the hot magmatic basalt, forming mineral precipitates and becoming acidic and enriched with metallic elements. Superheated hydrothermal solutions rise up through the ocean floor. When discharging, the hydrothermal fluids mix with cold sea water causing metallic sulphides and calcium sulphate to precipitate, so forming the smoker chimneys.

Depending of the amount of mixing with sea water, hydrothermal springs can have different physiochemical features, even within a single site. However, hydrothermal emissions are distinguished from ambient seawater by low pH, high concentrations of dissolved gases (hydrogen sulphide, methane, carbon monoxide, carbon dioxide, hydrogen) and minerals (silicon, iron, manganese, zinc, etc.). For several hydrothermal fluids sampled at 21°N, pH values ranged from 4.24 to 7.23, hydrogen sulphide concentrations were up to 4.3 mM, and iron manganese concentrations reached 1055 and 206 μM respectively [9]. Some of these compounds constitute energy sources used by chemoautotrophic bacteria to fix carbon dioxide. These redox reactions require electron acceptors (oxygen, NO_3^-, SO_4^{2-}) which are present in sea water. Almost all of the presently known types of bacterial metabolism have been reported or suggested to occur within vent ecosystems [108–110] on the basis of the availability of appropriate electron donors and acceptors. The bacterial communities in hydrothermal vents are distributed within three main niches: sea water, exposed surfaces and invertebrate tissues. As such, they are exposed to the main constraints of the environment which are (in addition to hydrostatic pressure) temperature gradients and high concentrations of toxic compounds. These different niches will be considered successively.

6.3.3 *Abundance and activity of bacteria in sea water*

Data concerning moderate-temperature ($< 50°C$) hydrothermal fluids and surrounding sea water will be considered in this section. The first bacterial counts of milky-whitish water samples at the Galapagos sites indicated high cell densities, up to 10^9 cells per ml [83]. Further counts of samples from 13°N or 21°N sites were lower [110], ranging from 6×10^4 to 10^6 cells per ml. These population densities are substantially greater than those normally found in deep-sea water, but are in the same range as those present in coastal waters [111]. The differences observed have been explained by variations in mixing between sea water and hydrothermal fluids and sampling at different distances from the vent [77]. Moreover, some fractions of surrounding bacterial mats, as suggested by the occurrence of bacterial aggregates, may be swept out by the venting fluids [108].

Several types of bacterial metabolism (especially autotrophic sulphur-oxidising and heterotrophic bacteria) have been reported in these water samples. Counts of autotrophs fluctuate within the range of 10 to 2.2×10^6 cells per ml [76, 78, 83, 112], while populations of heterotrophs range between 10^2 and 2×10^7 cells per ml [83, 112]. Biomass values corresponding to the bacterial counts have been calculated by converting bacterial numbers to dry weight of cell carbon, using standard coefficients [5, 112], or by ATP determinations. Using the latter method, it has been estimated [113] that at the Galapagos sites the concentration of bacterial carbon was in the range of 100–250 μg per litre of hydrothermal fluid, which contrasts sharply with the particulate organic carbon concentration in the deep sea, usually $< 10 \, \mu$g C/l.

Measurements of bacterial activity have also been determined at the Galapagos and 21°N sites to estimate total bacterial production and the role of the autotrophic and heterotrophic microbial communities [76, 78]. The data obtained indicate great variability, not only between vents but also within a single vent, owing to the method used and to the rapid dilution of the emanating fluid with sea water. Despite this variability, the results indicate that bacterial production in sea water was too low to account for the amount of invertebrate biomass present [76]. The respective roles of the autotrophic and heterotrophic communities in the total bacterial carbon production have been comprehensively reviewed, and led Karl [109] to conclude that 'it is probable that the role of free-living chemolithotrophic bacteria has been overestimated'.

One of the consequences of high-temperature vents is the production of a plume which is detectable several hundred kilometres away from the vent [114]. Such plumes, produced above the Juan de Fuca Ridge by warm and hot vents, have been sampled [114] and analysed for their microbiological content. Samples collected in the vicinity of the vents (6–50 m) contained bacterial cells showing the highest incorporation of [^{3}H] adenine at 30°C. Hydrocast surveys in the distant plume showed that bacteria growing near the vents were entrained in the plume, and originated mainly from moderate-temperature vent waters.

Minerals discharged by the vents are also entrained in the plume, particularly manganese, which is frequently used as a plume tracer [115]. Manganese can be oxidised by micro-organisms, and such metabolism has been reported for deep-sea bacteria [115], and particularly for those within hydrothermal vents, where it is present at high concentrations [117, 118]. The influence of bacteria transported by the plume in the scavenging and partitioning of manganese has been reported [119]. This is caused by bacterial cells binding manganese in their capsular polysaccharides [119]. In-situ manganese-binding experiments using these micro-organisms indicate that the process is enhanced by hydrostatic pressure and the hypothesis of 'a binding mechanism involving predominantly organic polymers produced by barophilic bacteria' has been postulated [115].

6.3.4   Bacterial communities on inert surfaces and bacterial mats

Colonisation of inert surfaces by deep-sea bacteria was first reported by scientists aboard submersibles such as *Alvin*. This colonisation is spectacular at the Guyamas site, where whitish bacterial mats are particularly abundant [90]. Scanning electron microscopy (SEM) of pieces of rock, chimneys and mollusc shells has revealed morphologically diverse bacterial communities [79, 83, 107]. Artificial substrates (glass, polycarbonate membranes) exposed for several days to 1 year were also observed to be colonised by similar communities [79]. Surfaces sampled from Galapagos sites [79] were covered by a biofilm 5–10 μm thick, composed mainly of coccoid cells (1 μm in diameter) arranged in multiple layers, with some filamentous cells interspersed amongst them. Many of the cells were surrounded by metal encrustations, principally of iron and manganese. TEM studies showed that the biofilm appeared to be composed of numerous cells with intracellular membranes, typical of nitrifying and methane-oxidising bacteria [120, 121]. Some filaments contained intracellular sulphur granules, and resembled *Beggiatoa*. Others, devoid of sulphur granules, were similar to *Leucothrix*, *Thiothrix* or *Leptothrix*, and were sometimes encrusted with metallic deposits. Morphological criteria are insufficient to give the observed filamentous cells a taxonomic position. Some similarities between the filamentous cells found in the vents and some cyanobacteria [79] are consistent with the hypothesis of Lewin about double genera, represented by both chemoheterotrophic and chemoautotrophic forms. In addition to the large filaments small prosthecate forms, resembling *Hyphomicrobium* or *Pedomicrobium*, have also been observed, with some strains being classified further as *Hyphomonas*. Several strains of this latest genus have been isolated from shellfish beds of the eastern Pacific vents, and assigned to the new species *H. hirschiana* and *H. jannaschiana* [122]. These two species have G + C contents of 57 and 60 mol% respectively, and optimum growth temperatures of 25–31°C and 37°C respectively.

Short-time surface colonisation experiments have been carried out at 13°N (D. Prieur and P. Fera, unpublished data) with glass and stainless-steel surfaces. Bacterial densities fluctuated from 2.3×10^4 to 5.2×10^5 cells per cm^2 and 1.3×10^5 to 5.2×10^7 cells per cm^2 for surfaces exposed for 3 and 10 days respectively. These values are greater than those obtained by the same authors in the Bay of Biscay at 3000 m depth, after 3 months of exposure, but comparable to colonisation rates observed in temperate coastal waters during spring [123]. In addition to the cell morphologies described above, surfaces exposed were very often found to be colonised by ring shaped (*Microcyclus*?) or twisted cells (*Gallionella*).

Although found at several hydrothermal vent sites, bacterial mats are particularly abundant in the Guaymas Basin [79], and are mainly composed of the large filamentous bacteria *Beggiatoa* and *Thiothrix*. The mats observed in the Guaymas Basin are one to several centimetres thick [124]. Samples collected by Nelson *et al.* [124] contained filamentous cells in different cell width classes: 24–32, 40–42 and 116–122 μm. At least two of the width classes showed Calvin cycle enzyme activities. The presence of internal S° globules and their gliding motility allowed the authors to assign these bacteria to the genus *Beggiatoa*.

6.3.5 Main features of mesophilic bacteria isolated from sea water and surfaces

Most of the mesophilic strains isolated from deep-sea vents and listed in previous reviews are sulphur-oxidising bacteria [87, 108, 109, 125, 126]. Some strains which are obligate chemolithotrophs have been assigned to the genus *Thiomicrospira*, and appear similar (morphologically and physiologically) to the type strain *T. pelophila*. They are comma- or spiral-shaped (0.3–0.4 μm × 0.7–2.0 μm) and have mostly been isolated from pieces of periostracum of the bivalve *Bathymodiolus thermophilus*, at the Galapagos site [87]. They grow optimally at pH 8 and 25°C, and require several cations ($2 M\ Na^+$. Ca^{2+}, Mg^{2+}). This bacterium is microaerophilic, and carbon dioxide incorporation is not inhibited by S^{2-} concentration of 300 μM. Carbon dioxide incorporation by *Thiomicrospira* strain L-12 has been studied as a function of hydrostatic pressure [87]. Under an *in-situ* pressure of 250 atmospheres, the rate of carbon dioxide incorporation was reduced by 25% compared with that recorded at atmospheric pressure, and was completely inhibited by 500 atmospheres. From the outer part of a tube of *Riftia pachyptila*, a different strain of *Thiomicrospira* has been isolated [126] and assigned to a new species, *T. crunogena*. This bacterium showed a higher growth rate and a G + C content of 41.8–42.6 mol% [126] against that of 43–44 mol% for *T. pelophila*.

However, most of the sulphur-oxidising bacteria isolated also have the capacity for heterotrophic metabolism [125], being mixotrophic or facultative chemoautotrophic. Those bacteria which oxidise thiosulphate to sulphate or polythiorates have been assigned to the genera *Thiobacillus* and *Pseudomonas* respectively. Mixotrophy has also been observed for a bacterium, presumed autotrophic, which oxidises manganese, with an inducible enzyme system, while bacteria associated with manganese nodules have constitutive enzymes [117]. These manganese oxidisers have been isolated from the periostracum of Mytilidae and on exposed glass slides at the Galapagos site.

Hydrothermal plumes have also been studied in the north Fiji Basin [127], and a *Thiobacterium* species has been isolated amongst other sulphur-oxidising autotrophs [128]. The cells of this bacterium exhibited some morphological changes during their growth phases, with short rods ($0.2 \times 1.0\ \mu m$) being dominant during the lag phase whilst during exponential phase the cells were larger ($0.5 \times 1.5\ \mu m$). Short chains of cells were observed during the late exponential growth phase, and progressively increased in number during the stationary phase; some cells also formed branching structures. At this stage, the cells appeared embedded in a gelatinous matrix which formed large particles, possibly available for filter feeders.

Although abundant, relatively few heterotrophic bacteria have been isolated and characterised [112]. All are Gram-negative motile rods, mostly oxidative and nitrate-respiring, utilising cysteine and producing hydrogen sulphide. In addition to the strains reported above which are strict aerobes or facultative anaerobes, a single anaerobic mesophilic bacterium, a spirochaete, has been isolated from a water sample [129].

6.3.6 *Invertebrate-associated bacteria*

In order to establish the significance of deep-sea vent ecosystems, three methodological approaches have been used: analysis of stable isotopes of carbon and nitrogen [130], a search for the presence of enzymes of the Calvin cycle [131] and examination of invertebrate tissues by SEM and TEM [80–82, 132, 133]. Vestimentiferan tubeworms, bivalves, gastropods and poly-chaetes have been intensively studied by these methods.

6.3.6.1 Riftia pachyptila *and the Vestimentifera.* *Riftia pachyptila* has been found at almost all the explored sites of the East Pacific rise where it constitutes 'bushes' up to 1.5 m high, with densities reaching 180–200 individuals per m^2 [7]. The worm lives inside a tube from which a well-irrigated and retractable gill emerges. Analysis of stable isotopes of carbon and $^{13}C/^{12}C$ ratio of *Riftia* tissues clearly indicate that the tube worm utilises a locally generated carbon source [134–136]. However, this species, in common with other Vestimentifera and Pogonophora, does not have a digestive tract. Instead these organisms have a trophosome, a well-irrigated organ which represents the main part of the animal trunk and is entirely full of Gram-negative prokaryotic cells [81] with a density up to 3.7×10^9 cells per gram wet weight. The prokaryotic cells are mostly spherical (3–5 μm in diameter), but also include rod-shaped and pleomorphic cells. The following relationships between bacterial and the host cells has been proposed [137, 138]. Trophosome cells are colonised by one or several rod-shaped bacteria which grow while bacteriocytes begin to degenerate. Bacterial cells then increase in size and also degenerate, after which the bacteria lyse and the bacteriocytes are disorganised. The products of bacterial lysis and bacteriocyte degeneration pass into the vestimentiferan worm circulatory system.

The metabolic type of the trophosome bacteria has been determined by analysis for enzymes characteristic of sulphide oxidation (rhodanese, APS reductase. ATP sulphurylase) and autotrophic fixation of carbon dioxide via the Calvin–Benson cycle (ribulose-1,5-biphosphate carboxylase, ribulose-5-phosphate kinase) [6, 131]. Sulphide oxidation by intracellular bacteria requires the presence of hydrogen sulphide, carbon dioxide and oxygen in the trophosome. It has been postulated that carbon dioxide could be incorporated at the gill level [139, 140] into a C_4 acid (probably malate) by carboxylases, transported in this form into the bloodstream, then decarboxy-lated in the trophosome to release carbon dioxide. Oxygen is transported by a haemoglobin-like pigment [141, 142]. Sulphide transport is effected by a specific blood protein able to fix sulphide reversibly. This protein has several roles [143]: it effects sulphide transport, protects cytochrome c oxidase from poisoning by sulphides [144], allows *Riftia* to concentrate sulphides from the outer environment and inhibits spontaneous oxidation of sulphides by oxygen [145]. This specific protein may correspond to a fraction of the haemoglobin-like pigment, but has a different function from that involved in oxygen transport [143].

Samples of DNA purified from trophosome tissues have been analysed for DNA base composition and genome size [146]. DNA base composition was determined to be 58 mol% G + C, while genome size was approximately 2.1×10^9 Da, a value typical of free-living bacteria. Moreover, the data strongly suggest that the trophosome tissues contained a single type of prokaryotic symbiont. Determination of 5S rRNA sequences for *Riftia* symbionts has led to the conclusion [147] that the symbionts are more related to the genus *Thiomicrospira* than *Thiobacillus*. However, DNA base composition of *Thiomicrospira* ranged from 36 to 44% G + C. 5S rRNA sequences were also determined for symbionts of *Calyptogena magnifica* and *Thiomicrospira* strain L-12, and used for a phylogenetic study [148]. All the strains belong to the gamma subdivision of the purple bacteria [149], which also includes the barophiles described previously [70]. The symbiont of *C. magnifica* appeared to be closer to *Thiomicrospira* L-12 than *Thiomicrospira* L-12 to *T. pelophila*. The symbiont of *R. pachyptila* was more related to *Thiobacillus ferrooxidans* strain M1, which are together near the root separating the gamma from the beta subdivision of purple bacteria.

North-west Pacific Vestimentifera have also been studied, and symbionts of similar morphology were found in different tube worms in this area. However, one or two types may exist within single individuals of *Ridgeia piscesae* and *R. phaesophila* [150]. All the bacteria have Gram-negative cell walls, and many of them showed internal membrane systems. The most common form in *R. piscesae* is spherical with a cell diameter ranging from 0.6 to 10 μm, and an average of 3 μm. The second form in *R. piscesae* was also found in *R. phaeophila*. It is a small bacterium with glycogen- like granules scattered in the cytoplasm. The second form in *R. phaeophila* is a round bacterium 5 μm in diameter. The authors of this work concluded that 'a given vestimentiferan species may contain bacteria differing in shape, size and states', these differences possibly being due to 'the age of bacterial cells and the physiological state of both host tissues and symbionts'. Molecular analysis of vestimentiferan symbionts [146, 147, 151] indicate a nearly or completely monospecific symbiosis, with little variation between individual worms.

In another study, four vestimentiferan tube worms typical of the north-east Pacific, *Ridgeia piscesae*, *R. phaeophila* and two unnamed types, were examined under TEM, and the size of their symbionts determined [152]. Except for those of *R. phaeophila*, all the bacteria examined had similar morphologies. More small bacteria were found in the anterior part than in the posterior part of the trophosome. Dividing cells, although rather few, were also found in the anterior part of the trophosome, which, moreover, contained larger amounts of sulphide and zinc. The small cells, sometimes dividing, may correspond to symbiotic bacteria in the exponential growth phase. Such a physiological gradient could be related to the external environmental gradient of sulphide and temperature, between the base and the top of the upward-growing tube worms.

During experimental fertilisation of *Riftia* [153], female gametes were examined under both SEM and TEM. These observations agree with previous data, and no bacteria were detected either on the egg surface or within the egg cells. The discovery that early juveniles of *R. pachyptila* lack symbiotic bacteria indicates an uptake of the symbiont during a later developmental stage. More precisely, a ciliated functional duct has been described [154] which occurs only in juvenile vestimentiferans. It has been proposed that this duct is the route of entry for bacteria in the trophosome.

6.3.6.2 *Bivalve molluscs. Calyptogena magnifica* belongs to the family Vesicomyidae, and is well represented at the Galapagos and 21°N sites (East Pacific Rise) [90]. This large species (30 cm in length) was first considered as a classical filter-feeder, ingesting particulate matter and bacteria [75]. However, a study of its functional anatomy indicated that the labial palps, organs usually involved in the transport of food particles, are small. Moreover, the digestive tract appears to be reduced, mostly empty and contains no recognisable material [85]. More recently [155], it has been shown that the gill of this species is only slightly adapted to the transport of particulate matter. Analysis of the stable isotopes of carbon and nitrogen in the tissues of *C. magnifica* indicate that the carbon sources are probably local [134]. The enzymes involved in the autotrophic fixation of carbon dioxide via the Calvin–Benson cycle have been found in the gill tissues [156], and prokaryotic cells have been observed in this organ by TEM [157]. These cells show the typical structure of Gram-negative bacteria and appear as cocci or short rods, with a diameter of 0.64 μm. The bacteria are located within the cells (bacteriocytes) of the gill lamella, and clustered in bacterial 'pockets' [132]. Numerous cells were found to be dividing, and no bacterial lysis was observed. Further observations revealed that the bacteriocytes show different stages, and the occurrence of a cyclic process has been proposed [155]: bacterial growth, reabsorption and lysis of bacteria, transfer of organic molecules through the circulatory system of the mollusc, and infestation of gill cells which then become bacteriocytes. It has been suggested [158] that the blood of *C. magnifica* has properties close to *Riftia* blood, and could affect the transport of hydrogen sulphide, carbon dioxide, and oxygen which .are required for the metabolism of the symbiotic bacteria.

Bathymodiolus thermophilus was found to be abundant at the Galapagos and 13°N sites [90]. The animal is 15–16 cm in length [86], and analysis of the stable isotopes of carbon [130] indicated that local bacteria constitute their trophic source. The nutritional process appears, however, to be different from that of *Calyptogena magnifica*, as *B. thermophilus* has well-developed labial palps. Although its digestive tract is simpler than that of coastal mytilid species, it is functional and the categories of digestive cells described previously in coastal bivalves have been found within the digestive gland [159]. The stomach is bulky and contains particles clearly identified as bacteria of different shapes, benthic foraminiferans, and also fragments of diatom frustules [82]. These results clearly indicate that *Bathymodiolus* feeds on suspended particles, some of them stemming from the euphotic zone. As a complement to this usual nutrition, particular cells of gill filaments contain Gram-negative bacteria, with a diameter of 0.5 μm, clustered in small vacuoles, located at the apical part of the bacteriocytes, and containing about 10 bacterial cells each [160]. Some bacteria seemed to be in a degenerating stage, and almost no cell division was observed.

It has been shown [161] that the gill tissues of *B. thermophilus*, and particularly the associated bacteria, are capable of carbon dioxide fixation, and that the gill cells which contain more bacteria incorporate more labelled carbon dioxide. These results indicate that the symbiotic bacteria have an autotrophic metabolism. Enzyme analysis [156], however, has indicated low levels of Calvin–Benson cycle enzymes, and no sulphide oxidation enzyme activity. Taking into account these observations, and those concerning the digestive tract, it is probable that *B. thermophilus* is a mixotrophic species [162]. This species would be less dependent on hydrothermal activity than *Calyptogena*, and this could explain why these invertebrates have also been observed on the fringe of active hydrothermal areas. The bacteria associated with the gills of *B. thermophilus* have not yet been cultivated, but some heterotrophic bacteria, probably from the alimentary groove, have been isolated and described [112, 163].

6.3.6.3 *Gastropods.*   Although they are not dominant in the eastern Pacific vent communities, gastropods exist in these areas, and *Neomphalus fretterae* is the only one among several gastropod vent species to have been described in detail [164]. It feeds by grazing bacteria, but also filter-feeds on particulate matter. A new species of Archaeogastropod has been collected at 1850 m on the Juan de Fuca Ridge, and its gill epithelium has been found to be colonised by filamentous bacteria [164]. The bacteria, tentatively assigned to the Actinomycetes, are located between and on the surfaces of the gill filaments. Viewed by TEM, the bacterial filaments appear 'embedded in the epithelial cell surfaces' and some cells were found to be endocytosed. Endocytosis is followed by degradation in lysosome-like organelles. The authors interpreted bacterial endocytosis 'as a result of a deliberate parasite invasion of the limpet cells, an invasion thwarted by immediate lysosomal breakdown of the bacteria, in other words, a defense mechanism' or 'an intermediate stage in the development of an intracellular symbiotic relationship'. No data are available about the metabolic type of the bacteria, which presumably provide food for the limpet. However, bacteria were also found within the digestive tract, which makes the interpretation of the limpet–bacteria association more complicated.

In the west and south-west Pacific, vent-associated macrofauna is dominated by the Gastropod *Alviniconcha hessleri* [94, 165]. The animal harbours in its gill filaments slender rod-shaped bacteria contained in bacteriocytes [133]. These bacteria are sulphur-oxidising and showed high activities of enzymes catalysing autotrophic carbon dioxide fixation [166]. TEM studies of those associations revealed another type of bacteria in the bacteriocytes of a specimen of *Alviniconcha*, possessing intracellular membranes, resembling those of type 1 methanotrophs. Methane-oxidising activity has not been detected [166], but this result could be due to the small quantity of tissues available for the enzymatic analysis. However, the possibility that those bacteria could be nitrifying bacteria cannot be excluded.

6.3.6.4 *Crustaceans.* The vent communities of the mid-Atlantic ridge are characterised by the absence of vestimentiferan tube worms, and the scarcity of bivalves [167]. The fauna is dominated by dense (approximately $1500 \, \mathrm{m}^{-2}$) and motile swarms of carridean shrimps belonging to a new genus: *Rimicaris exoculata* and *Rimicaris chacei* [168]. These swarms were observed on the surfaces of black smoker chimneys, in zones where temperatures were approximately 20–30°C. The shrimps have a normal digestive tract, and mineral particles have been found inside. No bacterial cells were observed, but lipopolysaccharide assays of stomach contents indicated the presence of a large amount of bacterial cell wall material (approximately 1 μg of bacterial carbon per mg wet weight of stomach content). Filamentous micro-organisms, probably prokaryotes, were found on particular appendages (maxilliped I, maxilla II, pereiopods III and IV). The animals appear to be heterotrophic, probably feeding on bacteria living on the surface of the chimneys. However, such a diet would require a particularly high biomass and productivity of chimney-associated bacteria, which still remains to be demonstrated.

6.3.6.5 *Bivalves from cold seeps and subduction zones.* Childress *et al.* [169] studied unidentified mussels living in the vicinity of hydrocarbon seeps in the Gulf of Mexico (Louisiana slope). Mussels were collected at depths of 600–700 m and used for oxygen and methane consumption experiments. The data indicated that methane was consumed by gill tissues exclusively and was associated with an increase of oxygen consumption and carbon dioxide production. TEM observation of the gill filaments showed bacteriocytes containing bacteria with stacked internal membranes, typical of type I methanotrophs.

Mussels from the Florida escarpment were studied by Cavanaugh *et al.* [170]. These mussels have large thick gills that contain Gram-negative bacteria of two morphological types. The largest (1.6 μm in diameter) are coccoid and contain internal membranes. The smallest (0.4 μm in diameter) are coccoid or rod-shaped, without internal membranes. No explanation on the role of the smaller bacteria was found, however the internal membranes of the large coccoid bacteria and the detection of hexulose phosphate synthetase (a key enzyme of one of the carbon assimilatory pathways of methane oxidisers) suggested that the larger mussel symbionts are methylotrophs. Examination of different bivalve species [171] and carbon dioxide fixation experiments indicated that the clams *Pseudomiltha* spp. (Lucinidae) and probably *Calyptogena ponderosa* and *Vesicomya cordata* contained autotrophic sulphur-oxidising bacteria. Analysis of carbon-stable isotopes confirmed that the mussel symbionts were methanotrophs, and analysis of stable isotopes of nitrogen indicated possibly bacterial nitrogen fixation.

Symbiotic associations of *Bathymodiolus thermophilus* have been compared with similar mussels from the Louisiana slope [172]. The seep mussel symbionts differ in their larger size, the occurrence of internal membranes, and also the small number of symbionts (three or less) within each vacuole. These authors confirmed the occurrence of methanotrophy for the seep mussel only. In addition, they looked for ribulose biphosphate (RuBP) carboxylase activity, and found low activity levels for the seep mussel and for *Bathymodiolus*. The absence of ATP sulphurylase and ATP reductase within the seep mussel gill tissues indicated that sulphur oxidation was not a major energy source for this animal. The low level of RuBP carboxylase could, however, be due to a contaminant or a minor gill-associated bacterium. In the case of *Bathymodiolus*, these results could indicate that the major symbiont was a thiosulphate-oxidising bacterium (perhaps mixotrophic), but not an autotrophic sulphur bacterium.

Two new species of the genus *Calyptogena* were collected and described from the Japan subduction zone. *C. laubieri* [173] was found between 3800 and 4020 m and *C. phaseoliformis* between 5130 and 5960 m. Both species have large thick gills in which bacteriocytes are abundant [174]. TEM observations have only been carried out on *C. phaseoliformis*. The bacterial cells within the bacteriocytes were $0.6-1 \mu m$ in diameter, and showed the typical cell wall of Gram-negative bacteria. Some dividing stages were observed. The bacteriocytes also contained lipid inclusions and lysozyme, which were involved in the reabsorption of the bacteria. The metabolic type of these endocellular bacteria has not yet been determined. These bivalves live in sulphide-enriched sediments, but the emitted fluids also contain methane. On the basis of geochemical analysis it has been proposed that the endosymbionts are methanotrophs [175]. These endocelluar bacteria did not, however, show any internal membranes, which are typical in methanotrophs [176], and the question of their energy source remains open.

One of the remaining questions concerning symbionts is the relationship between symbionts of different bivalve species. Classical microbiological methods (i.e. culture techniques) have not yet been successful in studying these organisms. Several heterotrophic bacterial strains have been isolated from *Bathymodiolus thermophilus* [163], but might have arisen from the alimentary groove. Isolation of strains from several coastal Lucinid species (*Lucinima borealis*, *Thyasira flexuosa*, *T. sarsi*, *T. obsoleta* and *Myrtea spinifera*) has been reported [177]. These bacteria are facultative methylotrophs, can oxidise thiosulphate to tetrathionate, and also exhibit a heterotrophic metabolism. However no data concerning the similarity between the supposed symbionts from the different bivalves analysed are available. More reliable information has been gathered using molecular biology techniques. It has been concluded [147] that the rRNA sequences of several invertebrate symbionts of *Calyptogena magnifica, Solemya velum* and the vestimentiferan *Riftia pachyptila* have a homology of at least 90%. These data support the conclusion that only a small number of free-living, sulphur-oxidising bacteria are likely to become symbionts [178], a hypothesis also proposed for sponge symbionts [179]. More extensive studies of the symbiotic associations presently reported with molecular biological techniques are required to clarify this issue.

The mode of acquisition of bacterial symbionts by the juveniles of vent invertebrates remains an important questions which has not yet been answered. Recent TEM examination of *Calyptogena soyoae*, collected at 1160 m from a seepage area of the Sagami Bay (Japan), revealed bacteria, resembling the gill endosymbionts of this species, in the follicle cells and the oocytes [180]. Endow and Ohta [180] reported the presence of '6 bacteria within a single primary oocyte, on a single ultra-thin section'. These data represent the first report of a mother–juvenile transmission of sulphur-oxidising gill symbionts in bivalves.

6.3.6.6 *The Polychaete annelid* Alvinella pompejana.   This annelid has been found within high-temperature environments at 21°N and 13°N, where it builds its tube on the outer walls of active sulphide chimneys and is exposed to temperatures in the range 20–40°C [89]. This species, first described as composed of two forms [88], has been further separated into two distinct species [181]: *A. pompejana* and *A. caudata*. Both species have a functional digestive tract in which bacteria and sulphide particles have been observed [182]. Analysis of stable isotopes of carbon and nitrogen [182] indicate that their food, probably bacterial, has a local origin.

However, *Alvinella* is also well known for an abundant epibiotic microflora covering the worm's teguments [80, 183]. Both species have, dispersed on their surfaces, three main morphological types of bacteria—rod, prosthecate and spiral-shaped form—the last two resembling *Hyphomicrobium* and *Spirillum*, respectively [80]. Clump-like communities have been also found located within the intersegmentary spaces of both species. These clumps are enmeshed in an organic matrix, and are composed of rods, cocci and filaments (both sheathed and not sheathed).

However, two *Alvinella* species are distinguished by specific bacterial associations. Macroscopic dorsal extensions are located in the intersegmentary spaces of *A. pompejana*, in which filamentous bacteria are located. These filaments (0.3–1 μm in diameter, 100–200 μm in length) are made up of cylindrical cells and are located in a thin sheath [80]. Rear parapods of *A. caudata* are transformed and bear filamentous bacteria reaching 600 μm in length with a diameter of 2.5 μm. The filaments are made of long cells (6 μm long) [80]. Finally, both species live in organomineral tubes [184], with the inner walls covered with various, mainly filamentous, types of bacteria [182].

The role of these bacterial communities has not been yet completely established. '*In-situ*' experiments using ^{14}C-labelled bicarbonate [185] showed that the polychaetes take up more bicarbonate in the absence of antibiotics. Epibiotic bacteria could thus play a trophic role, complementary to the food ingested through the digestive tract, and rich in bacteria produced in the smoker environment [107]. Moreover, detection of enzymes specifically involved in carbon dioxide fixation [110] indicates that at least some of these bacteria are autotrophic, and probably sulphur-oxidising. However, bicarbonate uptake in these experiments [185] was restricted to a few coccoid bacteria, suggesting metabolic diversity (autotrophy and heterotrophy) or physiological plasticity in response to environmental conditions. Heterotrophic bacteria isolated from various *Alvinella* samples [112] produced hydrogen sulphide from sulphur-containing amino-acids.

A more extensive survey of bacterial metabolic types associated with the worm's teguments and tubes was carried out more recently [186] and confirmed the high diversity of the worm's microflora. Sulphur-oxidising bacteria were commonly found, but also sulphate reducers, nitrifiers (ammonia and nitrite oxidisers), denitrifiers (nitrate-respiring and also nitrogen-producing bacteria), nitrogen-fixing strains and various non-specific heterotrophs, together with a few fermentative bacteria. The microflora associated with *A. pompejana* and *A. caudata* thus appears more complex than previously expected, and both aerobic and anaerobic processes, usually confined to different redox conditions, have been recorded. Two possible hypotheses have been proposed. The first is the existence of microanaerobic zones, within an aerobic environment, as a result of the high density of bacteria living on the worm and tube surfaces, and also probably producing exopolymers, as previously reported for several types of biofilms [187]. The second suggested that different metabolic processes could be active in response to successive contrasting environmental conditions arising from vent activity, and consequent fluctuations in the fluid temperatures. Such fluctuations have been reported (D. Desbruyères *et al.*, unpublished data) to occur in a range 10–20°C within a few seconds.

Another feature of the smoker environment, and particularly the *Alvinella* habitat, is its high toxicity due to sulphide production and the high concentrations of heavy metals in the hydrothermal fluids [5]. These observations, along with the detection of different minerals in the tissues of *Alvinella* [183], have led several workers to suggest a detoxifying role for some of these epibiotic bacteria [163,183]. The most abundant metals (sulphur, arsenic, zinc) have been detected not only within the worm's epidermic cells, but also within epibiotic bacteria [80]. Moreover, metallo-thionein-like proteins have been found in *Alvinella* tissues, with which these bacteria are associated [188]. Accumulated minerals have also been found in *Riftia* [188], *Calyptogena magnifica* [189] and *Bathymodiolus thermophilus* [190]. They are always located in bacteria-associated tissues, which suggests that bacterial detoxification is a common process in major hydrothermal vent invertebrate species.

Manganese is one of the minerals emitted by the active vents (see above), and manganese-oxidising bacteria have been found in many samples [191]. About 50 strains isolated from *Alvinella* have been studied [118]. All are Gram-negative and are related to the genera *Pseudomonas* and *Aeromonas*. Manganese oxidation appeared to occur during the stationary phase, when cultures were incubated at 40°C. At 20°C, oxidation was very slow or non-existent. Growth of the strains, which was faster at 40°C than at 20°C, did not seem to be affected by the presence or absence of manganese in the culture medium. Manganese has been reported to be toxic for several bacteria [192], and the result of oxidation could be the transformation into oxidised form, which is probably less toxic [185].

However, some other metals contained in vent fluids are considered to be even more toxic, and a search for heavy metal-resistant bacteria associated with *A. pompejana*, *A. caudata* and their tubes has been carried out recently [185, 193, 194]. Culture medium amended with copper, zinc, cadmium, arsenic and silver was used to select and isolate about 300 strains of heterotrophic bacteria from the polychaetes and their tubes. Minimal inhibitory concentrations (MIC) of metals were determined as $90 \mu g/ml$ cadmium, $70 \mu g/ml$ zinc, $20 \mu g/ml$ silver and $3500 \mu g/ml$ AsO_4^{3-}. Of these strains, 92.3% were resistant to at least one metal, and 57% were resistant to two metals or more. Some fluctuations were noted in the percentages of strains resistant to the different metals according to the sample from which they were isolated. However, the main feature was the resistance to arsenate, exhibited by 70% of the strains, independent of their origin. All the strains studied were Gram-negative rods, mostly aerobic, produced hydrogen sulphide from cysteine and were tentatively assigned to the genera *Pseudomonas*, *Alteromonas* and *Vibrio*. Screening of the 300 strains for their plasmid content indicated that about 25% carried between one and five plasmids, with sizes ranging from 25 to 134 kb [195].

On the basis of their plasmid contents (1–5) and the sizes of these plasmids, the strains have been classified into five groups [195]. Several strains, sometimes from different groups, carried plasmids of similar sizes, and homology has been demonstrated for two of them (45.3 and 87 kb). Six of those strains were selected for curing experiments, and two of them lost one of their plasmids (45.3 kb) after exposure to mitomycin [195]. The two mutants appeared to be sensitive to arsenite and arsenate, suggesting that arsenic resistance was encoded by the 45.3-kb plasmid. However, preliminary exposure of the mutants to arsenite or arsenate induced resistance, suggesting that remaining plasmid(s) or the chromosome could also play a role in the process. Nevertheless, these results represent the first record of arsenic resistance encoded by a plasmid for a pseudomonad, and moreover, the first for a deep-sea bacterium.

To test the hypothesis of a detoxifying role for at least some of the worm's epibionts, metal accumulation experiments have been carried out [195]. Most of the strains selected for those experiments appeared capable of metal accumulation. However, accumulation metals and metallothioneins have been found mainly in the dorsal epidermic cells of the worm, the dorsal epidermis also being colonised by epibiotic bacteria [183]. For these reasons, it has been suggested that some of the epibiotic bacteria which are able to accumulate metals could be partially responsible for the metal accumulation in the worm, as has been demonstrated in several other cases of epibiosis in plants [196] or crustaceans [197].

6.3.6.7 *Thermophilic bacterial communities.*   Since the discovery of the deep-sea vents, assays of microbiological cultures from samples of hot fluids have been tried, and the occurrence of bacterial communities producing methane, hydrogen and carbon monoxide at 100°C under atmospheric pressure have been reported since 1982 [9]. These workers proposed that part of the gases emitted by the vents could have a biological origin. This hypothesis required verification that bacterial communities able to grow under black-smoker conditions (250–300°C, 250 atmospheres) exist. Under these conditions, sea water remains liquid, and reports of mixed cultures growing at 265 atmospheres in the temperature range 150–250°C have been published [198]. Generation times observed were 8 hours at 150°C, 1.5 hours at 200°C and 40 minutes at 250°C. These surprising results were followed by a major controversy [199, 200], and presently, despite several attempts, these data have not been confirmed. Probably some artefacts occurred during the criticised experiments [199] but the same year, 1983, a new thermophilic archaeobacterium, *Pyrodictium*, with an optimum growth temperature of 105°C was reported [201]. Several other studies indicate clearly that thermophilic bacteria do exist within the hot-vent environments. Particulate material collected near a 21°N black smoker was found to contain viable micro-organisms which were active at atmospheric pressure at temperatures between 21 and 90°C, the highest temperature tested in these experiments [202]. Under those conditions carbon production has been estimated to be 19 μg C/h.

Analysis of black-smoker fluids (above 150°C) collected in the Guaymas Basin indicated that fluids contained no recognisable bacteria and had very low ATP contents [203]. Incorporation of [^{3}H]adenine in hot-water samples was < 1% of the maximum values obtained for samples collected at only 25 cm from the hot fluid discharge. These maximum values, obtained for temperatures between 25 and 45°C, are indicative of a mesophilic bacterial community in the plume. During the same cruise [203] two kinds of sample devices (vent cap and smoker poker) were used for *in-situ* collection and incubation experiments. No bacterial cells were found in the black smoker fluids, and the authors concluded that 'bacteria observed in hydrothermal fluids are not derived from the hot hydrothermal fluids but must originate in peripheral habitats'.

Black-smoker fluids have been also analysed for their DNA content [204]. It was found that 'most of the superheated (174 to 357°C) smoker fluid samples contained particulate DNA in concentrations too high (0.86 to 1.32 ng ml^{-1} in samples containing 15–16% maximum of seawater) to be attributable to entrained sea water'. The question is whether this DNA corresponds to intact bacteria or not. The use of the 'sterivex' method [204] only allows the collection of particulate DNA (associated with inorganic particles or nearly intact cells). Quantification of DNA after enzymatic lysis, with or without French press lysis, and correlation between DNA content and total cell counts favoured the association of DNA with intact cells [204].

In the Guaymas Basin, hydrothermal fluids are partially emitted through a deep layer of sediments, which form petroliferous material by thermal alteration [205]. The potential for some natural hydrocarbon molecules to serve as carbon source for a specific microflora has been investigated [205]. Aerobic bacteria, utilising hexadecane or naphthalene as sole carbon source were cultivated, and their growth occurred at 4 or 25°C, occasionally at 55°C, but never at 80°C. These results are consistent with previous work cited by these authors [205] and confirm that known hydrocarbon-utilising bacteria are only moderate thermophiles.

In the same hydrothermal area, sulphate reduction has been studied by radiotracer methods [206] and extremely thermophilic activity has been observed in subsurface sediments. Sulphate reduction in the 5–10 cm layer was higher at 50 or 70°C than at 20 or 35°C. Moreover, in the sediment slurry experiments, with sediment from 30 (*in-situ* temperature 70°C) or 45 cm depth (*in-situ* temperature 75°C), sulphate reduction was found to be optimal at 77°C (maximum 83°C) and 83°C (maximum 90°C) respectively. No significant sulphate reduction was found in a sample from 45 cm depth with an *in-situ* temperature of 110°C.

Isolation of ultrathermophilic bacteria in pure cultures from different vent sites has confirmed the existence of thermophiles adapted to these extreme environments. A new species of methanogen, *Methanococcus jannaschii* [207], has been isolated from material collected at the base of a white smoker at 21°N. The bacteria are irregular cocci, with a diameter of 1.5 μm. They are motile, with a complex flagellar apparatus (two bundles of numerous flagella inserted at the same pole), and occur singly or in pairs. *M. jannaschii* is a strict autotroph producing methane from hydrogen and carbon dioxide. Sulphides, with an optimum concentration of 1–3 mM, are required. The temperature range for growth is 50–86°C, with an optimum near 85°C. Sodium chloride and pH optima are 0.5 M Na$^+$ and 6 respectively, and the minimum doubling time measured was 26 minutes at pH 6 and 85°C. The strain has a DNA base composition of 31 mol% G + C. A new species has been created in the genus *Methanococcus* from nucleotide sequences of the 16S rRNA.

From sediment collected in the Guaymas Basin, another strain of thermophilic methanogen has been isolated [208] and has also been placed in the genus *Methanococcus*. This new strain (CS1) is an irregular motile coccus, with a diameter of 1 μm. This bacterium uses hydrogen and carbon dioxide or formate for growth and produces methane. The temperature range for growth is 48–94°C, with an optimum near 85°C. Optimal pH is 6 and optimal sodium chloride concentration is 0.5 M. At 85°C and pH 6, the minimum doubling time is 25 minutes. The DNA base composition was found to be 30.9 mol% G + C. As previously reported for *M. jannaschii*, CS1 contains a macrocyclic diether in its membrane lipid [209], but in addition a new series of C35 isoprenoids has been detected [210]. The authors noted that thermophilic methanococci are frequently isolated from deep-sea vent samples and are the 'predominant representatives of methanogenic Archaeobacteria' in these environments. Moreover, taking into account the doubling time and temperature optimum, the hypothesis of biological production of part of the methane contained in the vent fluids [9] was supported [207].

From the Guaymas Basin, another strain of methanogenic bacterium (AG 86) has been isolated [211]. On the basis of physiological data, G + C content (33%) and 16S rRNA sequences, this strain is considered to be related to *M. jannaschii*. AG 86 and *M. jannaschii* were screened for plasmid content. Strain AG 86 harbours a plasmid (pURB900) of 20 kb, and *M. jannaschii* two plasmids of 64 kb (pURB800) and 18 kb (pURB801). No homology was found between the plasmid of isolate AG 86 and those of *M. jannaschii*.

A new genus of archaeobacterium, *Staphylothermus marinus*, has been isolated from samples collected at 11°N on the East Pacific Rise (strain A12), and (strain F1) from geothermally heated, coastal marine sediments in Italy [212]. These bacteria are coccoid, non-motile, with a diameter of 0.5–1 μm, and appear singly, in short chains or in large clusters of up to 100 cells. Under certain culture conditions, and particularly with high concentrations of yeast extract, giant cells with diameter of 15 μm have been observed. The cells stain Gram-negative. *S. marinus* is a strictly anaerobic heterotroph and is obligately dependent on S°. Complex substrates such as peptone, yeast extract, meat extract or bacterial extract are metabolised. However, no growth occurred on casein, sugars, alcohols, organic acids or amino acids. Metabolism products are carbon dioxide, acetate, isovalerate and small amounts of hydrogen sulphide. Strain F1 grows in the range 65–98°C with an optimum at 92°C. No growth was observed at 60 or 100°C. The pH range for growth is 4.5–8.5 with an optimum at 6.5; optimum sodium chloride concentration is 1.5%. Minimum doubling time is 270 minutes at 92°C. DNA base composition of strains F1 and Al2 are 34.9–35.3 and 34.5 mol% G + C respectively, and homology between the two strains is 93–98%. The fact that the two strains of *S. marinus* were isolated from a Pacific deep-sea vent, and a Mediterranean shallow ven suggests a very wide distribution of this type of bacterium in oceanic hydrothermal systems [212].

The genus *Desulfurococcus* is also considered representative of thermophilic bacteria commonly found in different deep-sea vent ecosystems, and has been isolated from the eastern Pacific (11°N, Guaymas Basin) and the mid-Atlantic ridge [213]. Two deep-sea vent strains (S and SY) have been described [213] but not placed in a particular species, while terrestrial isolates from Icelandic hot springs have been described as *D. mobilis* and *D. mucosus* [214].

Strains S and SY have been isolated from the outer wall of a black smoker inhabited by alvinellid polychaetes. The cells are Gram-negative, irregular cocci with diameter of 0.8–1.2 μm. Strain SY is not motile, but strain S is motile and bears a bundle of flagella, similar to that of *Thermococcus celer* or *Desulfurococcus mobilis*. Both strains are heterotrophic and grow well on peptone, yeast extract or tryptone. No growth was observed on casamino acids, starch, single alcohols, carbohydrates or organic acids. The cells are strictly anaerobic, but at low temperatures which do not allow growth they are less sensitive to oxygen. Strains do not require S°, but elemental sulphur enhances growth rate by 5–5.5 times. DNA base composition of the strains S and SY is 52.01 ad 52.42 mol% G + C respectively. Temperature ranges for growth are 50–90°C (strain S) and 50–95°C (strain SY), with optimal values of 85 and 90°C respectively. At optimum temperatures, minimum doubling times are 34.5 and 33.8 minutes respectively. Optimal pH values are 7.5 for S and 7 for SY; optimal sodium chloride requirements are 65% (S) and 75% (SY) of seawater ionic strength. No growth was observed at 99°C under 1 atmosphere, and the cells do not survive autoclaving at 121°C for 15 minutes. The effect of hydrostatic pressure has been tested for both strains, and growth rates were 44–41% (S) and 27–35% (SY) lower than those at 1 atmosphere. Under these *in-situ* pressure conditions, no growth occurred at 100 or 110°C.

More recently, a new group of methanogenic bacteria has been discovered in hot sediments of the Guaymas Basin [215], and the genus *Methanopyrus* has been described. Members of the genus *Methanopyrus* are rod-shaped (0.5 × 8–10 μm) motile cells, staining Gram-positive, and showing the typical fluorescence of methanogens under the UV microscope at 420 nm. This bacterium is a strict autotroph which produces methane from hydrogen and carbon dioxide. Growth occurs in the range 84–110°C, with an optimum at 98°C and a doubling time of 50 minutes. At 110°C, the doubling time is 8 hours. No growth was observed at 80 or 115°C. The optima for pH and sodium chloride are 6.5 and 1.5% respectively. The DNA base composition is 60 mol% G + C. Lipid analysis revealed the presence of 1.1 M cyclic 2,3-diphosphoglycerate, found previously in lower amounts in methanogens, and presumably partially involved in enzyme thermostability [215]. The presence of such compounds, and the temperature range supporting growth of these new methanogens, are new data supporting the hypothesis of high-temperature biogenic methane production in the deep-sea environment.

Previous thermophilic deep-sea vent bacteria have been isolated from liquid or solid inert material, but a new unidentified Archaeobacterium (strain ES1) has been isolated from a macerated polychaete (*Paralvinella*) collected from an active smoker of the Juan de Fuca Ridge [216]. Isolate ES1 is a coccoid, non-motile, Gram-negative-staining bacterium (0.8–1.5 μm). The size varies according to culture conditions. It is an obligate heterotroph and is capable of reducing sulphur. Sulphur does not enhance growth, but does induce morphological changes, including production of giant cells up to 10 μm. Growth occurs on yeast extract plus peptone, or yeast extract, peptone or casein hydrolysates alone. Neither nitrate nor sulphate is reduced. No growth occurred on glucose, ribose, sucrose, glycine, glutamic acid, methanol, formate, acetate, lactate, glycerol or starch. Growth occurs in the range 50–91°C, with an optimum at 82°C, or 85°C when S° is present in the medium. The pH range for growth is 5.0–8.3, and optimum sea salt concentration is 32.3‰. Although isolated from a macerated polychaete, strain ES1 obviously does not grow within the polychaete for temperature reasons. It could be, however, part of the trophic resource of the worm.

The first ultrathermophilic, anaerobic Archaeobacteria described were either methanogens or sulphur-metabolising organisms. Discovery of the ultrathermophilic sulphur reducer *Archaeoglobus* [217] has been considered as a possible missing link between the two other groups of anaerobic thermophilic archaeobacteria [218]. The first representative of this genus, *Archaeoglobus fulgidus*, was isolated from a coastal, geothermally heated sediment in Italy [219]. Recently, similar bacteria were isolated from sediments and active chimney walls in the Guaymas Basin and assigned to the species *A. profundus* [220]. Cells are irregular cocci, single or in pairs, often showing a triangular shape (1.3 μm in width). Under the UV microscope, cells showed a characteristic blue–green fluorescene at 420 nm. Sulphate is used as an electron acceptor, but while *A. fulgidus* is a strict autotroph that uses only hydrogen as an electron donor, *A. profundus* is an obligate mixotroph, and requires hydrogen and organic carbon (acetate, lactate, pyruvate, yeast extract). Thiosulphate may also be used as electron acceptor. No growth was observed on S°, which can be an inhibitor in presence of sulphate. However, S° can be reduced but does not support growth. Growth occurs in the temperature range 65–90°C, with an optimum at 82°C. Optimum pH and sodium chloride levels are 6 and 1.8% respectively. In optimal conditions the minimum doubling time is 4 hours. DNA base composition is 41 mol% G + C. In addition to mixotrophy, lower G + C content (5 mol%), different lipid profile and weak DNA homology distinguished *A. profundus* from *A. fulgidus*.

All the ultrathermophilic archaeobacteria isolated from deep-sea vents have been isolated from various debris collected in the close vicinity of smokers, hot sediments or smoker fluids themselves. Recently, a new habitat for thermophiles, formed by trapped pools hot fluid 'trapped beneath densely colonized, sulfide-sulfate-silicate ledges, or flanges' has been discovered and an ultrathermophilic bacterium has been isolated from this habitat, but not yet described [221].

More generally however, the exact location, at a microbial scale, of those bacteria is not yet precisely known. The fact that both autotrophs and heterotrophs have been found in similar samples may indicate that metabolic exchange can occur between these dissimilar metabolic types. But the culture of a particular strain from a particular sample does not prove conclusively that the cultured bacteria were active at the place of collection. Several ultrathermophiles are known to be less sensitive to oxygen when exposed to temperatures lower than their lower temperature limit for growth [222]. Moreover, cultivation and identification of several ultrathermophilies from material resulting from the eruption of a submarine volcano and floating at the sea surface [222] has clearly demonstrated that these bacteria can survive in a well-oxygenated cold environment and are certainly dispersed by hydrothermal plumes.

As reported in section 6.2, barophily has only been demonstrated for heterotrophic psychrophilic deep-sea bacteria, and generally increased the upper temperature limit for growth of these bacteria [53]. The effect of hyperbaric pressures, the growth rate of *M. jannaschii* and methanogenesis were thermophilic bacteria. Formation of bacterial colonies at 120°C under a pressure of 265 atmospheres has been reported [223], and pressure was required for growth of this bacterium. However, pressure (200–300 atmospheres) appears to decrease growth rates of *Desulfurococcus* strains [213], whereas elevated pressure enhances growth of *Methanococcus thermolithotrophicus*, a coastal ultrathermophilic bacterium, at its optimum temperature [224], without extending its temperature range [225]. Cells of *M. thermolithotrophicus* exposed for 10 hours at 65°C and 50 MPa (1 atmosphere = 101.29 kPa) show alterations in their protein pattern [226] and the amino acid composition of the total cell wall hydrolysate, with formation of a series of proteins with a molecular mass ranging from 38 to 70 kDa in pressurised cells. The question of whether the observed alterations are caused by a perturbation in the balance of protein synthesis and turnover or by a pressure-induced synthesis of compounds analogous to heat-shock proteins remains unanswered.

The effects of hyperbaric pressure (up to 750 atmospheres) on *Methanococcus jannaschii* have been investigated [225]. When helium was used to produce hyperbaric pressures, the growth rate of *M. jannaschii* and methanogenesis were increased at 86 and 90°C, but the upper limit for growth was not extended. However, the temperature range for methanogensis was extended and methane formation occurred at 98°C and 250 atmospheres, while no methane was produced at 94°C and 7.8 atmospheres. The nature of the gas used for pressure appeared to be very important, and no methane was produced at 86°C when argon was substituted for helium. At 250 atmospheres pressure and 86°C, methanogenesis occurred when the culture was pressurised with hydrogen, but not with the commonly used mixture of hydrogen and carbon dioxide (80/20). The authors concluded that pressure had 'a stabilizing effect on an enzyme or enzymes crucial to the methane production' and that the results obtained 'illustrated that metabolic events other than growth rate are important to consider in studies of pressure effects on micro-organisms'.

Obviously, the question of hydrostatic pressure effects on deep-sea thermophilic bacteria remains open. A comprehensive study comparable to that done on psychrophilic organisms is now required. More intensive exploration of vents deeper than those actually studied, for instance on the Atlantic Ridge, could be of great interest for that purpose.

6.4 Biotechnology of deep-sea bacteria

The possible utilisation of aquatic and marine bacteria in biotechnological processes has been pointed out by several authors [227–229]. However, the number of bacterial species presently used in industrial processes is particularly low, and does not include any marine bacteria. Nevertheless, the marine environment is particularly rich in diversified ecosystems, each of them inhabited by adapted and specialised micro-organisms [230].

The deep sea represents the most extreme marine environment, and like other extreme environments is certainly an unexploited source of unique micro-organisms, including some potential candidates for biotechnological applications. However, there is presently no utilisation of deep-sea bacteria for any industrial purpose, and the following topics only constitute research projects.

Because they are particularly spectacular, deep-sea hydrothermal vents have aroused considerable interest amongst microbiologists as a potential source of new micro-organisms. From an ecological point of view, the main feature of deep-sea vent ecosystems is the food chain, based on bacterial chemosynthesis. The role of bacteria in the marine food chain has been studied by several authors and reviewed recently for bivalve molluscs [231]. If bacterial cells can be used as food by other species in the marine environment [232], the application of this property is obviously limited to some aquaculture processes, and particularly larvae nutrition. However, attempts to feed bivalve larvae with bacterial cells have not been very successful [231]. Bacteria used in such experiments were common marine or freshwater heterotrophic bacteria. If dead bacterial cells are used as food, the unconsumed cells are degraded by bacteria living in the culture tanks, and this process is generally harmful for larvae, because it results in a reduction in water quality. If live bacteria are used, they may proliferate in the culture tanks and give rise to the same problem as described above [233].

The bacteria mostly involved in the deep-sea vent food chains are autotrophic or mixotrophic sulphur oxidisers. If these bacteria, and particularly the true autotrophs, were released as food in culture tanks, they would not grow since no energy source would be available. However, they could remain alive (those which are not consumed) long enough to avoid degradation by heterotrophs. Such an aquaculture food chain based on sulphur-oxidising bacteria has been proposed [77], and these workers designed a model bacterial reactor to produce bacterial biomass on sulphide-rich residues.

An alternative approach would be to rear symbiont-bearing animals [234]. Berg and Alatalo [234] speculated 'that industrial sulphide wastes might be adapted for such a mariculture system, thus disposing of an abundant waste product while producing edible animal protein'. However, the feasibility of such a system has yet to be demonstrated.

The most typical feature of the deep sea is the elevated hydrostatic pressure. However, this feature does not seem to constitute, alone, a biotechnological potential. Another existing feature of the deep-sea vent is the existence of hot temperature environments, where sea water remains liquid at temperatures above 300°C because of the hydrostatic pressure. For these reasons, if bacteria able to grow at temperatures above 110°C, the highest so far unequivocally demonstrated, do exist, they may reside in deep-sea hydrothermal vents where the upper limit temperature for growth may lie between 110 and 150°C [235].

However, before such a unique micro-organism is discovered and cultured in the laboratory, several deep-sea ultrathermophiles are presently available for biotechnological applications. Although some new genera or species have been described, the deep-sea strains exhibit metabolic pathways which are well known, and many of them have shallow-vent or even terrestrial counterparts. Thermophilic bacteria, and particularly archaeobacteria, which are the most thermophilic, are particularly interesting because of the high stability of their macromolecules, and particularly their enzymes. The advantages of using thermostable enzymes have been considered by many authors [228, 236, 237] and two chapters in this book are concerned with thermophilic bacteria.

The particular advantage of thermophilic archaeobacteria isolated from deep-sea vents is that they originate from a unique ecosystem in which they probably develop unique biochemical and molecular adaptations. Currently, this is a hypothesis, and only the screening of newly isolated deep-sea strains for lipase, protease, RNA polymerase or other enzymes will verify whether or not these assumptions are correct. Such screening programmes are presently in progress in several laboratories.

A particular feature of the deep-sea vent environment is that, although it is extreme, for the reasons reported above it is inhabited by morphologically and metabolically diverse microbial communities. As a consequence, besides the properties described above, which are unique to the deep-sea vent environments, hydrothermal bacteria exhibit features also known in terrestrial or aquatic environments. Some of these features are metabolic properties (nitrification, denitrification, sulphate reduction), which are of great interest in bio-remediation processes. The expected advantage of using deep-sea vent bacteria, in addition to the possible but not yet verified higher activities, is that they are adapted to the constraints of this environment and particularly hydrostatic pressure, large spatial temperature gradients and elevated concentrations of toxic compounds.

In our laboratory, we have developed in cooperation with French and foreign colleagues, research programmes to investigate the feasibility of using deep-sea vent bacteria belonging to various metabolic types for biotechnological processes [191]. One of the topics studied is the screening of heterotrophic bacteria for exopolysaccharide production. Many aquatic and particularly marine bacteria are known to produce exopolysaccharides [230, 238], by which they often attach on inert or living surfaces. Several authors [79, 110] reported that natural or artificial surfaces exposed in vent waters are colonised by morphologically diverse bacteria.

Polysaccharides are essentially used in food or pharmaceutical industries as thickeners or gelling agents. They are extracted from terrestrial plants (starch, cellulose, pectin), from seaweeds (alginates, carrageenans) or crustacean shells (chitosan). Only a few are produced by microbial (fungi or bacteria) fermentation. Marine bacteria polysaccharides have also been suggested [227, 239] to be useful in oil recovery or cleansing of heavy metal-contaminated waters. In addition, utilisation of some compounds for medical applications cannot be excluded, as an antitumour polysaccharide has been isolated from a marine *Flavobacterium* strain [229].

Screening of about 500 deep-sea vent mesophilic heterotrophic bacteria has led to the selection of 17 strains which produce high-viscosity products in liquid media [240]. Some of these compounds have been chemically analysed and reveal a proportion of acidic sugars (uronic and galacturonic acids) up to 50% [241, 242]. Such a composition is in favour of heavy metal-binding capabilities for those compounds (G. Geesey, personal communication), and experiments in this field are presently in progress.

Although deep-sea micro-organisms were first demonstrated more than a century ago, deep-sea microbiology has remained only a small part of microbiology. However, in addition to its significance for biological oceanography, the discovery of deep-sea vents, and particularly the proposal that micro-organisms could grow there at temperatures higher than those generally admitted, has aroused a great interest in the deep sea, and as a consequence in the marine environment. In the same way, the development of molecular biology will certainly bring together microbiologists and molecular biologists interested in powerful molecular tools and new microbiological models. There is a general feeling that increasing exchange between the disciplines of microbiology will probably lead to significant biotechnological developments.

Acknowledgements

The author thanks G. Erauso and Dr C. Jeanthon for their help in gathering references and Dr S. de Goer and Dr C.C. Sommerville for helpful discussions and critical reading of the manuscript.

References

1. A. Certes, *C.R. Acad. Sci. Paris* **98** (1884) 690.
2. H.W. Jannasch and C.D. Taylor, *Ann. Rev. Microbiol.* **38** (1984) 487.
3. R.Y. Morita, *Bact. Rev.* **39** (1975) 144.
4. R.Y. Morita, *Sarsia* **64** (1979) 9.
5. J.B. Corliss, J. Dymond, L.I. Gordon, J.M. Edmond, R.P. Von Herzen, R.D. Ballard, K. Green, D. Williams, H. Bainbridge, K. Crane and T.H. Van Handel, *Science* **203** (1979) 1073.
6. H. Felbeck and G.N. Somero, *Trends Biochem. Sci.* **7** (1982) 201.
7. A. Fustec, D. Desbruyères and S.K. Juniper, *Biol. Oceanogr.* **4** (1987) 121.
8. J.M. Edmond and K. Von Damm, *Sci. Am.* **248** (1983) 78.
9. J.A. Baross, M.D. Lilley and L.I. Gordon, *Nature* **298** (1982) 366.
10. B. Fischer, *Zentrlbl. Baketriol.* **15** (1894) 657.
11. C.E. Zobell, *Marine Microbiology*, Chronica Botanica, Waltham (1946).
12. C.E. Zobell, *Bull. Misaki Mar. Biol. Inst.* **12** (1968) 77.
13. R.Y. Morita, in *Marine Ecology*, Wiley Interscience, New York (1972) 1361.
14. R.Y. Morita, in *The Survival of Vegetative Microbes*, Cambridge University Press, New York (1976) 279.
15. M.M. Quigley and R.R. Colwell, *J. Bact.* **95** (1968) 211.
16. H.W. Jannasch, K. Eimhjellen, C.O. Wirsen and A. Farmanfaian, *Science* **171** (1972) 672.
17. H.W. Jannasch and C.O. Wirsen, *Science* **180** (1973) 641.
18. J.R. Schwarz, J.D. Walker and R.R. Colwell, *Dev. Industrial Microb.* **15** (1974) 239.
19. H.W. Jannasch, C.O. Wirsen, C.D. Taylor, *Appl. Environ. Microbiol.* **32** (1976) 360.
20. H.W. Jannasch, *Oceanus* **21** (1978) 50.
21. H.W. Jannasch, *Bioscience* **29** (1979) 228.
22. H.W. Jannasch and C.O. Wirsen, in *Deep-Sea Biology*, Wiley, New York (1983) Chapter 6.
23. J.W. Deming, P.S. Tabor and R.R. Colwell, in *Advanced Concepts in Ocean Measurements for Marine Biology*, The Belle W. Baruch Library in Marine Science, Columbia (1980) 285.
24. J.R. Schwarz, J.D. Walker and R.R. Colwell, *Can. J. Microbiol.* **21** (1975) 682.
25. C.O. Wirsen and H.W. Jannasch, *Environ. Sci. Tech.* **10** (1976) 880.
26. J.R. Schwarz, A.A. Yayanos and R.R. Colwell, *Appl. Environ. Microbiol.* **31** (1976) 46.
27. K. Ohwada, P.S. Tabor and R.R. Colwell, *Appl. Environ. Microbiol.* **40** (1980) 746.
28. J.W. Deming and R.R. Colwell, *Appl. Environ. Microbiol.* **44** (1982) 1222.
29. P.S. Tabor, J.W. Deming, K. Ohwada and R.R. Colwell, *Appl. Environ. Microbiol.* **44** (1982) 413.
30. C.O. Wirsen and H.W. Jannasch, *Marine Biol.* **78** (1983) 69.
31. J.W. Deming and R.R. Colwell, *Appl. Environ. Microbiol.* **50** (1985) 1002.
32. J.W. Deming, *Mar. Ecol. Prog. Ser.* **25** (1985) 305.
33. C.O. Wirsen and H.W. Jannasch, *Marine Biol.* **91** (1986) 277.
34. K. Lochte and C.M. Turley, *Nature* **33** (1988) 67.
35. P.A. Marden, K. Tunlid, G. Malmcrona-Friberg, G. Odham and S. Kjelleberg. *Arch. Micobiol.* **142** (1985) 326.

36. J.A. Novitsky and R.Y. Morita, *Marine Biol.* **49** (1978) 7.
37. J.A. Novitsky and R.Y. Morita, *Appl. Environ. Microbiol.* **32** (1976) 617.
38. J.A. Novitsky and R.Y. Morita, *Appl. Environ. Microbiol.* **33** (1977) 635.
39. P.S. Amy, C. Pauling and R.Y. Morita, *Appl. Environ. Microbiol.* **45** (1983) 1041.
40. P.S. Amy and R.Y. Morita, *Appl. Environ. Microbiol.* **45** (1983) 1109.
41. P.S. Amy, C. Pauling and R.Y. Morita, *Appl. Environ. Microbiol.* **45** (1983) 1685.
42. P.S. Amy and R.Y. Morita, *Appl. Environ. Microbiol.* **45** (1983) 1748.
43. T.N. Nyström, N. Albertson and S. Kjelleberg, *J. Gen. Microbiol.* **134** (1988) 1645.
44. C.L. Moyer and R.Y. Morita, *Appl. Environ. Microbiol.* **55** (1989) 2710.
45. C.E. Zobell, and F.H. Johnson, *J. Bact.* **57** (1949) 179.
46. A.A. Yayanos, *Science* **200** (1978) 1056.
47. A.S. Dietz and A.A. Yayanos, *Appl. Environ. Microbiol.* **36** (1978) 966.
48. A.A. Yayanos, A.S. Dietz and R. Van Boxtel, *Science* **205** (1979) 808.
49. A.A. Yayanos, A.S. Dietz and R. Van Boxtel, *Proc. Natl. Acad. Sci. USA* **78** (1981) 5212.
50. A.A. Yayanos, A.S. Dietz and R. Van Boxtel, *Appl. Environ. Microbiol.* **44** (1982) 1356.
51. A.A. Yayanos, A.S. Dietz, *Appl. Environ. Microbiol.* **43** (1982) 1481.
52. A.A. Yayanos, R. Van Boxtel and A.S. Dietz, *Appl. Environ. Microbiol.* **48** (1984) 771.
53. A.A. Yayanos, in *Recent Advances in Microbial Ecology*, Japan Scientific Societies Press, (1989) Tokyo 38.
54. H.W. Jannasch and C.O. Wirsen, *Arch. Microbiol.* **139** (1984) 281.
55. A.A. Yayanos, *Proc. Natl. Acad. Sci. USA* **83** (1986) 9542.
56. J.A. Baross, F.J. Hanus and R.Y. Morita, *Appl. Microbiol.* **30** (1975) 309.
57. L.J. Albright and R.Y. Morita, *Limnol. Oceanogr.* **13** (1968) 637.
58. E. Pollard and P.K. Wellek, *Biochim. Biophys. Acta* **112** (1966) 573.
59. R.M. Arnold and L.J. Albright, *Biochem. Biophys. Acta* **238** (1971) 347.
60. D.H. Pope, W.P. Smith, R.W. Swartz and J.V. Landau, *J. Bact.* **121** (1975) 664.
61. H.W. Smith, D.H. Pope and J.V. Landau, *J. Bact.* **124** (1975).
62. J.V. Landau, W.P. Smith and D.H. Pope, *J. Bact.* **130** (1977) 154.
63. E.F. Delong and A.A. Yayanos, *Science* **228** (1985) 1101.
64. J.D. Oliver and R.R. Colwell, *Int. J. Syst. Bacteriol.* **23** (1973) 442.
65. E.F. Delong and A.A. Yayanos, *Appl. Environ. Microbiol.* **51** (1986) 730.
66. E.F. Delong and and A.A. Yayanos, *Appl. Environ. Microbiol.* **53** (1987) 527.
67. E. Helmke and H. Weyland, *Marine Biol.* **91** (1986) 91.
68. W.L. Straube, M. O'Brien, K. Davis and R.R. Colwell, *Appl. Environ. Microbiol.* **56** (1990) 812.
69. D. Bartlett, M. Wright, A.A. Yayanos and M. Silverman, *Nature* **342** (1989) 572.
70. J.W. Deming, H. Hada, R.R. Colwell and K.R. Luehrsen, *J. Gen. Microbiol.* **130** (1984) 1911.
71. M.T. MacDonell and R.R. Colwell, *Syst. Appl. Microbiol.* **6** (1985) 171.
72. J.W. Deming, L.K. Somers, W.L. Straube and M.T. McDonell, *Syst. Appl. Microbiol.* **10** (1988) 152.
73. A.T. Wortmann and R.R. Colwell, *Appl. Environ. Microbiol.* **54** (1988) 1284.
74. L. Laubier and D. Desbruyères, *La Recherche* **15** (1984) 1506.
75. P. Lonsdale, *Deep-Sea Research* **24** (1977) 857.
76. J.H. Tuttle, C.O. Wirsen and H.W. Jannasch, *Marine Biol.* **73** (1983) 293.
77. H.W. Jannasch and C.O. Wirsen, *Bioscience* **29** (1979) 592.
78. C.O. Wirsen, J.H. Tuttle and H.W. Jannasch, *Marine Biol.* **92** (1986) 449.
79. H.W. Jannasch and C.O. Wirsen, *Appl. Environ. Microbiol.* **41** (1981) 528.
80. F. Gaill, D. Desbruyères, D. Prieur and J.P. Gourret, *C.R. Acad. Sci. Paris* **298** (1984) 553.
81. C.M. Cavanaugh, S.L. Gardiner, M.L. Jones, H.W. Jannasch and J.B. Waterbury, *Science* **213** (1981) 340.
82. M. Le Pennec and D. Prieur. *C.R. Acad. Sci. Paris* **298** (1984) 493.
83. M.D. Lilley, J.A. Baross and L.I. Gordon, in *Hydrothermal Processes of Seafloor Spreading Centres*, Plenum Press, New York (1983) 411.
84. M.L. Jones, *Proc. Biol. Soc. Wash.* **93** (1981) 1295.
85. K.J. Boss and R.D. Turner, *Malacologia* **20** (1980) 161.
86. V.C. Kenk and B.R. Wilson, *Malacologia* **26** (1985) 253.
87. E.G. Ruby and H.W. Jannasch, *J. Bact.* **149** (1982) 161.
88. D. Desbruyères and L. Laubier, *Oceanologica Acta* **3** (1980) 267.

89. D. Desbruyères, P. Crassous, J. Grassle, A. Khripounoff, D. Reyss, M. Rio and M. Van Praet, *C.R. Acad. Sci. Paris* **295** (1982) 489.
90. J.F. Grassle, *Science* **229** (1985) 713.
91. V. Tunnicliffe, S.K. Juniper and M.E. DeBurgh, *Biol. Soc. Wash. Bull.* **6** (1985) 453.
92. J.L. Charlou, L. Dmitriev, H. Bougault and H.D. Needham, *Deep-Sea Research* **35** (1988) 121.
93. P.A. Rona, G. Klinkhammer, T.A. Nelsen, J.H. Trefry and H. Elderfield, *Nature* **321** (1986) 33.
94. R. Hessler, P. Lonsdale and J. Hawkins, *New Scientist* **117** (1988) 47.
95. J.M. Auzende, E. Honza and the scientific crew, *C.R. Acad. Sci. Paris* **306** (1988) 971.
96. J.M. Auzende, T. Urabe, C. Deplus, J.P. Eissen, D. Grimaud, P. Huchon, J. Ishibashi, M. Joshima, Y. Lagabrielle, C. Mével, J. Naka, E. Ruellan, T. Tanaka and M. Tanahashi, *C.R. Acad. Sci. Paris* **309** (1989) 1787.
97. Y. Fouquet and the 'Nautilau' group, *Ecos* **71** (1990) 678.
98. D.M. Karl, A.M. Brittain and B.D. Tilbrook, *Deep-Sea Research* **36** (1989) 1655.
99. E. Uchupi and R.D. Ballard, *Deep-Sea Research* **36** (1989) 1443.
100. C.K. Paull, B. Hecker, R. Commeau, R.P. Freeman-Lynde, C. Neumann, W.P. Corso, S. Golubic, J.E. Hook, E. Sikes and J. Curray, *Science* **226** (1984) 965.
101. M.C. Kennicut II, J.M. Brooks, R.R. Bidigare, R.R. Fay, T.L. Wade and T.J. MacDonald, *Nature* **317** (1985) 351.
102. L.D. Kulm, E. Suess, J.C. Moore, B. Carson, B.T. Lewis, D. Ritger, D.C. Kadko, T.M. Thornburg, R.W. Embley, W.D. Rugh, G.J. Massoth, M.G. Lanseth, G.R. Cochrane and R.L. Scamman, *Science* **231** (1986) 561.
103. L. Laubier, S. Ohta and M. Sibuet, *C.R. Acad. Sci. Paris.* **303** (1986) 25.
104. S.K. Juniper and M. Sibuet, *Mar. Ecol. Prog. Ser.* **40** (1987) 115.
105. J.M. Edmond, K.L. Van Damm, R.E. MacDuff and C.I. Measures, *Nature* **297** (1982) 187.
106. J.M. Edmond and K.L. Van Damm, *Pour la Science* **68** (1983) 34.
107. J.A. Baross and J.W. Deming, *Biol. Soc. Wash. Bull.* **6** (1985) 355.
108. H.W. Jannasch and M.J. Mottl, *Science* **229** (1985) 717.
109. D.M. Karl, in *Ecology of Microbial Communities*, Cambridge University Press Press, New York (1987).
110. D. Prieur, C. Jeanthon and E. Jacq, *Vie et Milieu* **37** (1987) 149.
111. R.L. Ferguson and P. Rublee, *Limnol. Oceanogr.* **21** (1976) 141.
112. D. Prieur, in *Proc. 21st EMBS*, Polish Academy of Science, Institute of Oceanology (1989) 393.
113. D.M. Karl, C.O. Wirsen and H.W. Jannasch, *Science* **207** (1980) 1345.
114. C.D. Winn, D.M. Karl and G.J. Massoth, *Nature* **320** (1986) 744.
115. J.P. Cowen, *Appl. Environ. Microbiol.* **55** (1989) 764.
116. H.L. Ehrlich, *Dev. Ind. Microbiol.* **7** (1966) 279.
117. H.L. Ehrlich, *Environmental Biogeochemistry, Ecol. Bull.* **35** (1983) 357.
118. P. Durand, D. Prieur, C. Jeanthon and E. Jacq, *C.R. Acad. Sci. Paris* **310** (1990) 273.
119. J.P. Cowen, G.J. Massoth and E.T. Baker, *Nature* **322** (1986) 38.
120. S.L. Davis and R. Whittenbury, *J. Gen. Microbiol.* **61** (1970) 227.
121. S.N. Watson and M. Mandel, *J. Bact.* **107** (1971) 563.
122. R.M. Weiner, R.A. Devine, D.M. Powell, L. Dagasan and R.L. Moore, *Intern. J. Syst. Bact.* **35** (1985) 237.
123. P. Fera and D. Prieur, in *Deuxième Colloque International de Bactériologie marine, CNRS, Brest*, IFREMER, Actes de colloque **3** (1986) 219.
124. D.C. Nelson, C.O. Wirsen and H.W. Jannasch, *Appl. Environ. Microbiol.* **55** (1989) 2909.
125. E.G. Ruby, C.O. Wirsen and H.W. Jannasch, *Appl. Environ. Microbiol.* **42** (1981) 317.
126. H.W. Jannasch, C.O. Wirsen, D.C. Nelson and L.A. Robertson, *Int. J. Syst. Bacteriol.* **35** (1985) 422.
127. T. Naganuma, A. Otsuki and H. Seki, *Deep-Sea Research* **36** (1989) 1379.
128. H. Seki and T. Naganuma, *Mar. Ecol. Prog. Ser.* **54** (1989) 199.
129. C.S. Harwood, H.W. Jannasch and E. Canale-Parola, *Appl. Environ. Microbiol.* **44** (1982) 234.
130. G.H. Rau and J.I. Hedges, *Science* **203** (1979) 648.
131. H. Felbeck, *Science* **213** (1981) 336.

132. A. Fiala-Médioni, *C.R. Acad. Sci. Paris* **298** (1984) 487.
133. K. Endow and S. Ohta, in *Recent Advances in Microbial Ecology* Japan Scientific Societies Press, Tokyo (1989) 28.
134. G.H. Rau, *Science* **213** (1981) 338.
135. G.H. Rau, *Biol. Soc. Wash. Bull.* **6** (1985) 243.
136. P.M. Williams, K.L. Smith, E.M. Druffel and T.W. Linick, *Nature* **292** (1981) 448.
137. C. Bosch and P.P. Grassé, *C.R. Acad. Sci. Paris* **299** (1984) 371.
138. C. Bosch and P.P. Grassé, *C.R. Acad. Sci. Paris* **299** (1984) 413.
139. H. Felbeck, *Physiol. Zool.* **58** (1985) 272.
140. H. Felbeck, M.A. Powell, S.C. Hand and G.N. Somero, *Biol. Soc. Wash. Bull.* **6** (1985) 261.
141. R.C. Terwilliger, N.B. Terwilliger and E. Schabtach, *Comp. Biochem. Physiol.* **65B** (1980) 531.
142. J.B. Wittenberg, R.J. Morris, Q.H. Gibson and M.L. Jones, *Biochim. Biophys. Acta* **670** (1981) 255.
143. J.J. Childress, A.J. Arp and C.R. Fisher, *Marine Biol.* **83** (1984) 109.
144. M.A. Powell, and G.N. Somero, *Science* **219** (1983) 297.
145. C.R. Fischer and J.J. Childress, *Mar. Biol. Lett.* **5** (1984) 171.
146. D.C. Nelson, J.B. Waterbury and H.W. Jannasch, *FEMS Microbiol. Lett.* **24** (1984) 267.
147. D.A. Stahl, D.J. Lane, G.J. Olsen and N.R. Pace, *Science* **224** (1984) 409.
148. D.J. Lane, D.A. Stahl, G.J. Olsen, D.J. Heller, and N.R. Pace, *J. Bact.* **163** (1985) 75.
149. C.R. Woese, *Microbiol. Rev.* **51** (1987) 221.
150. M.E. De Burgh, S.K. Juniper and C.L. Singla, *Marine Biol.* **101** (1989) 97.
151. D.L. Distel, D.J. Lane, G.J. Olsen, F.J. Giovannoni, N.R. Pace, D.A. Stahl and H. Felbeck, *J. Bact.* **170** (1988) 2506.
152. M.E. De Burgh, *Can. J. Zool.* **64** (1986) 1095.
153. S.C. Cary, H. Felbeck and N.D. Holland, *Mar. Ecol. Prog. Ser.* **52** (1989) 89.
154. M.L. Jones, *Oceanologica Acta* **8** (1988) 69.
155. A. Fiala-Médioni and C. Métivier, *Mar. Biol.* **90** (1986) 215.
156. H. Felbeck, J.J. Childress and G.N. Somero, *Nature* **239** (1981) 291.
157. C.M. Cavanaugh, *Nature* **302** (1983) 58.
158. A.J. Arp, J.J. Childress and C.R. Fischer, *Physiol. Zool.* **57** (1984) 648.
159. A. Hily, M. Le Pennec and M. Henry, *C.R. Acad. Sci. Paris* **302** (1986) 495.
160. M. Le Pennec, *Oceanologica Acta* **8** (1987) 181.
161. S. Belkin, D.C. Nelson and H.W. Jannasch, *Biol. Bull.* **170** (1986) 110.
162. M. Le Pennec, D. Prieur and A. Lucas, in *Proc. 19th EMBS*, Cambridge University Press, Cambridge (1985) 159.
163. D. Prieur and C. Jeanthon, *Symbiosis* **4** (1987) 87.
164. M.E. De Burgh and C.L. Singla, *Marine Biol.* **84** (1984) 1.
165. T. Okutani and S. Ohta, *Venus* **47** (1988) 1.
166. J.L. Stein, S.C. Cary, R.R. Hessler, S. Ohta, R.D. Vetter, J.J. Childress and H. Felbeck, *Biol. Bull.* **174** (1988) 373.
167. C.L. Van Dover, B. Fry, J.F. Grassle, S. Humphris and P.A. Rona, *Marine Biol.* **98** (1988) 209.
168. A.B. Williams and P.A. Rona, *J. Crustacean Biol.* **6** (1986) 446.
169. J.J. Childress, C.R. Fischer, J.M. Brooks, M.C. Kennicutt II, R. Bidigare and A.E. Anderson, *Science* **233** (1986) 1306.
170. C.M. Cavanaugh, P.R. Levering, J.S. Maki, R. Mitchell and M.E. Lidstrom, *Nature* **325** (1987) 346.
171. J.M. Brooks, M.C. Kennicutt II, C.R. Fisher, S.A. Macko, K. Cole, J.J. Childress, R.R. Bidigare and R.D. Vetter, *Science* **238** (1987) 1138.
172. C.R. Fisher, J.J. Childress, R.S. Oremland and R.R. Bidigare, *Marine Biol.* **96** (1987) 59.
173. T. Okutani and B. Métivier, *Venus* **45** (1986) 147.
174. A. Fiala-Médioni and M. Le Pennec, *Oceanologica Acta* **11** (1998) 185.
175. J. Bou ègue, E.L. Benedetti, D. Dron, A. Mariotti and R. Letolle, *Earth Planet Sci. Lett.* **83** (1983) 343.
176. R. Whittenbury and H. Dalton, in *The Prokaryotes*, Springer, Berlin (1981) 895.
177. A.P. Wood and D.P. Kelly, *J. Mar. Biol. Assoc. U.K.* **69** (1989) 165.
178. R.G.B. Reid and D.G. Brand, *Veliger* **29** (1986) 3.
179. C.R. Wilkinson, *Proc. R. Soc. Lond. B* **230** (1984) 79.
180. K. Endow and S. Ohta, *Mar. Ecol. Prog. Ser.* **64** (1990) 309.

181. D. Desbruyères and L. Laubier, *Can. J. Zool.* **64** (1986) 2227.
182. D. Desbruyères, F. Gaill, L. Laubier, D. Prieur and G. Rau, *Marine Biol.* **75** (1983) 201.
183. F. Gaill, D. Desbruyères and P. Prieur, *Microb. Ecol.* **13** (1987) 129.
184. J. Vovelle and F. Gaill, *Zoologica Scripta* **15** (1986) 33.
185. A.M. Alayse-Danet, F. Gaill and D. Desbruyères, in *Proc. 19th EMBS*, Cambridge University Press, Cambridge (1985) 167.
186. D. Prieur, S. Chamroux, P. Durand, G. Erauso, P. Fera, C. Jeanthon, L. Le Borgne, G. Mével and P. Vincent, *Marine Biol.* **106** (1990) 361.
187. J. McN. Sieburth, in *Microbial Seascapes*, University Park Press, Baltimore (1975).
188. M.A. Cosson-Mannevy, R. Cosson and F. Gaill, *C.R. Acad. Sci. Paris* **302** (1986) 347.
189. G. Roesijadi and E.A. Crecelius, *Marine Biol.* **83** (1984) 155.
190. C. Chassard-Bouchaud, A. Fiala-Médioni and P. Galle, *C.R. Acad. Sci. Paris* **302** (1986) 117.
191. D. Prieur, N. Benbouzid-Rollet, S. Chamroux, P. Durand, G. Erauso, E. Jacq, C. Jeanthon, G. Mével and P. Vincent, *Cah. Biol. Mar.* **30** (1989) 515.
192. H. Hajj and J. Makemson, *Appl. Environ. Microbiol.* **32** (1976) 699.
193. C. Jeanthon and D. Prieur, *Prog. Oceanogr.* **24** (1990) 81.
194. C. Jeanthon and D. Prieur, *Appl. Environ. Microbiol.* **56** (1990) 3308.
195. C. Jeanthon, *Thesis*, University of Brest (1991).
196. F.M. Patrick and M.W. Loutit, *Wat. Res.* **11** (1977) 699.
197. I. Johnson, N. Flower and M. Loutit, *Microb. Ecol.* **7** (1981) 245.
198. J.A. Baross and J.W. Deming, *Nature* **303** (1983) 423.
199. J.D. Trent, R.A. Chastain and A.A. Yayanos, *Nature* **307** (1984) 737.
200. R.H. White, *Nature* **310** (1984) 430.
201. K.O. Stetter, H. König and E. Stackebrandt, *Syst. Appl. Microbiol.* **4** (1983) 535.
202. D.M. Karl, D.J. Burns, K. Orrett and H.W. Jannasch, *Mar. Biol. Lett.* **5** (1984) 227.
203. D.M. Karl, G.T. Taylor, J.A. Novitsky, H.W. Jannasch, C.O. Wirsen, N.R. Pace, L.J. Lane, G.J. Olsen and S.J. Giovannoni, *Deep-Sea Research* **35** (1988) 777.
204. W.L. Straube, J.W. Deming, C.C. Somerville, R.R. Colwell and J.A. Baross, *Appl. Environ. Microbiol.* **56** (1990).
205. D.A. Bazylinski, C.O. Wirsen and H.W. Jannasch, *Appl. Environ. Microbiol.* **55** (1989) 2832.
206. B.B. Jorgensen, L.X. Zawacki and H.W. Jannasch, *Deep-Sea Research* **37** (1990) 695.
207. W.J. Jones, J.A. Leigh, F. Mayer, C.R. Woese and R.S. Wolfe, *Arch. Microbiol.* **136** (1983) 254.
208. W.J. Jones, C.E. Stugard and H.W. Jannasch, *Arch. Microbiol.* **151** (1989) 314.
209. P.B. Comita and R.B. Gagosian, *Science* **222** (1983) 1329.
210. G.U. Holzer, P.J. Kelly and W.J. Jones, *J. Microbiol. Meth.* **8** (1988) 161.
211. H. Zhao, A.G. Wood, F. Widdel and M.P. Bryant, *Arch. Microbiol.* **150** (1988) 178.
212. G. Fiala, K.O. Stetter, H.W. Jannasch, T.A. Langworthy and J. Madon, *Syst. Appl. Microbiol.* **8** (1986) 106.
213. H.W. Jannasch, C.O. Wirsen, S.J. Molyneaux and T.A. Langworthy, *Appl. Environ. Microbiol.* **54** (1988) 1203.
214. W. Zillig, K.O. Stetter, W. Schäfer, D. Janekovic, S. Wunderl, L. Holz and P. Palm, *Zbl. Bakt. Hyg., I. Abt. Orig.* **C2** (1981) 205.
215. R. Huber, M. Kurr, H.W. Jannasch and K.O. Stetter, *Nature* **342** (1989) 833.
216. R.J. Pledger and J.A. Baross, *Syst. Appl. Microbiol.* **12** (1989) 249.
217. K.O. Stetter, G. Lauerer, M. Thomm and A. Neuner, *Science* **236** (1987) 822.
218. L. Achenbach-Richter, K.O. Stetter and C.R. Woese, *Nature* (1987) 348.
219. K.O. Stetter, *System. Appl. Microbiol.* **10** (1988) 172.
220. S. Burggraf, H.W. Jannasch, B. Nicolaus and K.O. Stetter, *Syst. Appl. Microbiol.* **13** (1990) 24.
221. J.R. Delaney, R.E. McDuff, J.A. Baross, M.D. Lilley and M.S. Goldfarb, *Nature* (1990).
222. R. Huber, P. Stoffers, J.L. Cheminée, H.H. Richnow and K.O. Stetter, *Nature* **345** (1990) 179.
223. J.W. Deming and J.A. Baross, *Appl. Environ. Microbiol.* **51** (1986) 238.
224. G. Berhardt, R. Jaenicke, H.D. Ludemann, H. Konig and K.O. Stetter, *Appl. Environ. Microbiol.* **54** (1988) 1258.
225. J.F. Miller, N.N. Shah, C.M. Nelson, J.M. Ludlow and D.C. Clark, *Appl. Environ. Microbiol.* **54** (1988) 3039.
226. R. Jaenicke, G. Bernhardt, H.D. Lüdemann and K.O. Stetter, *Appl. Environ. Microbiol.* **54** (1988) 2375.

227. J.T. Staley and P.M. Stanley, *Microb. Ecol.* **12** (1986) 79.
228. J.W. Deming, *Microb. Ecol.* **12** (1986) 111.
229. Y. Ckami, *Microb. Ecol.* **12** (1986) 65.
230. J. McN. Sieburth, *Sea Microbes*, Oxford University Press, New York (1979).
231. D. Prieur, G. Mével, J.L. Nicolas, A. Plusquellec and M. Vigneulle, *Oceanogr. Mar. Biol. Ann. Rev.* **28** (1990) 277.
232. D. Prieur, *Kiel. Meeresforchung.* (submitted).
233. D. Prieur, *DSC thesis*, University of Brest (1981).
234. C.J. Berg Jr and P. Alatalo, *Aquaculture* **39** (1984) 165.
235. T.D. Brock, *La Recherche* **198** (1988) 476.
236. B. Sonnleitner and A. Fiechter, *Trends Biotechnol.* **1** (1983) 75.
237. M. Fossi, *Biofutur.* **53** (1987) 39.
238. J.W. Costerton, G.G. Geesey and K.J. Cheng, *Sci. Am.* **238** (1978) 86.
239. R.M Weiner, A. Segall and R.R. Colwell, *J. Appl. Environ. Microbiol.* **43** (1985) 83.
240. D. Prieur, C. Jeanthon, P. Vincent, F. Talmont and J. Guezennec, in *Current Topics in Marine Biotechnology*, The Japanese Society of Microbiology, Tokyo (1989).
241. P. Vincent, F. Talmont, D. Prieur, B. Fournet, P. Pignet and J. Guezennec, *Kieles. Meeresforschung.* (submitted).
242. F. Talmont, P. Vincent, T. Fontaine, J. Guezennec, D. Prieur and B. Fournet, *Food Hydrocolloids* (in press).

7 Physiology and molecular biology of psychrophilic micro-organisms

N.J. RUSSELL

7.1 What are psychrophiles and psychrotrophs?

It might seem perverse to begin a discussion of psychrophiles by immediately introducing the second term psychrotroph, but is important to establish from the outset what is meant by these terms which have engendered debate since the first use of the word psychrophile by Schmidt-Nielsen in 1902 to describe bacteria capable of growth as well as survival at 0°C [1]. Within a year the terminology was challenged by Muller, who argued that many bacteria capable of growth at zero had optimum growth temperatures that were really quite high, in the range 20–30°C, i.e. well into the growth temperature range of mesophiles, and so they should be regarded as being cold-tolerant [2]. Soon after Ekelof found that in Antarctic soils warmed by solar radiation the bacteria had optimum growth temperatures of 17.5°C [3] and so would be psychrotrophic according to modern definition. This was the prelude to a debate, particularly during the 1960s, about the distinctions between psychrophiles ('cold-loving') and psychrotrophs (psychrotolerant or 'cold-tolerant'); the terms obligate and facultative psychrophiles have also been used [4].

Such debate was far from being pedantic, because it sought to define the essential nature of these organisms, which is that they are all characterised by their ability to grow at or close to zero (i.e. degrees Celsius). Indeed, the ability to produce visible colonies on solid media within a specified time period (usually 1 or 2 weeks) was one of the early criteria used to define psychrophilic micro-organisms [4, 5]. Whilst this is no longer used as a definitive criterion, it exemplifies the fact that, unlike mesophiles and thermophiles, which are defined by their optimum and maximum growth temperatures respectively, psychrophiles (and psychrotrophs) are defined by their minimum growth temperature, i.e. their ability to grow at low temperatures. Nowadays, the most universally accepted definition of psychrophilic micro-organisms is that of Morita, who in 1975 proposed that this term should be reserved for those micro-organisms whose cardinal (i.e. minimum, optimum and maximum) growth temperatures are < 0, < 15 and < 20°C respectively [6]. In comparison, those micro-organisms which have cardinal temperatures of 0–5, > 15 and > 20°C should be regarded as being psychrotrophic. Inevitably there are micro-organisms which fall between these definitions, but despite the fact the some modern textbooks of microbiology fail to distinguish the two groups [7] it is a useful distinction to make because, as discussed below, there do seem to be differences in their ecological distribution and physiological adaptation mechanisms to growth at low temperatures.

Some of the lowest optimum and upper growth temperatures reported are those of the so-called 'snow algae' [8,9]. These algae are distributed in snow-covered habitats ranging from alpine areas of the temperate zone in the northern hemisphere to the Antarctic where they inhabit the upper 1 cm layer of snow, giving it a red, green or yellow colouration; others can give a grey colouration [10]. Their optimum growth temperatures are usually < 10°C and some have an extremely narrow growth temperature range, for instance *Chlainomonas rubra* loses its flagella above 4°C and suffers membrane damage at − 1°C.

The definitions of psychrophiles and psychrotrophs given above imply that the former are more likely to grow at subzero temperatures, and this raises the question 'What is the lowest temperature at which microbial growth occurs?' Minimum growth temperatures are notoriously difficult to determine, especially for psychrophiles and psychrotrophs, because of their very slow growth rates at such low temperatures; in addition, the composition of the growth medium can alter the minimum growth temperature, e.g. by supplying a specific nutrient whose synthesis would otherwise become limiting at a low temperature. The authoritative opinion of Ingraham is that the most reliable estimate is the reported value of − 12°C for bacteria, whilst the claims for fungal and yeast growth at temperatures as low as − 34°C need to be substantiated [11]. The value of − 12°C is consistent with the temperature at which the intracellular formation of ice occurs, accompanied by increases in solute concentration, both of which would have drastic consequences for metabolism and thus prevent growth [12,13].

Psychrophiles usually grow over a more restricted growth temperature range than psychrotrophs are capable of doing: psychrophiles could be regarded as specialists in low-temperature growth, whereas psychrotrophs are more versatile. Recently, Wiegel has suggested that the term psychrotrophic (which means 'cold-eating') is better replaced by the term 'eurythermal mesophile' [14], this being consistent with the use of the expression 'eurythermal' for those organisms which are capable of growing over a wide thermal range, as put forward by Brock [15]. Baross and Morita have pointed out that a number of soil bacteria originally considered as being mesophiles should more correctly be designated as being psychrotrophic [16]; the same is true of the pathogenic food-spoilage bacterium *Listeria monocytogenes* [17]. The iron-oxidising *Thiobacillus ferrooxidans* is a mesophilic bacterium with an optimum growth temperature varying from 25 to 35°C, depending on the strain and the pH of growth, and an upper limit of 42–43°C [18]. Iron oxidation occurs at temperatures as low as 4°C, and some strains have been considered as being psychrotrophic [19]. This is relevant to the occurrence of these bacteria in underground mines where temperatures typically range between 5 and 15°C, whereas surface temperatures in heap leaching vary more widely—from freezing up to 50°C or higher.

7.2 Microbial types of psychrophiles

Psychrophily, the ability to grow at low temperatures, is widespread amongst microbial flora. Representatives are found amongst the eubacteria and cyanobacteria, yeasts, fungi and algae, but to date no archaeobacterial psychrophiles are known (although it is this author's opinion that it is only a matter of time before they are discovered, most likely among the methanogens). It seems probable that, like the ability to carry out photo-synthesis, the ability to grow at low temperatures has evolved independently in several groups of micro-organisms representing a wide diversity of taxonomic types and general metabolic characteristics. Moreover, it is not infrequently observed that there are species within a single genus which are psychrophilic (or psychrotrophic), mesophilic and thermophilic: for example, this is true of the genera *Clostridium* and *Bacillus* amongst the eubacteria, and of the cyanobacterial genus *Phormidium*. Thus, even amongst very closely related species it is possible for there to be biochemical and physiological adaptations which enable them to colonise both extremes of temperature. In fact, such closely related species with vastly different thermal tolerances have been useful in defining some of these adaptive strategies (see below).

Although this review will consider examples from each of the major microbial groups, the emphasis will be on eubacteria. Moreover, no attempt is made to be encylopaedic; instead the intention is to demonstrate the essential features of psychrophiles, their special physiological and biochemical adaptations for low-temperature growth, their contribution to microbial activity in natural environments and finally how these abilities might be put to future biotechnological use.

Within the eubacteria there are psychrophiles which are autotrophs or heterotrophs; aerobes or anaerobes; phototrophs or non-phototrophs; protein degraders or cellulose degraders. Despite this metabolic diversity, there is a curious preponderance of Gram-negative relative to Gram-positive eubacterial psychrophiles (and psychrotrophs). For example, of the 144 eubacterial species isolated from a marine Antarctic environment by Tanner and Herbert, about 50% were *Vibrio* and *Aeromonas*, and 25% were *Alcaligenes* plus some *Pseudomonas* and *Flavobacterium* spp. [20]. The dominant species in surface waters of the Beaufort Sea in the sub-Arctic were found to be *Flavobacterium* and *Microcyclus* species, with *Vibrio* and other

Gram-negative rods, but *Pseudomonas* were not major contributors to the flora [21]. In soils on Signy Island in the Antarctic peninsula it is Gram-negative rods (*Pseudomonas*, *Achromobacter* and *Alcaligenes*) which predominate [22]. A number of explanations, such as the use of inappropriate isolation media or the greater protection afforded by the complex Gram-negative cell envelope structure, have been proposed over the years, but none offers molecular insights into the phenomenon. In apparent contradiction of the general trend, in the Antarctic dry valleys, which are among the most hostile permanently cold environments in the world, the Gram-positive genera *Planococcus* and *Micrococcus* are the dominant eubacteria (for example, see refs. 23 and 24). Undoubtedly there are ecological factors besides temperature which influence the species distribution of eubacteria in cold Antarctic soils, and the same will be true of other microbial types as well as other habitats. For example, in sea-ice microbial communities (SIMCOs) the species distribution of the free-living bacteria (mostly Gram-negative rods and cocci) differ from the attached populations (on microalgae and diatoms), which include a wider range of species and morphological types [25].

In some cold environments it is cyanobacteria which are the dominant micro-organisms. For instance, in Antarctic freshwater streams, ponds and lakes oscillatorians of the genera *Phormidium*, *Oscillatoria*, *Lyngbya*, *Micrococeus* and *Schizothrix* are the most common mat-formers, where they form substantial cohesive mats from a few millimetres up to 1–2 cm thick over the bottom sediments, sometimes lifting off to give spectacular pink, grey–green or orange rafts on the water surface, while *Nostoc* and *Gleocapsa* form black mucilaginous mats over stone and sediment surfaces [26].

7.3 Ecology of psychrophiles and psychrotrophs

In the broadest terms psychrophiles and psychrotrophs are found in two situations, namely in foodstuffs stored at low (chill) temperatures and in cold terrestrial or aquatic environments. Within each of these categories there are habitats varying considerably in their properties, which is reflected in the species diversity of the inhabitants.

7.3.1 *Food*

The microbial flora of a food will reflect both its natural origin and the final composition if it is processed or preserved in any way. Temperature is an important factor in determining the shelf-life of preserved food, because the number of microbial types which are able to multiply and hence cause spoilage is considerably reduced at typical chill-cabinet temperatures. Micro-organisms cannot grow in properly stored frozen foodstuffs, i.e. nominally at a temperature of $-18°C$ or below. Psychrotrophic micro-organisms tend to predominate over psychrophilic ones in foodstuffs, and they include both food-poisoning and food-spoilage organisms (see refs. 17 and 27 for reviews). The most common psychrotrophic/psychrophilic food-spoilage bacteria are Gram-negative genera, mainly *Pseudomonas*, with some *Flavobacterium*, *Achromobacter*, *Alcaligenes*, *Escherichia* and *Aerobacter*. Psychrotrophic yeasts are usually *Candida* and *Rhodotorula*. Because of their ability to grow at low water activity, fungi are more likely to spoil dried or partially dehydrated foodstuffs.

Although the majority of psychrotrophic/psychrophilic micro-organisms which contaminate food and cause low-temperature spoilage do not pose a health risk to humans or animals, there are some notable exceptions. Recently, there has been considerable public concern expressed about outbreaks of listeriosis caused by *Listeria monocytogenes*, which will grow at temperatures close to 10°C [28], particularly in chilled raw milk or soft cheeses and in processed poultry [29]. This organism is perhaps unusual in the present context of thermal properties of micro-organisms because, despite its psychrotrophic properties, it is rather heat-resistant compared with other *Listeria* spp. and non-spore-forming bacteria [28, 30]. This has prompted a re-examination of the thermal processing procedures and a search for augmented methods of killing [31], although some workers have questioned the necessity for such measures [30]. Certain serotypes of *Yersinia enterocolitica* can cause severe but rare infections in children from ingestion of contaminated milk products or fresh spring-water; the bacterium is capable of growing, albeit slowly, at temperatures as low as 1°C [17]. Among the fungi which cause low-temperature deterioration of nuts and grain, the presence of psychrotrophic strains of *Aspergillus flavus* and *A. parasiticus* is a particular concern, because they produce highly toxic aflatoxins that survive the heat treatments used to kill the vegetative fungi [32].

Amongst the food-spoilage bacteria, several Gram-negative rods, including *Alcaligenes*, *Alteromonas*, *Flavobacterium*, *Moraxella* and *Pseudomonas*, that typically spoil proteinaceous foods are psychrotrophic, growing well at chill-cabinet temperatures. When such spoilage bacteria grow on the moist surfaces of meat, they produce individual colonies which coalesce to form a slime and malodours from breakdown products of the amino acids. *Pseudomonas* are the main slime formers on meats stored at chill temperatures. The fungus *Cladosporium herbarum* can grow as black spots on chilled meat. In the deeper tissue anaerobic spoilage by clostridia can occur: some non-proteolytic strains of *Clostridium botulinum* are psychrotrophic, capable of growth down to 3–4°C, and consequently have been of concern in the storage of foodstuffs at chill temperatures, especially because of the relative heat resistance of their spores. The growth of psychrotrophic lactobacilli (*Lactobacillus* spp.) and lactic acid bacteria (i.e. *Pediococcus*, *Leuconostoc* and *Enterococcus* spp., and *Brochothrix thermosphacta*) is favoured in packed meat and meat products, because of their tolerance of anaerobiosis, low pH and curing salts, as well as the chill-storage temperatures [33].

The microbial flora of marine fish from temperate zones are largely psychrotrophs and psychrophiles, mainly *Pseudomonas*, *Alteromonas* and *Moraxella* [34]. Salting tends to decrease the Gram-negative flora at the expense of the Gram-positive population. Some of the psychrophiles from fish have been reported as growing at temperatures as low as $-6°C$ (C. Harrison-Church, unpublished data cited in ref. 34).

Milk is a sterile product, but during its collection from the cow and transfer to storage vessels it becomes contaminated with a wide range of micro-organisms that are derived from the teat and udder surfaces of the cow, and the machinery, vessels and personnel in the milking parlour. Few of the micro-organisms, including most lactobacilli, which contaminate milk are capable of growth at its normal storage temperature of 4–7°C. However, during storage the psychrotrophic population becomes dominant, unless the milk has been pasteurised or subjected to any of the other heat treatments used to preserve the keeping qualities and safety of raw milk and milk products.

In line with the general pattern for food spoilage, the major psychrotrophic bacterial contaminants in milk are Gram-negative, most frequently the *Pseudomonas* spp. *Ps. fluorescens*, *Ps. putida* and *Ps. fragi*, together with *Acinetobacter* and *Aeromonas* species [35]. In addition, raw milk may also contain psychrotrophic spore-forming rods, mainly *Bacillus cereus*, *B. sphaericus* and *B. circulans*, with some *Clostridia*, which are slower growing but which have heat-resistant spores. The Gram-negative bacteria are the main agents of spoilage, because although their vegetative cells are generally killed by pasteurisation they all produce lipolytic and proteolytic enzymes [35, 36]. Some of these enzymes, despite being derived from psychrotrophic organisms, are relatively heat-stable and may cause rancidity and off-flavours in cheese made from pasteurised or otherwise heat-treated milk.

Many pigmented species of *Erwinia*, which cause plant diseases such as wilts and soft-rot, are psychrotrophic. Of the Gram-positive rods, several species of *Corynebacterium* found on vegetables are psychrophilic; some *Bacillus* species are also psychrophilic, but the relative heat sensitivity of their vegetative cells as well as the spores means that they seldom cause spoilage problems.

7.3.2 *Terrestrial and aquatic ecosystems*

It has been suggested that because of their different growth temperature ranges psychrophiles are more likely to be found in those places which are permanently cold, whereas psychrotrophs are more likely to be found in cold environments that undergo thermal fluctuations. This distinction has been extended to their distribution in aquatic (particularly marine) and terrestrial environments, since the low temperatures of the former are less likely to undergo periodical fluctuations. Ellis-Evans and Wynn-Williams [37] in a study of the lakes on Signy Island in the Antarctic peninsula found that most aquatic eubacteria had growth temperature optima below 20°C and upper limits of 21–24°C, whereas those from terrestrial sources had higher optima above 18°C and upper limits of 23–33°C. In contrast, most of the fungi in the lake water were psychrotrophic, with optima of 15–20°C and maxima above 30°C [38]. Thus, perhaps all of the microbial populations in the lake would be growing at ambient temperatures that are considerably below the optimum for growth, again reflecting the fact that ecological factors other than temperature influence the species distribution even in permanently cold ecosystems; these will include irradiance, humidity, nutrient availability and grazing by predators. For photosynthetic micro-organisms, light is predictably often the critical factor, as shown for the development of algal blooms in sea ice in the Canadian Arctic [39].

There are reports of very high proportions of psychrophiles from marine environments in the Southern Ocean. For example, Wiebe and Hendricks found that although the population densities were low, up to 80% of the eubacterial isolates did not grow above 10°C, with the percentages of psychrophiles being greatest in the deeper water samples compared with surface waters [40]. Ruger reported that most of the *Alteromonas* and *Vibrio* species isolated from the central part of the north-west African upwelling region were psychrophiles, including 12% with optimum growth temperatures that were less than 6°C and upper limits below 12°C, whereas in the northern part of this upwelling the dominant bacteria were psychrotrophic *Bacillus* species; the difference was believed to be due to the upwelling of colder waters in the central region, supported by the observation that the distribution of psychrophilic bacteria was also dependent on water depth and hence temperature [41]. Arctic bacterial populations are generally less tolerant of thermal fluctuations than are those from the sub-Arctic [21], i.e. the former are more likely to contain greater proportions of psychrophiles. However, the consensus of opinion is that although it is true that psychrophiles are more likely to be isolated from permanently cold environments, it is generally psychrotrophs which are the predominant flora in terms of microbial numbers. For example, Upton and Nedwell [42] and Delille and Perret [43] found this to be so in samples from marine and terrestrial Antarctic sources respectively. Atlas and Morita cited unpublished data from H. Weyland that psychrotrophs dominate in polar waters, but that the absolute numbers of psychrophiles per unit volume of water is greater in Antarctic than in Arctic waters [21].

The temperature of most Arctic and Antarctic soils does not rise above 5–10°C, except for short periods during the summer. The average temperature of the polar seas is −1.8°C and the bulk of oceanic water is < 5°C. Therefore, apart from the most extreme psychrophiles, the ambient temperatures are considerably below the optimum growth temperatures of the indigenous microbial flora. Consequently, if the temperature rises there is the capacity for greatly enhanced metabolic activity, particularly on the part of the psychrotrophs which are well suited to take advantage of the altered thermal conditions; the presence of a proportion of psychrophiles in the population will ensure that metabolic activity can be maintained at the lowest temperatures around zero. A demonstration of this phenomenon in Arctic soils was reported recently by Nadelhoffer *et al.* [44], who showed that there was little increase in carbon and nitrogen mineralisation rates when the temperature was raised from 3 to 9°C, but that the rates doubled with a rise to 15°C [44]. The physiological and biochemical bases of the adaptations which contribute to the ability of psychrophiles and psychrotrophs to function at low temperatures are discussed in the next section.

Antarctic stream and pond cyanobacterial mats freeze in winter; water flow stops and the mats remain viable down to temperatures of − 60°C as dry crusts in which photosynthesis and respiration are restored within minutes of rehydration and warming when summer approaches [45]; soon after rehydration they are capable of incorporating metabolic precursors into all types of cellular macromolecules [26]. Yet these cyanobacteria are largely psychrotrophic, able to metabolise at temperatures up to 30°C, thus demonstrating their enormous thermal flexibility (N.J. Russell, W. Vincent and C. Howard-Williams, unpublished data). Davey also found that *Phormidium* mats on Signy Island recovered rapidly from desiccation or freezing, with 85% restoration of photosynthetic activity after 48 hours, whereas the alga *Prasiola* was less resilient [46].

In contrast, for soils in the Antarctic dry valleys where the environment is so consistently harsh, Vishniac and Klinger have argued that although eubacteria may be more numerous it is psychrophilic yeasts, particularly *Cryptococcus* spp., which are of primary importance in relation to their physiological adaptation and contribution to soil fertility [47]. The species *Cryptococcus friedmannii* is found nowhere else in the world [48].

One of the most startling habitats for cold-adapted micro-organisms is within the air spaces a few centimetres below the surfaces of exposed rocks in Antarctic dry valleys. These cryptoendolithic microbial communities have been studied by Friedmann's group [49], who have shown that the microflora is predominantly eukaryotic or cyanobacterial depending on the type of rock being colonised [50]. The lichen communities include the free-living green alga *Hemichloris antarctica*, which like the yeast above is found in no other environment. The endoliths are protected from the harshest features of the climate of the dry valleys; for example, the magnitude of the temperature fluctuations is less and the small amounts of moisture are more likely to be retained. Microbial activity is regulated primarily by temperature [51] and continuous monitoring by satellite has shown that the communities become active in mid-March when the temperatures rise above − 10°C and there is enough solar radiation and humidity within the air spaces under the rock surface [52].

7.4   Molecular mechanisms of adaptation to low temperature

7.4.1   *Lipids, membranes and nutrient uptake*

It is well established that the lipid composition of membranes must provide the correct conditions of fluidity and phase structure in order that the integral proteins can carry out such essential functions as ion or nutrient uptake and electron transfer in an efficient manner. Temperature is a critical parameter in this respect because it has a major influence on both the fluidity and the phase behaviour of membrane lipids. Micro-organisms are too small to insulate themselves from the effects of thermal change and generally are unable to adapt by avoidance mechanisms through moving to an area of more acceptable temperature. Therefore, micro-organisms must regulate their lipid composition within certain limits if they are to maintain the activity of their membrane-associated proteins, and although this has been demonstrated directly in very few micro-organisms there are myriad reports of the effects of temperature fatty acid composition which are assumed to be responsible for fluidity regulation.

With the acceptance of the fluid-mosaic model for membrane structure and emphasis on the necessity for a liquid-crystalline lipid phase, it was believed that such lipid changes were a mechanism for preserving membrane fluidity. This idea was formalised by Sinensky who coined the term 'homeoviscous adaptation', which postulated that the fluidity of the membrane was maintained at a relatively constant viscosity (i.e fluidity) throughout the growth temperature range [53]. This was supported by the observation that 15–20% of the fatty acids in the acyl lipids of *E. coli* grown at 37°C had to be unsaturated for optimum growth rate and that this percentage increased at lower temperatures [54]. There were conflicting data for the relatively few organisms in which fluidity was measured directly. The lipid phase in membranes of the psychrotroph *Micrococcus cryophilus* remains fluid at temperatures down to − 30°C [55]. The lichen-dominated endolithic communities discussed above have a lipid composition that remains fluid down to − 20°C; they also have unusual hydration properties which could be related to the uniquely stressful cold environment which they inhabit [56]. In the thermophile *Bacillus stearothermophilus* the fatty acid composition altered so that the upper end of the gel-to-liquid-crystalline phase transition was always just below the growth temperature, i.e. the fluidity was kept at a constant level [57], but later studies on a different strain of the same organism suggested that a constant fluidity was not maintained [58]. In psychrotrophic bacteria it has been shown that membrane fluidity is maintained at a constant level over the growth temperature range in some, e.g. *Pseudomonas fluorescens* [59], but not others, e.g. *Micrococcus cryophilus* [60].

The notion that a range of membrane fluidity could be tolerated was demonstrated clearly in a series of experiments by a number of research groups working with the wall-less bacterium *Acholeplasma laidlawii*, which requires lipid supplementation for growth, and unsaturated fatty auxotrophic mutants of *E. coli*, in both of which the lipid fatty acid composition can be manipulated within large limits. Using these experimental approaches, it was shown that *E. coli* could grow with up to 50% and *A. laidlawii* with up to 80–90% of the lipid in their membranes in the gel phase [61, 62]. The studies with *A. laidlawii* in particular led to an alternative theory of 'homeophasic adaptation' being proposed [63, 64], which emphasised the necessity to maintain the membrane lipids in a bilayer phase, i.e. that it was not so important to keep an exact level of membrane fluidity so long as the correct lipid phase was preserved.

The experiments with *A. laidlawii* and mutants of *E. coli* demonstrated that, although a wide range of fatty acid compositions could be tolerated, there were clear upper and lower limits of lipid fluidity [62, 64]. By supplementing with high-melting-point fatty acids it is possible to raise the minimum growth temperature, but the optimum and maximum growth temperatures are much less affected by supplementation with low-melting-point fatty acids. Recently, Murata's group in Japan has shown that the cold tolerance of the normally chill-sensitive cyanobacterium *Anacystis nidulans* can be enhanced genetically by increasing its capacity to synthesise unsaturated fatty acids through the introduction of the cloned gene for a Δ12 desaturase [65]. Native microbial membranes normally contain a variety of fatty acids, although studies with model lipid systems show that a requisite level of fluidity could be achieved with a single type of fatty acyl chain in the lipids, and it was suggested that such mixtures serve to extend the range of temperature over which a micro-organism can grow [64].

Since psychrophiles and psychrotrophs live at low temperatures, one might expect them to have fatty acid compositions that are somehow distinct; fluidity might be achieved by a number of different combinations of fatty acid types, but they would be distinguishable from those of mesophiles and thermophiles, for example by having greater proportions of unsaturated fatty acids which give lower melting point lipids than the comparable saturated lipids [66]. Psychrophilic and psychrotrophic micro-organisms do tend to have relatively large proportions of (poly)unsaturated or short-chain fatty acids in their lipids. However, the trend is by no means clear-cut and, for example, many mesophilic eubacteria contain high or higher proportions of unsaturated fatty acids than psychrophilic/psychrotrophic counterparts. The relative lack of branched-chain fatty acids amongst the psychrophiles and psychrotrophs compared with thermophiles seems to be more a reflection of generic differences in the distribution of these two groups [67]. As so often in science, the simplistic prediction is not borne out in practice, and even when psychrophilic, mesophilic and thermophilic species of the same genus are compared they do not necessarily have fatty acid compositions that follow the predicted trend [68].

Such comparisons fail to take into account other genotypic differences in lipid composition, and a more reliable approach to determining the mechanisms of lipid fluidity maintenance is to investigate the phenotypic modification of fatty acid composition in a particular organism grown at different temperatures. The response will depend on the fatty acid composition of each individual species and its metabolic capability, but by comparing the responses in different types of micro-organisms general patterns can be deduced [69]. The most frequently observed temperature-dependent alteration is in lipid un-saturation: in eubacteria and one group of cyanobacteria the change following a decrease in growth temperature is from saturated to monounsaturated because, with few exceptions, they lack the capability for synthesising polyunsaturated fatty acids [70]; in yeasts, fungi, algae and the other two groups of cyanobacteria, all of which contain polyunsaturated fatty acids, the unsaturation changes are more complex and may involve increases in both the ratio of unsaturated to saturated and the proportion of polyunsaturated fatty acids. In all the microbial groupings the double bonds in unsaturated fatty acids are formed by desaturase enzymes, which modify the (saturated) products of fatty acid synthetase, usually after they have been incorporated into acyl lipids (see below); some eubacteria utilise an alternative mechanism, termed the anaerobic pathway, in which the double bond is introduced as part of the fatty acid biosynthetic mechanism [70]. Until recently it was believed that the two mechanisms were mutually exclusive, but both appear to be present in a psychrotrophic pseudomonad [71].

A number of other responses to thermal shifts are found amongst micro-organisms [69]. Representatives of all the groupings may alter the acyl chain length of their lipids, a decrease in growth temperature favouring the formation of shorter, and therefore lower-melting-point, fatty acids. Eubacteria, particularly many Gram-positives, may alter the amount of methyl-branched relative to straight-chain fatty acids and/or the relative proportions of *iso*-and *anteiso*-branched fatty acids, lower temperatures generally favouring the formation of *anteiso*-branched fatty acids. The effects of temperature appears to be on the primer specificity of fatty acid synthetase, which is the stage at which the methyl branch is introduced, but the molecular basis of this effect has not been elucidated [72].

The significance of the almost universal occurrence of the desaturase mechanism for increasing the proportion of unsaturated lipid following a drop in temperature may be the fact that the substrate for desaturation seems to be the acyl chains of membrane lipids (at least in the relatively few organisms in which it has been investigated). In ecological terms this provides a mechanism for modifying large proportions of the membrane lipid to restore its fluidity after a sudden temperature drop, such as an overnight frost, both rapidly and at a lower energy cost than if the complete fatty acid had to be made *de novo* [69]. Russell termed this mode of lipid change 'modification synthesis', in comparison with 'addition synthesis' which does require fatty acid biosynthesis [69]. All other types of temperature-dependent changes in fatty acid chain length and branching are mediated by addition synthesis, and take correspondingly longer. This is illustrated particularly clearly in the response of some bacilli to thermal shifts: the initial change after temperature shift-down is the induction of a desaturase which alters the unsaturation of membrane lipids; this is followed by a much slower response in which the fatty acid synthetase alters its pattern of products with differences in both the chain length and methyl branching of the fatty acids [73]. It would be interesting to determine whether, following a temperature shift-down, there was a temporal sequence in the utilisation of the desaturase and the anaerobic mode of unsaturated fatty acid biosynthesis in the psychrotrophic pseudomonad which unusually contains both mechanisms for making unsaturated fatty acid (see above).

One of the major functions of membranes is the uptake of nutrients and the regulation of intracellular ionic composition, processes which are performed by carrier systems and ion pumps. The activities of these integral membrane proteins are influenced by membrane phase and fluidity [66] and so it is not surprising that nutrient uptake at low temperatures has often been mooted as one of the determinants of psychrophily. For example, Morton et al., in a study of psychrophilic, mesophilic and thermophilic Torulopsis yeast species, found that only the psychrophilic T. psychrophila transported glucose at 2°C [74], although in a later paper it is stated that the mesophile (sic) T. candida can grow from 0 to 36°C [75], which would make it a psychrotroph.

There are two reports of studies in which sugar uptake/utilisation has been compared in psychrophilic and psychrotrophic bacteria. Herbert and co-workers studied some psychrophilic Vibrio spp. and psychrotrophic Pseudomonas spp. (reviewed in ref. 76), finding that the uptake and utilisation of glucose and lactose in one of the psychrophiles was greatest at 0°C and decreased up to 15–20°C, whereas the psychrotrophs showed the converse with a maximum at approximately 20°C. The fatty acid and phospholipid compositions of most of the psychrophiles but none of the psychrotrophs altered with growth temperature; in addition, the phospholipid content of the former was maximal at zero and decreased up to 15°C. It was postulated that these changes reflected not only the larger size and volume of cells grown at 0°C but also their increased transport activity. It was concluded from a statistical analysis of the data from one psychrophile that temperature affected the activity of the uptake system; however, it was not established whether this was a direct effect or one mediated through lipid fluidity changes. The suggestion was made that psychrotrophs but not psychrophiles might be able to modify their fatty acid composition and thus extend the upper limit of growth through maintenance of solute uptake.

In the second study, Ellis-Evans and Wynn-Williams found that glucose affinity in a psychrotroph from a permanently cold lake environment was maximal at zero and fell at higher temperatures, whereas the reverse was true of a psychrotroph from a terrestrial environment undergoing regular thermal fluctuations [37].

In both of these studies no distinction was made between sugar uptake and its subsequent utilisation. In order to investigate whether there was a difference in the solute uptake capability of cold-adapted bacteria, Fukunaga and Russell investigated glucose uptake *per se* in two psychrotrophic bacteria isolated from Antarctic lake sediment; one isolate was regarded as being more 'psychrophilic' than the other in having optimum and maximum growth temperatures of 9.7 and 26°C, compared with 20.9 and 32.0°C for the other [77]. These authors found that although both isolates could grow at 0°C and adapted their fatty acid composition in response to temperature changes, the glucose uptake system in the one with more psychrophilic characteristics was better adapted to function at 0°C. There was no evidence that the collapse of transport ability was the reason for the upper growth temperature limit. More detailed studies are needed to resolve the question of whether solute uptake ability at low temperatures is a feature that truly distinguishes psychrophiles and mesophiles, i.e. a determinant of psychrophily [13].

7.4.2   *Proteins and protein synthesis*

Thermostability of proteins is invariably discussed in terms of their stability and activity at elevated temperatures. Mesophilic and especially thermophilic proteins are cold-labile, losing their activity at lower (psychrophilic) temperatures. Therefore, psychrophilic proteins must be adapted to maintain structural integrity and catalytic function at low temperatures. The paucity of information about their structure means that conclusions about psychrophilic proteins must be extrapolated from data on mesophilic and thermophilic proteins.

This chapter is not the appropriate place for an in-depth discussion of thermostability and the essentials only will be covered. Two experimental approaches have been used to investigate this problem. In the first approach, the native structure of a protein has been compared with that of natural, chemically-derived or site-directed mutants in which selected residues have been altered [78, 79]. In the second approach, the structure of proteins such as lactate dehydrogenase from mesophilic and thermophilic species of the same genus (chosen so as to minimise other genetic differences) have been compared [80]. The structures of the same protein, e.g. ferredoxin, having different thermostabilities in unrelated organisms have also been compared [81].

These experimental approaches have yielded essentially the same answer, namely that the stability of proteins is nearly always an intrinsic property of the polypeptide, being the balance between relatively large stabilising and destabilising forces with a net free energy of stabilisation that is of the order of only tens of kilojoules per mole of protein [78]. As a result, it requires relatively few changes in amino acid composition to alter thermostability significantly. The intrinsic thermostability of thermophilic proteins is generally brought about by a few additional hydrophobic bonds within the interior or extra hydrophilic bonds on the exterior of the protein. These extra bonds prevent the protein from adopting too loose a conformation at elevated temperatures; as the temperature falls, the protein 'tightens up' and activity is lost, thus explaining why the minimum growth temperatures of thermophiles are higher than those of mesophiles. One would predict, therefore, that psychrophilic proteins would have the opposite kind of adaptations in their structure. Since hydrophobic bonds are stronger at higher temperatures, whereas ionic interactions are stronger at lower temperatures, it seems more likely that hydrogen bonds or salt linkages will contribute to the stability of psychrophilic proteins. Recently, the gene for a psychrophilic lactate dehydrogenase has been isolated and cloned [82] and the initial results indicate that the supposition is indeed correct, i.e. one can predict that the flexibility of psychrophilic proteins at low temperatures will approximate that of thermophilic proteins at high temperatures. Just how psychrophilic proteins maintain their flexibility at or close to 0°C is not understood, but this is a key problem to be resolved if a molecular understanding of the basis of psychrophily is to be achieved.

Another way in which thermostability can be improved is by enhancing the area of (hydrophobic) surface contacts such as found in subunit enzymes [83]. This could be the explanation why the dimeric form of isocitrate dehydrogenase in the psychrophilic bacterium *Vibrio* sp. strain ABE-1 is more thermostable than the monomeric form [84]. The monomeric form of the enzyme appears to be highly cold-adapted in that it is inactivated above 15°C but is rapidly reactivated by chilling at 0°C. However, the presence of multiple subunit forms of the same enzyme is unusual in bacteria and thus cannot be considered a general property of psychrophiles.

Differences in the amino acid sequences and three-dimensional structures between psychrophilic and other proteins will be imprinted at the genetic level in the gene base sequences which have evolved over long periods of time and many cell generations. However, there are more immediate effects of temperature on cellular proteins, affecting either the total amount synthesised or the relative abundance of certain proteins by influencing, respectively, either the overall translation rate on the ribosomes or the synthesis, initiation or translation rate of specific mRNAs. Since psychrophiles can grow at 0°C, it is to be expected that their ribosomes would have special thermal qualities. Krajewska and Szer showed that *in vitro* protein synthesis in a psychrophilic pseudomonad had a very low miscoding rate compared with that of mesophiles and thermophiles tested at the same low temperatures [85]. It could be argued, however, that a better comparison would be the error frequencies at the respective optimum growth temperatures. That the ability to synthesise efficiently at 0°C was a property of the ribosomal proteins *per se* rather than accessory soluble factors was demonstrated by cross-mixing experiments with ribosomes and soluble factors from the psychrophile and *E. coli*. Ribosomes from the psychrophile were not abnormally heat-labile, but the Mg^{2+} requirements were different at 2 and 45°C, indicating that perhaps there were temperature-dependent changes in ribosome stability. Subsequently, Szer isolated a protein from ribosomes which conferred the ability to translate at 0°C [86]. A different result was found by Araki working with a psychrophilic *Vibrio* sp. strain ANT-300 [87]. In this bacterium the rates of protein synthesis *in vitro* at 0 and 13°C were regulated by soluble factors: the levels of 24 proteins varied with an alteration in temperature, but their individual functions have not been elucidated except that overall there appears to be a change in the activation energy of translation.

In mesophiles the slowing of protein synthesis at 0°C [88] appears to be due to a block in the formation of the initiation complex [89–91]. It should be noted that protein synthesis in some mesophilic bacteria can also be inhibited by low temperatures *in vivo*, not by direct thermal effects on the translation machinery but secondarily by leakage of amino acids through a damaged cell membrane [92]. In contrast, cold-adapted micro-organisms have membranes that are adapted so that they do not leak metabolites at low temperatures (see above).

In the psychrotroph *Micrococcus cryophilus*, the primary lesion responsible for the upper growth temperature limit is the presence of temperature-sensitive aminoacyl-tRNA synthetase enzymes for histidine, glutamic acid and proline; the active tetramers dissociate into inactive monomers above 25°C [93,94]. A temperature-resistant mutant of *M. cryophilus* contained glutamyl-tRNA synthetase which remained in the active tetrameric state at 30°C [95,96]. Similar thermolabile aminoacyl-tRNA synthetases have been identified in yeast [97]. Thus it appears that one of the adaptations in psychrophiles is in their translation machinery, specifically in its ability to maintain efficient initiation at low temperatures, something which mesophilic and thermophilic systems cannot do.

There are sporadic reports of increased protein content in cold-adapted micro-organisms grown near their lower cut-off temperatures, for instance in a psychrophilic pseudomonad [98] and a psychrophilic vibrio [76] grown at 0°C compared with their optimum growth temperatures of 14 and 13°C. However, apart from the study of Araki [87] mentioned above, there is no information on whether the protein composition of psychrophiles/psychrotrophs alters as zero temperature is approached, although there are some relevant data obtained using the mesophile *E. coli* reported by Neidhardt's group, which determined the levels of separate cellular proteins during growth at different temperatures by gel electrophoretic separation of radiolabelled cellular proteins [99]. They found that at growth temperatures near the upper or lower limits there were some differences in protein composition, which was otherwise constant over the central part of the growth range. Three proteins did show large changes in their levels with an increase in growth temperature from 13.5 to 46°C: the amount of two proteins decreased in a linear fashion by approximately nine- and 13-fold, whilst a third increased 2.5-fold. The levels of 10 other proteins also altered over the growth temperature range, but by less than 1.5-fold. At both extremes of growth temperature there were reduced levels of ribosomal proteins and other translational and transcriptional proteins; such changes are, however, characteristic of restricted growth in general and not specifically related to thermal limitation.

It should be pointed out that the proteins under discussion here are quite distinct from those of the heat-shock response, in which the levels of a specific set of proteins increase transiently following temperature up-shock [100]. Recently, heat-shock proteins have been discovered in psychrophilic and psychrotrophic bacteria [101, 102] and yeasts [103]. In the psychrophilic *Aquaspirillum* sp. Res-10 and the psychrotrophic *Bacillus psychrophilus* homologous sequences to the *dnaK* gene of the major heat-shock protein in *E. coli* were identified [102]. This homology is to be expected in view of the universal occurrence and close similarities between the heat-shock response in not only micro-organisms but higher organisms too [104].

A different set of proteins is induced by cold shock in *E. coli* [105]. The major protein, called CS7.4 after its molecular weight of 7.4 kDa, has been identified, and it has been hypothesised that in view of the large amounts made after a thermal shift-down and its small size (it contains only 70 amino acid residues) it could act as an antifreeze protein in protecting cells from damage due to ice formation [106]. The gene encoding the protein has been cloned and sequenced by Goldstein *et al.* [106]. A comparison of the derived amino acid sequence with those in the GenBank v62 database by Wistow [107] revealed that it has striking similarities with those of some human DNA-binding proteins, which would be relevant to the suggestion that the CS7.4 protein is autoregulatory in that it interacts with its own gene promoter [106]. This finding also raises the possibility that aspects of the cold-shock response may be conserved in all organisms in a similar fashion to the heat-shock response (see above).

In the natural environment, psychrotrophs in particular must be adapted to withstand large and often sudden falls in temperature. Thus it is no surprise that the cold-shock response has been demonstrated in the psychrotrophic yeast *Trichosporum pullulans* [108]. Up to 26 different cold-shock proteins were induced following thermal shift-down, the number and pattern depending on the magnitude of the temperature shift. The levels of cold-shock proteins were maximal 12 hours after the shift-down, which is longer than the time to reach peak levels of heat-shock proteins in the same organism (see refs. 103 and 108). The explanation is probably the slower rate of protein synthesis at the lower temperatures after a shift-down.

Gounot and co-workers have studied protein turnover in a psychrotrophic

Arthrobacter species isolated from an Arctic glacier where surface temperatures vary quite widely and the organism can grow from -5 to $+32°C$ [109]. Both the protein content and the rate of proteolysis *in vivo* are higher at the upper growth temperature limit compared with lower temperatures; when cultures were shifted up in temperature there was a rapid increase in protein synthesis, including heat-shock proteins, and in proteolysis [110–112]. It is speculated that these changes might be related to the thermal characteristics of growth by helping the bacterium to cope with sudden changes in growth temperature through the provision of more enzymes and other proteins; the decline in efficiency of the proteolytic system above 30°C could contribute to setting the upper growth limit by allowing defective proteins to persist inside the cell [112].

7.5 Biotechnological uses and potential of psychrophiles

The majority of industrial processes are operated at elevated temperatures, either because they are exothermic or by artificial heating, in order to speed the rate of reaction and thereby shorten the throughput time. Therefore, in the context of extremophiles, it is not surprising that emphasis has been placed on the search for thermophilic rather than psychrophilic enzymes and other products. Yet, most biotransformations occurring in the natural environment proceed at relatively low temperatures. The indigenous psychrophilic/psychro-trophic flora represents a potential source of enzymes that could function more efficiently than their mesophilic/thermophilic counterparts in biotechnological processes run at low temperatures. Although such a process might take longer, this disadvantage could be more than outweighed by the cost saving on energy requirements (or the necessity for expensive cooling facilities) in situations where there is less pressure on space and rapid turnover times. Psychrophilic processors could operate in the open, being less influenced by ambient temperatures, including overnight frosts. There would be no necessity for complex or expensive heating, cooling or recycling facilities, thus making such open-air psychrophilic reactors suitable for developing countries with limited financial and engineering resources. They are also appropriate to developed countries with the current emphasis on energy conservation.

Reichardt screened > 600 psychrophilic aerobic bacteria from Antarctic marine sediments for their cold-adapted enzymic activities towards proteins and polysaccharides [113]. He found that, although intracellular enzymes and physiological systems were cold-adapted, the catalytic activity of most extra-cellular enzymes was optimal at temperatures considerably above the optimum for growth [114]. However, their synthesis and secretion was cold-adapted and some hydrolase enzymes did have temperature optima as low as 25–28°C. Morita also reported that some hydrolytic enzymes from psychrophiles have activity optima at the relatively low temperatures of 20–25°C [6].

It would seem worthwhile to search for enzymes with requisite catalytic activity for specified biodegradative tasks to deal with, for example, industrial wastes. Kohring et al. observed the anaerobic biodegradation of 2,4-dichloro-phenol in freshwater lake sediments at temperatures between 5 and 50°C [115]. This compound is used in the production of wood preservatives and herbicides, and is highly toxic. Thus it serves as a model for many compounds which could be degraded in anaerobic digesters working at low temperatures.

Besides their use in biodegradation, psychrophiles also have potential in biosynthetic processes, including the production of foodstuffs and low-temperature fermentations. For example, cultivation of edible basidiomycete fungi, not only for their direct use as food but also for the conversion of industrial and agricultural lignocellulose waste into feed, has been highlighted as being suitable for helping the development of Third-World countries [116]. Some progress has been made in defining the genetic control of fruiting in 'low-temperature' strains of the oyster mushroom Pleurotus ostreatus, which have better organoleptic qualities than 'temperature-tolerant' strains [117]. Fungal mycelia have also been considered for the low-temperature production of medically and nutritionally valuable ω-3 fatty acids such as eicosapentaenoic acid (20:5 Δ 5, 5, 8, 11, 14, 17 all cis), one of the polyunsaturated fatty acids thought to help prevent atherosclerosis and thrombosis [118]. The synthesis of this acid by Mortierella spp. is switched on at 12°C [119]. The diatom Phaeodactylum tricornutum (formerly Nitzschia closterium) also produces large amounts of this acid with optimum production at about 22°C [120]. Lower cultural temperatures also enhance the eicosapentaenoic acid content of the alga Chlorella minutissima [121], reflecting a trend often seen for temperature effects on fatty acid composition (see section 7.4.1). These micro-organisms represent alternative sources of the acid, which conventionally is extracted from fish and fish oils; however, fish is unlikely to meet future demand and it is difficult to separate the eicosapentaenoic and docosahexaenoic acid from fish on a processing scale [122].

Recently, the potential for two rather unusual and unexpected bio-technological uses of some psychrotrophic phytopathogenic bacteria has become apparent. Strains of Pseudomonas fluorescens, together with some pathovars of Ps. syringae, Ps. viridiflava, Erwinia herbicola and Xanthomonas campestris are important plant pathogens, because they induce frost damage of sensitive crop plants. These bacteria trigger ice crystal formation on leaf and flower surfaces by producing ice nucleation proteins at subzero temperatures when the water would would otherwise remain supercooled [123, 124]. The bacterial proteins are related, containing internally repeated sequences of eight amino acids [125], and models for a number of structures have been put forward to explain how they might align water molecules, thereby initiating ice formation [124, 126].

The discovery of ice nucleation-active (Ice$^+$) bacteria provides the bio-technological potential for their use, such as in ice-cream manufacture or making synthetic snow, since ice crystal formation in the presence of Ice$^+$ bacteria can occur at -3 to $-5°C$ instead of the normal temperatures of -8 to $-10°C$ [127]. Another and potentially more beneficial biotechnological use of these bacteria arises from the finding that ice nucleation-deficient (Ice$^-$) strains, both naturally occurring ones and those in which the ice nucleation gene had been deleted, could antagonise the activity of Ice$^+$ strains on plant surfaces [128]. Controlled experiments in the field have confirmed the economic potential for crop protection by the deliberate introduction of Ice$^-$ bacteria [129].

A molecular genetic use of psychrophiles and psychrotrophs could be as cloning vehicles for the production of commercially or medically useful molecules. For instance, it has been pointed out that lower temperatures stabilise α_2-interferon against proteolysis and so might be a candidate for such a cloning strategy [130]. The disadvantage of a longer process time could be offset by the greater yield of a less degraded product of higher quality.

Compared with thermophiles there has been disappointingly little progress in the molecular genetics of psychrophiles. Recently, Kimura and Horikoshi reported the nucleotide sequence of an α-amylase gene from an alkalo-psychrotrophic species of *Micrococcus* [131]. At low temperatures and high pH this enzyme converts starch into maltotetrose, one of several malto-oligosaccharides that are used as food additives and clinical substrates. Psychrophilic enzymes that are also resistant to alkali should also have potential for use in cold-water washing processes.

References

1. S. Schmidt-Nielsen, *Zentralbl. Bakteriol. Parasitenkd. Infektionskr. Hyg.* **9** (1902) 145.
2. M. Muller, *Arch. Hygiene* **47** (1903) 127.
3. E. Ekelof, in *Wissenschaftliche Ergebnisse der Schwedischen Sudpolar-Expedition 1901–1903*, ed. O. Nordenskjold, Lithographic Institute Generalstabs, Stockholm (1908).
4. J.L. Ingraham and J.L. Stokes, *Bacteriol. Rev.* **23** (1959) 97.
5. J.L. Stokes, *Recent Prog. Microbiol.* **8** (1963) 187.
6. R.Y. Morita, *Bacteriol. Rev.* **39** (1975) 144.
7. F.C. Niedhardt, J.L. Ingraham and M. Schaechter, *Physiology of the Bacterial Cell. A Molecular Approach*, Sinauer Associates, Sunderland, USA, Chapter 8.
8. R.W. Hoham, *Arctic Alpine Res.* **7** (1975) 13.
9. R.W. Hoham, in *Phytoflagellates*, ed. E.R. Cox, Elsevier, New York (1980) 61.
10. H.U. Ling and R.D. Seppelt, *Antarctic Sci.* **2** (1990) 143.
11. J.L. Ingraham, *Cryobiology* **6** (1969) 188.
12. P. Mazur, *Orig. Life* **10** (1980) 137.
13. N.J. Russell, *Phil. Trans. Roy. Soc.* **B326** (1990) 595.
14. J. Wiegel, *FEMS Microbiol. Rev.* **75** (1990) 155.
15. T.D. Brock, in *Thermophiles: General, Molecular, and Applied Microbiology*, ed. T.D. Brock, Wiley, New York (1987) Chapter 1.
16. J.A. Baross and R.Y. Morita, in *Microbial Life in Extreme Environments*, ed. D.J. Kushner, Academic Press, London (1978) 9.
17. G.W. Gould and N.J. Russell, in *Food Preservatives*, eds. N.J. Russell and G.W. Gould, Blackie, Glasgow (1991) Chapter 1.

18. L. Ahonen and O.H. Tuovinen, *Appl. Environ. Microbiol.* **55** (1989) 312.
19. G.D. Ferroni, L.G. Leduc and M. Todd, *J. Gen. Appl. Microbiol.* **32** (1986) 169.
20. A.C. Tanner and R.A. Herbert, *Kiel Meeresforsch. Sonderh.* **5** (1981) 390.
21. R.M. Atlas and R.Y. Morita, in *Perspectives in Microbial Ecology*, eds. F. Megusar and M. Gantar, Slovene Society for Microbiology, Ljubljana (1986) 185.
22. D.D. Wynn-Williams, in *Advances in Microbial Ecology*, Vol. 11, ed. K.C. Marshall, Plenum Press, New York (1990) 71.
23. R.E. Cameron, F.A. Morelli and R.M. Johnson, *Ant. J. United States* **7** (1972) 187.
24. K.L. Miller, S.B. Leschine and R.L. Huguenin, *Ant. J. United States* **18** (1983) 222.
25. S. Grossi, S.T. Kottmeier and C.W. Sullivan, *Microb. Ecol.* **10** (1984) 231.
26. W. Vincent and C. Howard-Williams, in *Perspectives in Microbial Ecology*, eds. F. Megusar and M. Gantar, Slovene Society for Microbiology, Ljubljana (1986) 201.
27. N.J. Russell and G.W. Gould, in *Food Preservatives*, eds. N.J. Russell and G.W. Gould, Blackie, Glasgow (1991) Chapter 2.
28. P.O. Wilkins, R. Bourgeois and R.G.E. Murray, *Can. J. Microbiol.* **18** (1972) 543.
29. K.A. Glass and M.P. Doyle, *Appl. Environ. Microbiol.* **55** (1989) 1565.
30. J.G. Bradshaw, J.T. Peeler and R.M. Twedt, *J. Food Prot.* **54** (1991) 12.
31. D.N. Kamau, S. Doores and K.M. Pruitt, *Appl. Environ. Microbiol.* **56** (1990) 2711.
32. L.R. Beuchat, in *Psychrotrophic Microorganisms in Spoilage and Pathogenicity*, eds. T.A. Roberts, G. Hobbs, J.H.B. Christian and N. Skovgaard, Academic Press, London (1981) Chapter 30.
33. G. Reuter, in *Psychrotrophic Microorganisms in Spoilage and Pathogenicity*, eds. T.A. Roberts, G. Hobbs, J.H.B. Christian and N. Skovgaard, Academic Press, London (1981) Chapter 23.
34. J.M. Shewan and C.K. Murray, in *Cold Tolerant Microbes in Spoilage and the Environment, SAB Tech. Ser. No. 13*, eds. A.D. Russell and R. Fuller, Academic Press, London (1979) 117.
35. B.A. Law, C.M. Cousins, M.E. Sharpe and F.L. Davies, in *Cold Tolerant Microbes in Spoilage and the Environment, SAB Tech. Ser. No. 13*, eds. A.D. Russell and R. Fuller, Academic Press, London (1979) 137.
36. M.A. Cousin, in *Psychrotrophic Microorganisms in Spoilage and Pathogenicity*, eds. T.A. Roberts, G. Hobbs, J.H.B. Christian and N. Skovgaard, Academic Press, London (1981) Chapter 6.
37. J.C. Ellis-Evans and D.D. Wynn-Williams, in *Antarctic Nutrient Cycles and Food Webs*, eds. W.R. Siegfried, P.R. Condy and R.M. Laws, Springer, Berlin (1985) 662.
38. J.C. Ellis-Evans, *Biologist* **32** (1985) 171.
39. H.E. Welch and M.A. Bergmann, *Can. J. Fish. Aquat. Sci.* **46** (1989) 1793.
40. W.J. Wiebe and C.W. Hendricks, in *Effects of the Ocean Environment on Microbial Activity*, eds. R.R. Colwell and R.Y. Morita, University Park Press, Baltimore (1974) 524.
41. H-J. Ruger, *Mar. Ecol. Prog. Ser.* **57** (1989) 45.
42. A.C. Upton and D.B. Nedwell, in *University Research in Antarctica*, ed. R.B. Heywood, British Antarctic Survey, Cambridge, England (1989) 97.
43. D. Delille and E. Perret, *Microb. Ecol.* **18** (1989) 117.
44. K.J. Nadelhoffer, A.E. Giblin, G.R. Shaver and J.A. Laundre, *Ecology* **72** (1991) 242.
45. W. Vincent and C. Howard-Williams, *Freshwater Biol.* **16** (1986) 219.
46. M.C. Davey, *Polar Biol.* **10** (1989) 29.
47. H.S. Vishniac and J. Klinger, in *Perspectives in Microbial Ecology*, eds. F. Megusar and M. Gantar, Slovene Society for Microbiology, Ljubljana (1986) 46.
48. H.S. Vishniac, *Mycologia* **77** (1985) 149.
49. E.I Friedmann, *Science* **215** (1982) 1045.
50. E.I. Friedmann, M. Hua and R. Ocampo-Friedmann, *Polarforschung* **58** (1988) 251.
51. J.A. Nienow, C.P. McKay and E.I. Friedmann, *Microb. Ecol.* **16** (1988) 253.
52. E.I. Friedmann, C.P. McKay and J.A. Nienow, *Polar Biol.* **7** (1987) 273.
53. M. Sinensky, *Proc. Natl. Acad. Sci. USA* **71** (1974) 522.
54. J.E. Cronan Jr and E.P. Gelmann, *J. Biol. Chem.* **248** (1973) 1188.
55. L. McGibbon, A.R. Cossins, P.J. Quinn and N.J. Russell, *Biochim. Biophys. Acta* **820** (1984) 115.
56. L. Finegold, M.A. Singer, T.W. Federle and J.R. Vestal, *Appl. Environ. Microbiol.* **56** (1990) 1191.
57. R.N. McElhaney and K.A. Souza, *Biochim. Biophys. Acta* **443** (1976) 348.
58. J. Reizer, N. Grossowicz and Y. Barenholz, *Biochim. Biophys. Acta* **815** (1985) 268.

59. J. Cullen, M.C. Phillips and G.G. Shipley, *Biochem. J.* **125** (1971) 733.
60. M. Foot, R. Jeffcoat, M.D. Barratt and N.J. Russell, *Arch. Biochem. Biophys.* **224** (1983) 718.
61. M.B. Jackson and J.E. Cronan, Jr, *Biochim. Biophys. Acta* **512** (1978) 472.
62. D.L. Melchior, *Curr. Top. Membr. Transp.* **17** (1982) 263.
63. J.R. Silvius, N. Mak and R.N. McElhaney, in *Membrane Fluidity: Biophysical Techniques and Cellular Regulation*, eds. M. Kates and A. Kuksis, Humana Press, New Jersey (1980) 213.
64. R.N. McElhaney, *Curr. Top. Membr. Transp.* **17** (1982) 317.
65. H. Wada, Z. Gombos and N. Murata, *Nature* **347** (1990) 200.
66. N.J. Russell, in *Microbial Lipids*, Vol. 2, eds. C. Ratledge and S.C. Wilkinson, Academic Press, London (1989) Chapter 5.
67. N.J. Russell and N. Fukunaga, *FEMS Microbiol. Rev.* **75** (1990) 171.
68. M. Chan, R.H. Himes and J.M. Agaki, *J. Bacteriol.* **106** (1971) 876.
69. N.J. Russell, *Trends Biochem. Sci.* **9** (1984) 108.
70. J.L. Harwood and N.J. Russell, *Lipids in Plants and Microbes*, George Allen & Unwin, London (1984).
71. M. Wada, N. Fukunaga and S. Sasaki, *J. Bacteriol.* **171** (1989) 4267.
72. T. Kaneda, *Bacteriol. Rev.* **41** (1977) 391.
73. A.J. Fulco and D.K. Fujii, in *Membrane Fluidity: Biophysical Techniques and Cellular Regulation*, eds. M. Kates and A. Kuksis, Humana Press, New Jersey (1980) 77.
74. H. Morton, K. Watson and M. Streamer, *FEMS Microbiol. Lett.* **4** (1978) 291.
75. C.A.J. Thorne and K. Watson, *FEMS Microbiol. Lett.* **10** (1981) 137.
76. R.A. Herbert and M. Bhakoo, in *Psychrotrophic Microorganisms in Spoilage and Pathogenicity*, ed. T.A. Roberts, Academic Press, London (1959) 1.
77. N. Fukunaga and N.J. Russell, *J. Gen. Microbiol.* **136** (1990) 1669.
78. R.M. Daniel, in *Protein Structure, Folding, and Design*, ed. D.L. Oxender, Alan Liss, New York (1986) 291.
79. B.W. Matthews, *Biochemistry* **26** (1987) 6885.
80. H. Zuber, *Biophys. Chem.* **29** (1988) 171.
81. M.F. Perutz and H. Raidt, *Nature* **255** (1975) 259.
82. D. Schlatter, O. Kriech, F. Suter and H. Zuber, *Biol. Chem. Hoppe-Seyler* **368** (1987) 1435.
83. E. Stellwagen and H. Wilgus, *Nature* **275** (1978) 342.
84. T. Ochiai, N. Fukunaga and S. Sasaki, *J. Gen. Appl. Microbiol.* **30** (1984) 479.
85. E. Krajewska and W. Szer, *Eur. J. Biochem.* **2** (1967) 250.
86. W. Szer, *Biochim. Biophys. Acta* **213** (1970) 159.
87. T. Araki, *J. Gen. Microbiol.* **137** (1991) 817.
88. H.K. Das and A. Goldstein, *J. Mol. Biol.* **31** (1968) 209.
89. H. Friedman, P. Lu and A. Rich, *Nature* **223** (1969) 909.
90. H. Friedman, P. Lu and A. Rich, *J. Mol. Biol.* **61** (1971) 105.
91. R.J. Broeze, C.J. Solomon and D.H. Pope, *J. Bacteriol.* **134** (1978) 861.
92. H. Sarayama, N. Fukunaga and S. Sasaki, *Plant Cell Physiol.* **26** (1985) 1345.
93. N.L. Malcolm, *Biochim. Biophys. Acta* **157** (1968) 493.
94. N.L. Malcolm, *Nature* **221** (1969) 1031.
95. N.L. Malcolm, *Biochim. Biophys. Acta* **190** (1969) 337.
96. N.L. Malcolm, *Biochim. Biophys. Acta* **190** (1969) 347.
97. C.H. Nash, D.W. Grant and W.A. Sinclair, *Can. J. Microbiol.* **15** (1969) 339.
98. W. Harder and H. Veldkamp, *Archiv. Mikrobiol.* **59** (1967) 123.
99. S.L. Herendeen, R.A. VanBogelen and F.C. Neidhardt, *J. Bacteriol.* **139** (1979) 185.
100. S. Linquist and E.A. Craig, *Ann. Rev. Genet.* **22** (1988) 631.
101. K.L. McCallum, J.J. Keikkula and W.E. Inniss, *Can. J. Microbiol.* **32** (1986) 516.
102. K.L. McCallum, B.J. Butler and W.E. Inniss, *Arch. Microbiol.* **152** (1989) 148.
103. G.R. Berg, W.E. Inniss and J.J. Keikkula, *Can. J. Microbiol.* **33** (1987) 383.
104. M.J. Schlesinger, M. Ashburner and A. Tissières (eds.) *Heat Shock: From Bacteria to Man*, Cold Spring Harbor, New York (1982).
105. P.G. Jones, R.A. VanBogelen and F.C. Neidhardt, *J. Bacteriol.* **169** (1987) 2092.
106. J. Goldstein, N.S. Pollitt and M. Inouye, *Proc. Natl. Acad. Sci. USA* **87** (1990) 283.
107. G. Wistow, *Nature* **344** (1990) 823.
108. C.R. Julseth and W.E. Inniss, *Can. J. Microbiol.* **36** (1990) 519.
109. A.M. Gounot, *Can. J. Microbiol.* **22** (1976) 357.
110. P. Potier, A.R. Hipkiss and D.J. Kushner, *Arch. Microbiol.* **142** (1985) 28.

111. P. Potier, P. Drevet, A.M. Gounot and A.R. Hipkiss, *FEMS Microbiol. Lett.* **44** (1987) 267.
112. P. Potier, P. Drevet, A-M. Gounot and A.R. Hipkiss, *J. Gen. Microbiol.* **136** (1990) 283.
113. W. Reichardt, in *FEMS Symposium No. 49, Microbiology of Extreme Environments and Its Potential for Biotechnology,* eds. M.S. Da Costa, J.C. Duarte and R.A.D. Williams, Elsevier, London and New York (1989) 405.
114. W. Reichardt, *Mar. Ecol. Prog. Ser.* **40** (1987) 127.
115. G-W. Kohring, J.E. Rogers and J. Wiegel, *Appl. Environ. Microbiol.* **55** (1989) 348.
116. National Research Council (1979) *Microbial Processes: Promising Technologies for Developing Countries,* National Academy of Sciences, Washington DC.
117 S.F. Li and G. Eger-Hummel, in *Advances in Biotechnology,* Vol. 1, ed. M. Moo-Young, Pergamon Press, Toronto (1981) 101.
118. J. Dyerberg, *Nutr. Rev.* **44** (1986) 125.
119. S. Shimizu, Y. Shinmen, H. Kawashima, K. Akimoto and H. Yamada, *Biochem. Biophys. Res. Commun.* **150** (1988) 335.
120. W. Yongmanitchai and O.P. Ward, *Appl. Environ. Microbiol.* **57** (1991) 419.
121. A. Seto, H.L. Wong and C.W. Hesseltine, *J. Am. Oil Chem. Soc.* **61** (1984) 892.
122. T. Fujita and M. Makuta, March 1983, US Patent 4,377,526.
123. S.E. Lindow, *Annu. Rev. Phytopathol.* **21** (1983) 363.
124. P. Wolber and G. Warren, *Trends Biochem. Sci.* **14** (1989) 179.
125. R.L. Green and G.J. Warren, *Nature* **317** (1985) 645.
126. M.A. Turner, F. Arellano and L.M. Kozloff, *J. Bacteriol.* **172** (1990) 2521.
127. L.R. Maki, E.L. Galyan, M. Chang-Chien and D.R. Caldwell, *Appl. Microbiol.* **28** (1974) 456.
128. S.E. Lindow, *Plant Dis.* **67** (1983) 327.
129. T. Suslow, *Trends Biochem. Sci.* **14** (1989) 180.
130. J.A. Chessyre and A.R. Hipkiss, *Appl. Microbiol. Biotechnol.* **31** (1989) 158.
131. T. Kimura and K. Horikoshi, *FEMS Microbiol. Lett.* **71** (1990) 35.

8 Molecular biology and biotechnology of microbial interactions with organic and inorganic heavy metal compounds

G.M. GADD

8.1 Introduction

A general feature of many heavy metals and their compounds, inorganic and organic, is toxicity towards microbial and other life forms [1]. However, many metal ions are essential at low concentrations for normal growth and metabolism, e.g. copper, zinc, iron, manganese and cobalt, and micro-organisms utilise mechanisms of varying specificity for their accumulation from the environment [2]. In contrast, many other metals have no essential biological functions, e.g. lead, aluminium, cadmium, tin, and mercury, but may still be accumulated [3]. An organometallic compound may simply be defined as a compound containing at least one metal–carbon bond, and these often exhibit enhanced microbial toxicity, hence their frequent exploitation as biocides. Where these compounds contain 'metalloid' elements such as germanium, arsenic, selenium or tellurium, which are not considered to be true metals by chemists, the term 'organometalloid' is often used [4].

Micro-organisms have encountered toxic metals in the environment throughout their evolutionary history which, for bacteria, may have begun around $3-4 \times 10^9$ years ago [5]. Microbial mechanisms of metal resistance have therefore had considerable time to evolve, and there is now increasing interest in the molecular and genetic basis of such attributes, in pro- and eukaryotic micro-organisms. Although average abundances of metals are generally low in the environment, and much is biologically unavailable because of sequestration in sediments, soils and minerals [6], elevated metal levels may occur naturally, albeit in rather specific locations, e.g. deep-sea vents, hot springs and volcanic soils [7]. However, it is now mainly as a result of anthropogenic activities that ecosystems are increasingly subject to pollution by heavy metals, organometal(loid)s and radionuclides [8]. Almost every industrial activity may lead to altered and/or increased metal distribution in natural habitats with fossil fuel combustion, mining and ore processing, nuclear waste, sewage treatment effluents and sludge, brewery and distillery wastes, fungicides, biocides and disinfectants being major offenders, as well as industrial effluents and discharges in a general context [6, 8].

Major mechanisms of metal toxicity are generally a consequence of the strong coordinating abilities of metal ions. Because of the wide variety of ligands found in living cells, and the range of potentially harmful interactions that may occur, almost every index of microbial activity may be affected under conditions of metal pollution [8]. These include primary productivity, nitrogen fixation, biogeochemical cycling of carbon, nitrogen, sulphur, phosphorus and other elements, litter decomposition, enzyme synthesis and activity in aquatic and terrestrial habitats [9]. Organometallic compounds can arise in the environment by natural processes, including those mediated by the biota, or by accidental or deliberate introduction, e.g. effluents and biocides [4, 10]. Because of the fundamental importance of micro-organisms in the environment, including plant productivity and symbiotic associations, pollution by heavy metals, organometal(loid)s and radionuclides can have serious immediate and long-term effects, ultimately posing a threat to humans by accumulation and transfer through food chains [1, 9].

In contrast to well-defined 'extreme' environments which may be characterised by a relatively restricted microflora, a range of micro-organisms from all major groups may be found in metal-polluted habitats, and the ability to survive and grow in the presence of potentially toxic concentrations is a common attribute [8]. It must be stressed that in a polluted environment, other adverse components may affect the microflora, e.g. extremes of pH, nutrient limitation and high salinity. More generally, the physicochemical characteristics of a given habitat determine metal speciation and biological availability. When factors such as pH, E_h, other anions and cations, particulate and soluble organic matter, clay minerals and salinity decrease the biological availability of metals, toxicity may also be reduced or even eliminated [1, 6]. Thus, the analysis of natural microbial population responses towards heavy metals is highly complicated and subject to misinterpretation [8]. Despite this, it is evident that elevated concentrations of heavy metals, and related compounds, can affect the qualitative and quantitative composition of microbial populations in the environment, while an extensive array of direct and indirect responses have been documented in laboratory studies [8, 10–12]. This chapter highlights the molecular basis of some of the best-characterised resistance mechanisms encountered in micro-organisms, and further explores possible ways by which metal-microbe interactions can be utilised in a biotechnological context.

8.2   Physiology of metal–microbe interactions

Microbial survival in metal-contaminated habitats may depend on genetic and/or physiological adaptation and, in a given population, both may occur to varying extents [8]. Terms such as 'tolerance' and 'resistance' are arbitrary and may be based on criteria such as the ability to grow on a certain metal concentration in the laboratory [13]. It is probably more appropriate to use 'resistance' to describe a direct response resulting from metal exposure, although this is not an easy distinction to make because of the multiplicity of interactions that metals can have with cellular components and the significance of environmental factors in modifying metal toxicity [1, 8, 14, 15]. The ability to grow in the presence of high metal concentrations is found in a range of micro-organisms, including those from 'unpolluted' sites, and not always is adaptation necessary. Tolerance may result from intrinsic properties of the organism, e.g. possession of an impermeable cell wall, extracellular mucilage or polysaccharide or lack of a transport system [1, 8, 12]. In other cases, environmental factors may dramatically determine toxicity, and these include pH, E_h, aeration, inorganic anions and cations, particulate and soluble organic matter, clay minerals and salinity [1, 9, 16]. Where such factors decrease biological availability of metal ions, toxicity may be reduced or even eliminated. Thus, considerable difficulties may be encountered in practical analysis of microbial metal responses, particularly in relation to natural environments. Laboratory studies under defined conditions have allowed detailed investigation of the mechanisms involved in metal resistance, but these can only be extrapolated to the field situation with caution.

Extracellular complexation, precipitation and crystallisation of heavy metal species exterior to cells can result in detoxification. Externally produced polysaccharides, organic acids, proteins, pigments and other metabolites can bind and/or convert metal ions to innocuous forms [3, 17–19], while iron-chelating siderophores may chelate other metals and conceivably confer protection from metal toxicity (see later sections). Microbial sulphide production leads to heavy metal removal from solution, and insoluble metal sulphide granules may be deposited in and around cell surfaces [2, 20, 21]. Sulphide production is also important in biological mercury cycling [2, 10]. Reduction of sulphate, for example by *Desulfovibrio* under anaerobic conditions, can lead to disproportionation of organomercury and organolead compounds by hydrogen sulphide to more volatile products and insoluble sulphides, e.g.

$$2\,CH_3Hg^+ + H_2S \rightarrow (CH_3)_2Hg + HgS$$

$$2(CH_3)_3Pb^+ + H_2S \rightarrow (CH_3)_4Pb \rightarrow (CH_3)_2PbS$$

Many other examples of metal crystallisation and precipitation on microbial surfaces are known, some of great significance in biogeochemical cycling, e.g. microfossil formation and iron and manganese deposition [7, 22–25] and silver and uranium mineralisation [26–28].

General 'biosorption' of metal species to cell walls can reduce metal ion concentrations in solution, but this is largely outside metabolic control and dependent on factors such as cell wall composition, metal and biomass concentration, and abiotic factors [29]. Biosorption may be significant in certain survival structures, e.g. melanised fungal chlamydospores which exhibit higher tolerance than hyaline cell types to inorganic and organic metal compounds, the wall acting as a permeability barrier [30–32].

Since toxicity may result when metal species interact with intracellular components, it is not surprising that decreased transport, impermeability or the occurrence of metal efflux systems constitute resistance mechanisms in many organisms [5, 8, 33]. Plasmid-mediated bacterial resistances to cadmium and arsenic, which can involve efflux, are dealt with later. In yeasts, decreased influx of cadmium, copper, cobalt and zinc may occur in resistant strains [32, 34] while some data have demonstrated the existence of efflux systems for toxic monovalent cations, e.g. Li^+, as well as Sr^{2+} and Mn^{2+} [35]. Some resistant strains take up more metal than sensitive strains because of the possession of additional or alternative resistance mechanisms such as more efficient internal detoxification [8]. Impermeability may be a result of general biosorption, as mentioned previously, but can also result from changes in membrane proteins [13, 33] or the accumulation, through synthesis, of high intracellular concentrations of glycerol [36]. Those abiotic factors which can reduce metal transport into cells, including ion competition and pH, may also enable microbial survival [1, 6, 8].

Once inside cells, metal ions may be compartmentalised and/or converted to less toxic forms, and organisms expressing such mechanisms may be able to accumulate metals to high internal concentrations. A variety of electron-dense bodies, including polyphosphate, have been implicated in metal sequestration in bacteria, cyanobacteria, algae and fungi [37–41]. In eukaryotes, e.g. algae, yeasts and fungi, there may be compartmentation of metals within certain organelles. In yeasts, a majority of accumulated Co^{2+}, Zn^{2+}, Mn^{2+} and Mg^{2+} is located in the vacuole, where it may be present in ionic form and/or bound to low molecular weight polyphosphates [34, 35, 42, 43]. An alternative response, particularly for metals like copper and cadmium, is the synthesis of metal-binding proteins, and these have been recorded in all microbial groups examined [3, 38, 39, 44–48]. Metal-binding proteins of yeasts (dealt with later) have been the subject of intensive research at the molecular and genetic level.

Micro-organisms may affect chemical transformations of heavy metals and their compounds, e.g. oxidation, reduction, methylation and demethylation, which are relevant to biogeochemical cycling and may also constitute resistance mechanisms. Plasmid-mediated mercury resistance in bacteria by means of reduction of Hg^{2+} to Hg^o is dealt with subsequently. As well as this, a variety of bacteria, algae, yeasts and fungi have been shown to be able to reduce Au^{3+} to elemental Au^o [49] and also Ag^+ to metallic Ag^o which may result in silver deposition on glass surfaces and in and around growing colonies [50, 51]. Methylation of mercury can be carried out by a variety of bacteria, fungi and yeasts as well as by abiotic mechanisms [52]. This may be viewed as a detoxification mechanism for mercury, and other metal(loid)s, since methylated species are usually more volatile and may be lost from the environment [2, 52]. There is general agreement that Hg^{2+} methylation, by direct and indirect microbial action, involves methylcobalamin (vitamin B_{12}; CH_3CoB_{12})[2, 4, 10, 53, 54]. Two products can form, CH_3Hg^+ (methylmercury) and $(CH_3)_2Hg$ (dimethylmercury):

$$CH_3CoB_{12} + Hg^{2+} \xrightarrow{H_2O} CH_3Hg^+ + H_2OCoB_{12}^+$$

$$CH_3CoB_{12} + CH_3Hg^+ \xrightarrow{H_2O} (CH_3)_2Hg + H_2OCoB_{12}^+$$

The latter step is usually slower than the first so methylmercury usually predominates.

Methylcobalamin may also be involved in biomethylation of lead, tin, palladium, platinum, gold and thallium, whereas another biological methylating agent, S-adenosylmethionine (SAM), is involved in biomethylation of arsenic and selenium [4, 10]. Other biomolecules, e.g. iodomethane (methyl iodide, CH_3I), betaine and humic acids, may also oxidise and/or methylate metal species [4, 10, 54]. It should be noted that considerable amounts of methylated metal compounds are introduced into aquatic and terrestrial ecosystems as a result of anthropogenic activities, some of which can act as methylating agents, e.g. $(CH_3)_3 S^+I^-$ and $(CH_3)_3Sn^+$ [10, 55].

Arsenic biomethylation, by transfer of carbonium ions (CH_3^+) from SAM, has been demonstrated in bacteria, fungi and yeasts, and evidence indicates that arsenic biomethylation is a significant process in the environment [4, 53]. It should be noted, however, that plasmid-mediated arsenate resistance in bacteria depends on efflux and not methylation [56]. Selenium biomethylation appears to be similar to that of arsenic, and the production of dimethylselenide, dimethyldiselenide and dimethylselenone from soils and sediment samples amended with selenium compounds has been demonstrated [54]. Other methyl-accumulating reactions for metals such as lead, thallium, palladium, platinum, gold, chromium, germanium and antimony have been described, though the relative roles of biotic/abiotic action are often unclear, while results for other metals, e.g. cadmium, silver and copper are based on little evidence [57]. Other bacterial metal-detoxifying enzymes of importance include arsenite oxidase catalysing $As^{3+} \rightarrow As^{5+}$ and chromate reductase catalysing $Cr^{6+} \rightarrow Cr^{3+}$ [58–61].

Organomercurials may be detoxified by organomercurial lyase (see later sections), while organotin degradation involves sequential removal of organic groups from the tin atom:

$$R_4Sn \rightarrow R_3SnX \rightarrow R_2SnX_2 \rightarrow RSnX_3 \rightarrow SnX_4$$

This generally results in a reduction of toxicity [62, 63]. It seems to be well established that micro-organisms are capable of organotin degradation, including bacteria, algae and fungi [55]. However, there are few pure culture studies, and a dearth of information still exists relating to biotic organotin degradation, including the relationship between resistance and degradative ability, the influence of the anionic radical on breakdown, and the relative importance of biotic and abiotic degradation in natural habitats [55, 62, 63]. Abiotic degradation is an important consideration for organometal(loid) compounds. Organotins and organoleads may be subject to photolytic decay whereas Hg(II) salts can demethylate other compounds, e.g. $(CH_3)_3SnCl$ [4]. Redistribution reactions with sulphide have been discussed previously.

8.3 Molecular biology of heavy metal tolerance

Only a few metal–microbe interactions have received detailed attention at the molecular and genetic level. These include bacterial iron transport and assimilation (outside the scope of this review but see refs. 64–66), plasmid-mediated bacterial heavy metal resistances (mainly mercury, cadmium and arsenic) and the synthesis and regulation of metal-binding proteins (metallo-thioneins and metal γ-glutamyl peptides) in yeasts. It is these two subject areas that will receive prominence here. It should be pointed out that other genetic studies do exist, though these are often confined to mutant isolation and subsequent physiological studies [67–69]. Furthermore, many organisms, particularly those of environmental significance, are not as amenable to molecular and genetical analysis as, for example, *Escherichia coli* or *Saccharomyces cerevisiae*, while knowledge of the basic physiology and biochemistry of metal responses is limited in many cases, which clearly hampers progress at the molecular level.

8.3.1 *Plasmid-mediated bacterial heavy metal resistance*

Many bacterial mechanisms of resistance to heavy metals and metalloids are encoded by plasmids and not by chromosomal genes [5, 56, 70, 71]. The range of metals and metal species for which bacterial resistance has been documented includes Hg^{2+} (and organomercurials), Ag^+, AsO_2^-, AsO_4^{3-}, Cd^{2+}, Co^{2+}, CrO_4^{2-}, Cu^{2+}, Ni^{2+}, TeO_3^{2-}, Tl^+ and Zn^{2+} though not all of these have received detailed genetical attention. Most detailed information available is concerned with mercury, cadmium and arsenic species, and readers are directed to the wealth of excellent reviews dealing with the subject area [5, 56, 58, 70, 72–84].

8.3.1.1 *Mercury resistance.* Mercury resistance is a common property of many Gram-positive and Gram-negative bacteria isolated from the environment or clinical sources [56, 77, 82, 84, 85]. The best-characterised mechanism of mercury resistance is the enzymatic reduction of Hg^{2+} by cytoplasmic mercuric reductase to elemental Hg^o, which is less toxic than Hg^{2+}, volatile and rapidly lost from the immediate environment of resistant cells [5, 58, 76, 77]. Mercury-resistant bacteria differ in their sensitivity to organomercurials, with 'narrow-spectrum' resistance being limited to a few compounds, e.g. merbromin and fluorescein mercuric acetate; 'broad-spectrum' resistance is exhibited to a wider range of compounds [77]. In broad-spectrum resistance, organomercurials are enzymatically detoxified by an organo-mercurial lyase which cleaves the mercury–carbon bond of, for example, methyl, ethyl and phenyl mercury, to form Hg^{2+} and methane, ethane and benzene respectively [77, 86, 87]. The resultant Hg^{2+} may then be reduced by mercuric reductase. Narrow-spectrum resistance to organomercurials may be a result of permeability barriers [77]. In most cases, mercury and organomercurial resistance is inducible, with constitutive resistance being demonstrated in only a few organisms [77].

The genes encoding mercury resistance are primarily extrachromosomal [5, 70, 77], and two narrow-spectrum mercury resistance determinants from Gram-negative bacteria have provided most information to date. These are the determinants of the multiple drug resistance plasmid R100, originally isolated in *Shigella* [88] and the determinant of transposon Tn*501*, originally isolated on plasmid pVS1 from *Pseudomonas aeruginosa* [73].

Genetic analysis and determination of nucleotide sequence have enabled identification of the mercury genes of Tn*501* and R100 [56, 77, 89–91]. The *merR* gene ecodes a *trans*-acting regulator responsible for the induction of expression of the *mer* operon. The operon contains the structural genes for mercuric reductase (*merA*) and mercury transport (*merT* and *merP*) as well as *merD* of unknown function. In the R100 determinant, an additional gene, *merC*, may be involved in mercury uptake [56, 76, 77].

There is close homology between the predicted structures of the gene products of Tn*501* and R100, and a model for the mechanism of mercury resistance has been described [56, 76, 77]. FAD-containing mercuric reductase is a flavoprotein and structurally and mechanistically related to glutathione reductase and lipoamide dehydrogenase [72, 92]. NADPH is the preferred reducing co-factor for mercuric reductase (and organomercurial lyase) and the enzyme requires an excess of thiols for activity. The thiols prevent formation of $NADPH-Hg^{2+}$ complexes and ensure that Hg^{2+} is present as a dimercaptide. The reaction can be represented:

$$RS-Hg-SR + NADPH + H^+ \rightarrow Hg^\circ + NADP + 2RSH$$

The postulated mechanism of action is that electrons are transferred from NADPH via FAD to reduce the active-site cystine, converting it to two cysteine residues with titratable SH groups. One Cys residue forms a charge transfer complex with FAD. The active-site cysteines then reduce Hg^{2+}, bound to the C-terminal cysteines, to Hg° [56, 58, 72, 76]. It seems likely that mercuric reductase has a dimeric structure *in vivo* with a subunit molecular weight of between 54 000 and 69 000 depending on the source [58]. In the model of mercury resistance in Gram-negative bacteria, the *merP* gene product scavenges the periplasm for Hg^{2+}, which binds to a cysteine pair, releasing $2H^+$. The Hg^{2+} is then transferred to the outermost cysteine pair of the $merT$ gene product, also with $2H^+$ release. The Hg^{2+} is then transferred to the innermost cysteine pair on the $merT$ gene product. Mercuric reductase may transiently associate with the $merT$ gene product in the inner membrane. After Hg^{2+} reduction, elemental $Hg°$ passes out through the cell envelope by simple diffusion, and since it is volatile will be rapidly lost to the external environment [77] (Figure 8.1). It is possible that additional gene(s) of the mer operon, for which no function have been ascribed, may facilitate $Hg°$ loss from cells [77]. It is presumed that organomercurial compounds cross the cytoplasmic membrane via the $merP$, $merT$ system in a similar manner to Hg^{2+} [73]. Once the organomercurial is transferred into the cytoplasm, it may be cleaved by the organomercurial lyase ($merB$ gene product) prior to Hg^{2+} reduction to $Hg°$ [77].

The genetic organisation of the determinants of broad-spectrum mercury resistance is less well described than those of narrow-spectrum resistance [77]. The broad-spectrum mer operon from the *Staphylococcus aureus* plasmid pI258 has been sequenced. Two genes ($merA$ and $merB$) were identified by their homology with equivalent genes in Gram-negative systems [56]. Work on other Gram-positive bacteria is limited, though there are now data pertaining to DNA sequences for a *Bacillus* sp. in which the mercury resistance determinant is chromosomal and not plasmid-located [93] and for *Streptomyces lividans* [5, 56].

Plasmid-encoded bacterial resistance to Hg^{2+} and other metals is of considerable environmental significance [85]. It is now known that in the

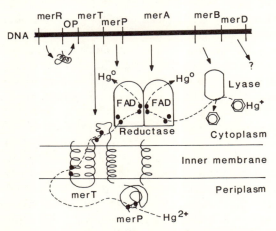

Figure 8.1 Model of the genetic determination of Hg^{2+} detoxification in Gram-negative bacteria. The top line shows the gene order on the DNA. The rest of the diagram shows the postulated locations of the gene products relative to the inner membrane of the bacterial cell. Solid circles represent the paired cysteine residues that bind Hg^{2+}. The mercuric reductase protein is represented as a dimer. Adapted from refs. 5 and 76 with permission.

aquatic environment there is widespread plasmid transfer and genetic rearrangement in Hg^{2+}-resistant strains [94], and experiments with a *Pseudomonas putida* recipient have shown that intact epilithic bacterial communities could transfer mercury resistance plasmids *in situ* at frequencies of up to 3.75×10^{-6} per recipient [95]. It has also been shown that even metal resistance genes incorporated into non-mobilisable plasmids can be exchanged between two different genera [96]. DNA probes have been developed for detection of resistance genes for mercury [97], cadmium, cobalt and zinc [98], and a correlation between soil heavy metal content and the presence of metal-resistant megaplasmid-bearing *Alcaligenes eutrophus* strains has been demonstrated [85, 98].

In contrast to the reduction of Hg^{2+} to $Hg°$, other mechanisms of mercury detoxification have received less attention at the genetic level. These include methylation of Hg^{2+}, extracellular precipitation by hydrogen sulphide and decreased uptake or impermeability (see previous sections).

8.3.1.2 *Cadmium resistance.* A variety of cadmium resistance mechanisms have been reported in bacteria, though only the *cadA* system has been cloned and DNA-sequenced [56, 99]. Plasmid-determined cadmium resistance has been found in *Staphylococcus aureus*. The separate genes *cadA* and *cadB* encode for energy-dependent Cd^{2+} efflux [100, 101] and increased binding respectively [102]. *CadA* confers both Cd^{2+} and Zn^{2+} resistance, whereas another resistance system in *S. aureus* confers Cd^{2+} resistance only [99]. Cd^{2+} may enter bacterial cells via the Mn^{2+} transport system in Gram-positive bacteria [102] or the Zn^{2+} transport system in Gram-negative bacteria [103], whereupon it is rapidly effluxed from resistant cells. It was originally proposed that the mechanism was a $Cd^{2+}/2H^{+}$ antiport [101], and DNA sequence analysis of the *cadA* determinant has now confirmed the inclusion of a cation-translocating ATPase [56] (Figure 8.2). The *cadA* DNA sequence potentially encodes a polypeptide of 727 amino acids, and sequence analysis shows that this protein is a member of the family of E1, E2 ATPases which include the bacterial K^{+}-ATPases, fungal H^{+}-ATPases, as well as those from higher eukaryotic systems [56, 75, 104]. The protein is low in cysteine content (4 out of 727 residues), though these are positioned strategically (Figure 8.2). The first two cysteines (positions 23 and 26) show homology to the Hg^{2+}-binding regions in mercuric reductase described previously and may function as soft metal-binding residues [5, 70, 75]. An approximate 186-residue domain constitutes a 'funnel' accepting Cd^{2+} from the cysteine pair and guiding it to the membrane channel. The second cysteine pair occurs at positions 371 and 373 bounding a proline residue. The following ATPase domain, of about 250 amino acids, includes an aspartate residue at position 415 which begins a seven amino acid sequence which is unaltered in all the ATPases examined. Lysine (position 489) functions in ATP binding prior to phosphorylation of the aspartate residue. The end of the ATPase domain and

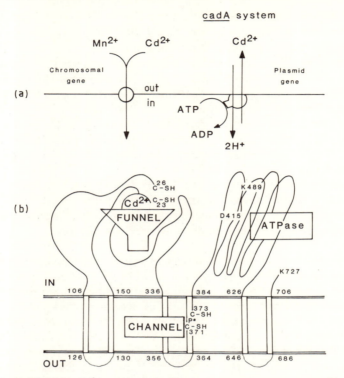

Figure 8.2 (a) Model for Cd^{2+} uptake and efflux in Gram-positive bacteria. Cd^{2+} uptake is *via* the Mn^{2+} transport system and efflux is *via* the *cadA* Cd^{2+} efflux ATPase. (b) Postulated model of the bacterial Cd^{2+}-ATPase from the amino acid sequence and homology to E1, E2 ATPases. Adapted from refs. 5, 70 and 75 with permission.

the next potential membrane-spanning segment are the most conserved sequences in this family of proteins [5, 70, 75]. It should be noted that this current model (Figure 8.2) is based on DNA sequence analysis alone, and supporting biochemical data are lacking [70, 75].

Other bacterial cadmium resistance mechanisms are not as well characterised. In *Bacillus subtilis*, a chromosomal mutation results in a change in the Mn^{2+} transport system, which transports both Mn^{2+} and Cd^{2+}, so that Cd^{2+} is no longer accumulated [105, 106]. Other resistance mechanisms include synthesis of polythiol-containing Cd^{2+}-binding proteins, analogous to metallothioneins [38, 39, 107, 108], and a system found with a 9.1-kb *Alcaligenes eutrophus* plasmid DNA fragment that confers resistance to Cd^{2+}, Zn^{2+} and Co^{2+} [109, 110].

8.3.1.3 *Arsenic and other resistances.* Arsenate, arsenite and antimony(III) resistance are all coded for by an inducible operon-like system on plasmids of *S. aureus* and *E. coli*, each ion inducing all three resistances [56, 74]. Arsenate accumulation is normally via transport systems for phosphate, and the reduced net arsenate uptake by resistant cells is a result of rapid efflux of AsO_4^{3-} mediated by an ATPase transport system [5, 56, 70, 74, 75]. DNA sequence analysis of the plasmid system shows the presence of three structural genes with products appearing to be a loosely membrane-associated ATPase (ArsA), a large (429 amino acid) membrane-embedded protein (ArsB) and a small soluble protein (ArsC) which determines substrate specificity [70, 75, 111, 112]. If the *arsC* gene is deleted, arsenate resistance is lost whereas arsenite and antimony(III) resistances remain [70, 75]. Silver *et al.* [75] and Tisa and Rosen [74] have proposed the current model (Figure 8.3) which has an analogous structure to the mammalian multidrug resistance efflux ATPase [75]. Whether these analogous structures indicate similar biochemical processes will not be clear until adequate biochemical analyses have been carried out [75].

There are other examples of plasmid-mediated bacterial heavy metal resistances [13, 33, 85]. Tellurite and tellurate resistances have been detected from plasmid-mediated determinants [113, 114] though the mechanism of resistance has not been established. The mechanism of plasmid-mediated chromate resistance in *Pseudomonas fluorescens* and *Alcaligenes eutrophus* is reduced uptake [85, 115–117]. Plasmid-determined Cu^{2+} resistance has been found on an antibiotic resistance plasmid from *E. coli* [118], in *E. coli* exposed to copper feed supplements [119] and in copper-exposed *Pseudomonas syringae* [120–122]. The genes determining Cu^{2+} resistance in *P. syringae* have been cloned and transferred into *E. coli* [121]. It is possible that cellular copper sequestration may be the mechanism of resistance [123]. Plasmid-encoded copper resistance, by means of copper precipitation as copper sulphide, has been demonstrated in *Mycobacterium scrofulaceum* [124] and *Alcaligenes eutrophus* [85].

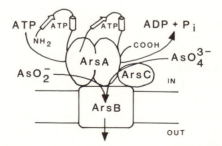

Figure 8.3   Model of the bacterial arsenic efflux ATPase. The integral membrane protein ArsB acts as an anion channel. The ArsA and ArsB protein complexes comprise an ATP-coupled pump for arsenite (or antimonite). Interaction of the ArsC protein with the complex allows arsenate efflux. Adapted from refs. 74 and 75 with permission.

In *E. coli*, plasmid pRJ1004 mediates copper resistance and there is evidence for two copper resistance determinants, *cdr*, which encodes for 'low-level' resistance and DNA repair, and the *pco* region (5.8 kb), consisting of four linked genes (termed *pco*ARBC) that are responsible for major resistance. The *pcc*R gene ecodes a DNA-binding regulatory protein involved in induction of copper resistance; PcoC is a cytoplasmic transport/storage protein, while PcoA and PcoB are responsible for chemical modification of copper and its export from cells [85, 125]. Copper is deposited around resistant colonies. Chromosomal gene products, CutC and CutD, are also required in conjuction with PcoA and PcoB for copper efflux. In addition, other chromosomal genes are involved in copper uptake (*cut*A), control of transport/storage proteins (*cut*R, *cut*E, *cut*F) and copper and zinc uptake (*cut*B). This system enables 'sensing' and adaptation to low (*cut*) and high (*pco*) levels of extracellular copper [85, 125]. Plasmid-determined silver resistance has been described in several bacteria, e.g. *Pseudomonas stutzeri* AG259 [126–128]. Plasmid pKK1, which encodes for Ag^+ resistance in *P. stutzeri* AG259, was introduced into *Pseudomonas putida* by high-voltage electroporation, which resulted in acquisition of Ag^+ resistance by the latter organism [129]. *E. coli* R1 is Ag^+-resistant and contains two large plasmids of 83 kb (pJT1) and 77 kb (pJT2). A silver-sensitive derivative was obtained on curing this strain of the 83-kb plasmid which indicated that Ag^+ resistance in *E. coli* R1 may be plasmid-encoded [130]. The level of silver resistance can markedly depend on the Cl^- concentration in the medium [128]; it is possible that resistant cells do not compete successfully with silver chloride precipitates for Ag^+ in contrast to sensitive cells [131]. More recently, plasmid-encoded resistance to zinc, cadmium, cobalt and nickel in *Alcaligenes eutrophus* has been shown to depend on inducible, energy-dependent cation efflux systems [85, 132, 133]. In *A. eutrophus*, determinants for heavy metal resistance are carried by large plasmids (165–250 kb); strain CH 34 carries pMOL28 and pMOL30, which carry resistance to nine metals [85].

At this point, it is appropriate to point out that metal ions are essential co-factors in several *trans*-acting bacterial gene regulators. On metal binding, the receptor proteins either repress gene expression or, in other systems, activate transcription. To date, two DNA-binding proteins (Fur, involved in iron metabolism in *E. coli*, which blocks transcription of iron-regulated genes [64, 134], and MerR, involved in Hg^{2+} resistance and discussed previously) have been shown by direct evidence to regulate their target genes in a metal-dependent manner. There is increasing evidence for other genetic control circuits regulated by metal ions obtained by indirect means, e.g. NifA proteins of rhizobia [134]. The key roles of metalloregulatory proteins in genetic control mechanisms are likely to receive considerable attention in the future, particularly because repression or activation of the appropriate genes enables bacteria to adapt and change in response not only to toxic metals, but to other environmental changes, e.g. light and oxygen [134].

8.3.2 *Metal-binding proteins of fungi*

8.3.2.1 Yeast metallothionein. Metallothioneins are small, cysteine-rich polypeptides that can bind essential metals, e.g. copper and zinc, as well as inessential metals such as cadmium. Copper resistance in *Saccharomyces cerevisiae* can be mediated by the induction of a 6573-dalton cysteine-rich protein, copper-metallothionein (Cu-MT) [135–137]. This is a protein which normally functions to maintain low levels of intracellular copper ions and thus prevent futile transcription of the *CUP1* structural gene [138]. Although yeast metallothionein can also bind cadmium and zinc *in vitro*, it is not transcriptionally induced by these ions and does not protect against them [139, 140].

S. *cerevisiae* contains a single Cu-MT gene which is present in the *CUP1* locus located on chromosome VIII [141, 142]. Copper-resistant S. *cerevisiae* strains (CUP1^r) contain 2–14 or more copies of the *CUP1* locus, which are tandemly repeated, and may grow on media containing $\geqslant 2\,\text{mM}$ copper [142, 143]. Copper-sensitive strains [cup1^s) do not grow $\geqslant 0.15\,\text{mM}$ [136]. Continuous subculture of S. *cerevisiae* in the presence of increasing copper concentrations can select for hyper-resistant strains [34] which appear to be disomic for chromosome VIII. The disomic chromosomes exhibit differential *CUP1* gene amplification patterns, which indicates that the copper resistance mechanism involves not only amplification of the *CUP1* locus on the chromosomes but also disomy or aneuploidy of chromosome VIII [136, 143]. The DNA fragments of the *CUP1* locus that confer copper resistance have been cloned [139, 141, 144], and evidence that the *CUP1* locus was present as multiple copies in CUP1^r strains was obtained after restriction enzyme analysis of the cloned DNA [136, 137]. The basic repeating unit '*CUP1* locus' is composed of 2.0-kb DNA fragments containing a unique *Xba*I site, two sites for *Kpn*I, and *Sau*3A restriction enzymes. The full nucleotide sequence of the *CUP1* locus has been determined [142].

Copper resistance in S. *cerevisiae* depends on amplification of the *CUP1* locus [136, 137]. In simple terms, the gene amplification model suggests that because of the proposed homology of DNA sequences at the junction of *CUP1* repeats, one or more units are looped out. If the loop is replicated, the copy number increases (amplification), but if the loop is degraded, the copy number decreases (deamplification) [136, 137, 143]. For more precise accounts of this model, and the regulation of MT gene expression, readers should consult the detailed works available [135–137, 141, 143, 145, 146]. It should be noted that copper MT is not directly involved in Cu^{2+} uptake in yeast [147].

The mechanism by which copper ions induce Cu-MT gene transcription is now receiving detailed attention, and recent progress has been rapid [136, 146, 148, 149, 150–153]. It is now known that a *trans*-acting regulatory protein, encoded by the *ACE1* locus [148, 149] and also known as the *CUP2* locus [152, 154], activates transcription of the Cu-MT gene in response to excess copper (or silver) ions [153]. Although the *ACE1* gene is constitutively expressed in the absence or presence of copper, the apoprotein cannot bind DNA [149]. However, in the presence of Cu(I) [or Ag(I)], the amino-terminal portion of ACE1 protein (amino acids 1–122) undergoes a conformational change such that it can specifically bind to the *CUP1* upstream activator sequence (UAS) [149, 155]. This metal-dependent DNA-binding domain is rich in cysteines and basic amino acids and bears structural similarities to yeast Cu-MT itself [149, 150, 153]. It is proposed that the cysteine residues of ACE1 form a polynuclear Cu(I)–S core in which cysteine thiolates coordinate multiple Cu(I) ions and Cu(I) ions are bound by multiple sulphur bonds [153]. Wrapping the protein around this Cu(I)–S scaffold would lead to the formation of positively charged loop structures that are postulated to bind DNA specifically and activate transcription [149]. It therefore appears that the role of copper in yeast Cu-MT regulation is to allow ACE1 to bind to the metallothionein gene control sequences [153].

It should be emphasised that this metal-regulated system provides a powerful tool in biotechnology. Metal-regulated DNA sequences are efficient elements for heterologous gene expression, and a variety of proteins have been expressed in the *CUP1* expression system, including human serum albumin, human ubiquitin gene, *E. coli galK* gene and human hepatitis virus antigen [136, 156]. Yeast Cu-MT may also be of potential in metal recovery since it can bind other metals besides copper, e.g. cadmium, zinc, silver, cobalt and gold, although these metals do not induce MT synthesis [136]. The Cu-MT gene has also been used to transform brewing strains of *S. cerevisiae*, which are not very amenable to classical genetic studies, and copper selection has proved a useful tool for the genetic manipulation of such strains and improvement of the product [136, 144, 157] as well as studying the mechanisms of protein turnover [158].

Genetic analyses of metal-regulated genes from some other fungi are available. The gene for *Neurospora crassa* Cu-MT has been cloned and the nucleotide sequence determined; this gene codes for a 26 amino acid protein [159, 160]. The *N. crassa* protein contains seven different amino acids and 28% is composed of cysteine; it appears that copper is present in cuprous [Cu(I)] form and all the cysteines are ligated to form Cu(I) complexes with six copper molecules per mole of protein [161]. The isolated protein from *S. cerevisiae* contains 8 mol of copper [Cu(I)] ligated to 12 cysteines per mole of protein [140, 162].

It should be mentioned that a cadmium-resistant strain of *S. cerevisiae* synthesised a cytoplasmic cadmium-binding protein when grown in cadmium-containing medium [163]. More recently, it was shown that this protein (9 kDa) exhibited the characteristics of a metallothionein, being rich in cysteine (18 mol%) and having a high cadmium content (63 μg of cadmium per mg of protein). There was a high similarity in amino acid composition between *S. cerevisiae* Cd-MT and Cu-MT [164].

8.3.2.2 *Metal γ-glutamyl peptides*. In addition to the synthesis of cysteine-rich polypeptides in the metallothionein family, another group of metal-binding molecules are short, cysteine-containing γ-glutamyl peptides, encompassed by the trivial name 'phytochelatins' [165, 166]. These γ-glutamyl peptides are the main metal detoxification mechanism in algae and plants [167, 168] as well as a range of fungi and yeasts and are of general formula (γGlu–Cys)$_n$–Gly [(γ-EC)$_n$G] [169]. Peptides of $n = 2$ to $n = 5$ are most common [170] and metal ions are bound within a metal–thiolate cluster composed of an oligomer of peptides [165]. Phytochelatins are not synthesised by *S. cerevisiae*, but *Schizosaccharomyces pombe* is now known to synthesise at least seven different homologous phytochelatins in response to metal exposure, probably by consumption of glutathione or its biosynthetic precursor [169]. The originally discovered two γ-glutamyl peptides in *S. pombe*, cadystin A and B ($n = 2$ and 3 respectively) were synthesised in response to cadmium [171, 172] though other metals including Cu^{2+}, Pb^{2+}, Zn^{2+} and Ag^+ may also be effective in *S. pombe* [169, 170, 173], a situation analogous to that in higher plants [166, 175]. However, the greatest amount of peptide synthesis occurs with cadmium [175] and it seems that larger peptides bind cadmium more strongly than smaller peptides [173, 175]. When *S. pombe* is exposed to Cu(II), a Cu-γ-glutamyl peptide is synthesised similar to structure to the Cd-γ-glutamyl peptide [170]. However, only one Cu-γ-glutamyl peptide complex was observed which did not contain labile sulphur, and with the metal being bound as Cu(I) [170].

Cadmium-induced (but not copper-induced) γ-glutamyl peptides contain labile sulphur. Cadmium stimulates the production of S^{2-} in *S. pombe* and *Candida glabrata*, and a portion of this may be incorporated within the Cd-γ-glutamyl peptide complexes [165, 176, 177]. The sulphide content is dependent on growth conditions and incubation time and S^{2-} levels ranging from 0.1 to 1.5 mol per mol of peptide have been demonstrated [165]. Sulphide incorporation enhances the stability and metal content of the peptide complex (the molar sulphide–cadmium ratios range from < 0.06 in low-sulphide forms to 0.8 in high-sulphide forms) and the enhanced stability may increase the efficacy of the complex in maintaining a low intracellular concentration of Cd^{2+} [165]. Complexes with a sulphur–cadmium ratio of 0.7 contain a 20-Å diameter cadmium sulphide crystallite coated with the γ-glutamyl peptides, which are quantum particles exhibiting size-dependent electronic states [178]. This is a feature of relevance to solid-state physics, since the crystallites exhibit properties analogous to those of semiconductor clusters [178–180]. The extracellular and cell-associated cadmium sulphide colloid formed by interaction of Cd^{2+} with S^{2-} also contributes to cadmium tolerance (see previous sections).

Candida (*Torulopsis*) *glabrata* is unique in expressing both metallothionein and γ-glutamyl peptide synthesis for metal detoxification in a metal-specific manner [176]. *C. glabrata* synthesises two distinct MT polypeptides in response to copper which exhibit repeats of the Cys–Xaa–Cys sequence, common in mammalian MT, and coordinates copper in Cu(I)–thiolate clusters. The cysteine content of the two molecules is 30 mol% (*S. cerevisiae* Cu–MT = 22 mol%; *N. crassa* Cu-MT = 28 mol%) with the copper content being between 10 and 13 mol per mol protein for each [176]. Two metallothionein genes (MT-I and MT-II) from *C. glabrata* have been cloned and sequenced. *C. glabrata* metallothioneins constitute a multigene family consisting of two classes of genes with multiple isoforms of one class. The MT-I gene is present as a single copy but multiple [3–9] and tandemly arranged copies of one MT-II gene were present in different strains. *C. glabrata* strains hyper-resistant to copper showed further stable chromosomal amplification (> 30 copies) of MT-II gene [182]. Both genes are inducible by silver but not by cadmium, the latter reducing levels of both MT-I and MT-II mRNAs [181]. The Cd-γ-glutamyl peptide of *C. glabrata* has $n = 2$, whereas peptides with $n = 3$ and 4 are more common in *S. pombe* and plants [170, 183]. Thus, *C. glabrata* can utilise two types of regulatory metal-binding molecules and the affinity and/or coordination preference of the molecules may determine whether MT or γ-glutamyl peptides are synthesised [176].

8.4 Biotechnological aspects of metal-microbe interactions

8.4.1 *Microbial removal and recovery of heavy metals and radionuclides*

Micro-organisms, including cyanobacteria, actinomycetes, and other bacteria, algae, fungi and yeasts, can accumulate heavy metals and radionuclides from their external environment [3, 7, 18, 24, 29, 184–200] (Table 8.1). Amounts accumulated can be large, and a multiplicity of physical, chemical and biological mechanisms may be involved, including adsorption, precipitation, complexation and transport. Living and dead cells are capable of metal/radionuclide removal from solution as well as products excreted by, or derived from, microbial cells, e.g. cell wall constituents, pigments, polysaccharides, metal-binding proteins and siderophores [3, 18, 201]. The removal of radionuclides, metal or metalloid species, compounds and particulates from solution by biological material, particularly by non-directed physicochemical interactions [185], is now frequently termed 'biosorption'. This is a general term which takes no account of mechanistic details, and although virtually all biological material has a significant biosorptive ability [18, 201] most work to date has been directed towards microbial systems.

Biosorption is an expanding area of biotechnology because the removal of potentially toxic and/or valuable metals and radionuclides from aqueous effluents can result in detoxification and therefore safe environmental discharge [3, 29, 195]. Furthermore, appropriate treatment of loaded biomass can enable recovery of valuable elements for recycling or further containment of highly toxic and/or radioactive species [198, 199, 202] (Figure 8.4). With

Table 8.1 Some examples of heavy metal and radionuclide accumulation by micro-organisms*

Organism	Element	Uptake (% dry weight)
Bacteria		
Streptomyces spp.	Uranium	2–14
S. viridochromogenes	Uranium	30
Thiobacillus ferrooxidans	Silver	25
Bacillus cereus	Cadmium	4–9
Zoogloea spp.	Cobalt	25
	Copper	34
	Nickel	13
	Uranium	44
Citrobacter spp.	Lead‡	34–40
	Cadmium‡	40
	Cadmium§	170
	Uranium**	900
Pseudomonas aeruginosa	Uranium	15
Mixed culture	Copper	30
Mixed culture	Silver	32
Algae		
Chlorella vulgaris‡	Gold	10
Chlorella regularis†	Uranium	15
C. regularis	Uranium	0.4
	Manganese	0.8
Fungi		
Phoma spp.	Silver	2
Penicillium spp.	Uranium	8–17
Rhizopus arrhizus	Copper	1.6
	Cadmium	3
	Lead	10.4
	Uranium	19.5
	Thorium	18.5
	Thorium††	11.6
	Silver	5.4
	Mercury	5.8
Aspergillus niger	Thorium	18.5
	Thorium††	3.8
	Uranium	21.5
Yeasts		
Saccharomyces cerevisiae	Uranium	10–15
	Thorium††	12
	Zinc	0.5
Yeasts (14 strains)	Silver	0.05–1
Yeasts (4 strains)	Copper	0.05–0.2

*Data adapted from refs. 3 and 189 and references therein. Consultation of original work is recommended for full experimental details, particularly metal and biomass concentration, suspension medium and time of incubation.

† Immobilised cells [97].

‡Enzymatic (phosphatase)-mediated metal removal from solution, batch-grown cells [195].

§Enzymatic (phosphatase)-mediated metal removal from solution, cells continuously grown under carbon limitation [95].

**Immobilised cells: enzymatic (phosphatase)-mediated uranium precipitation as UO_2HPO_4 (195, 213].

††Incubation in 1 M nitric acid [193, 212].

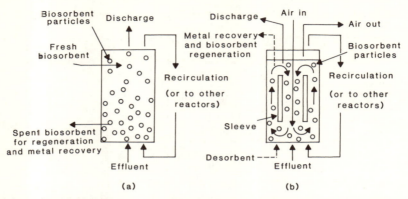

Figure 8.4 Simplified diagrams of (a) a fluidised-bed and (b) an air-lift bioreactor for metal/radionuclide biosorption by particulate or immobilized microbial biomass. Adapted from refs. 191 and 212 with permission.

accelerating depletion of natural mineral sources, there is a greater need for recycling, while more efficient means of effluent detoxification are essential for environmental protection [203, 204].

8.4.1.1 *Specific examples of microbial metal removal*

(a) *Living organisms.* Because of heavy metal toxicity, inactivation of living cells may be a problem, though it may be feasible to separate microbial growth from the metal removal phase or to use strains tolerant to the operating conditions encountered. Not surprisingly, most living cell systems exploited to date have been used for decontamination of effluents that contain metals at concentrations below toxic levels [184, 186]. Streams and ponds containing cyanobacteria and algae, sometimes encouraged by addition of sewage effluent can reduce levels of lead, copper, zinc and manganese in mining effluents, while a series of algal ponds has successfully removed a variety of elements, including uranium, molybdenum and radium from uranium mine waters [184]. Where there is death of algae, after exposure to toxic metal concentrations, decomposition of settled biomass can result in hydrogen sulphide production by sulphate-reducing bacteria and metal precipitation as sulphides. The formation of insoluble metal sulphides is virtually irreversible if reducing conditions are maintained, and this appears to be a process that may be exploitable by industry [186].

Activated sludge microbial biomass has received applied attention, particularly where there is extensive production of extracellular polymers by bacterial species such as *Zoogloea* and *Arthrobacter* [205–208]. A variety of batch and continuous processes have been described, while other examples of microbial consortia have been shown to achieve efficient removal of, for example, cadmium, silver and copper [209–211].

(b) *Immobilised cell systems*. Freely suspended microbial biomass has disadvantages that include small particle size and low mechanical strength, while a similar density to the effluent may complicate biomass/effluent separation [198]. Immobilised biomass appears to have greater potential in packed- or fluidised-bed reactors (Figure 8.4) with benefits including control of particle size, better regeneration of biomass, easy separation of biomass and effluent, high biomass loadings and minimal clogging under continuous flow [195, 198, 199, 212]. Furthermore, such systems may be mathematically defined by reference to such parameters as flow rate, metal concentration and loading capacity [195, 213]. Immobilised cells are particularly amenable to non-destructive recovery, and after loading the metal may be concentrated in a small volume of solid material or desorbed into a small volume of eluant for recovery, disposal or containment.

Derived products, as well as living or dead cells, can be used in immobilised forms, though toxicity and nutrient supply are important considerations for living cells. However, living cells may allow possibilities for longer term continuous processes, with biomass replenishment, and the exploitation of metal-removal mechanisms only expressed by living cells. Furthermore, living cells may provide an additional capacity for removal of other pollutants including hydrocarbons, pesticides and nitrates [185, 195, 214].

Metal-removal efficiency may be influenced by the method of immobilisation. Electrostatic adsorption of cells to surfaces is weak and affected by pH while chemical coupling may result in toxic symptoms [195]. Entrapment of cells within alginates, polyacrylamide and silica gels may be highly efficient for small-scale systems, though diffusional limitations may be a problem [195, 213]. However, small-scale systems may be adequate for low-volume waste containing commercially valuable element(s), e.g. gold [187]. Almost all microbial groups have been used successfully in gel-immobilised systems with efficient regeneration and prolonged function over extended time periods [3, 195, 215–217]. Selective recovery of UO_2^{2+}, Au^{3+}, Cu^{2+}, Hg^{2+} and Zn^{2+} from immobilised algae has been described [197] as well as other granulated systems for metal and radionuclide removal [198, 199, 214]. Some processes have been developed for commercial use and exhibit high metal loadings. Metal removal from dilute solutions (10–100 mg/1) can exceed 99% and granules can be used in fixed- or fluidised-bed bioreactors [214].

Other methods of effluent treatment rely on cells immobilised as a biofilm on inert supports. Ideal supports have a large surface area but are sufficiently porous to enable high flow rates and minimal clogging. Materials include those with planar surfaces (glass, metals, plastics) uneven surfaces (wood shavings, clays, sand, coke) and porous materials (foams, sponges) [195]. Biofilms of *Citrobacter* spp. exhibit metal removal capacities comparable to gel-immobilised cells [195, 213]. A large-scale commercial process (5.5×10^6 gallons of effluent per day from gold mining/milling) employs biofilm-covered rotating disc contactors for biosorptive removal of metals as well as degradation of cyanide, thiocyanate and ammonia [214].

(c) *Growth-decoupled enzymic metal removal.* A means of avoiding metal toxicity with living cell systems is to separate microbial propagation from the metal-contacting phase. This has been achieved by exploitation of an enzyme responsible for metal deposition around cells. Metal accumulation in immobilised *Citrobacter* spp. is catalysed by a phosphatase which releases HPO_4^{2-} from a supplied substrate, glycerol 2-phosphate, which precipitates divalent cations (M^{2+}) as metal phosphate at cell surfaces [18, 195, 213, 218, 219]. Such a process may be used continuously with up to 9 g metal per gram bacterial dry weight being accumulated. The process is non-specific and depends on the insolubility of the particular metal phosphate involved; species such as Cd^{2+}, Pb^{2+}, Cu^{2+} and UO_2^{2+} can be precipitated singly or in combination [220]. With time, metal phosphate formation may limit further metal accumulation because of column blockage, while the need for a phosphate donor, glycerol 2-phosphate, involves another economic consideration. The possibility of using a cheap or industrial waste product such as tributyl phosphate, which is a by-product of nuclear fuel reprocessing, may make this process attractive [18, 195, 213].

8.4.1.2 *Metal removal by induced, excreted or derived biomolecules.* Virtually all biological material has a high affinity for heavy metals/radionuclides, and a huge array of biomolecules may be involved in metal sequestration [18, 201]. Some biomolecules may be induced by the presence of certain metals, e.g. metallothioneins and other metal-binding proteins (see previous sections) or by deprivation of an essential metal ion, e.g. Fe^{3+}, leading to siderophore production [221]. Other molecules with significant metal-binding abilities may be overproduced as a result of exposure to potentially toxic metal concentrations, e.g. fungal melanins [3, 32]. However, the majority of metal-binding biomolecules found in microbial cells are synthesised or excreted as a result of 'normal' growth and/or are important structural components, particularly in cell walls [18, 24]. Metal binding by such compounds may be fortuitous, and relative efficiencies may largely depend on the metal species present and the chemical nature and reactivity of the metal binding ligands present [12, 29, 199]. It should be noted that the macromolecular composition of a given microbial species can be altered by cultural and genetic manipulations, and there is a wide diversity in chemical composition, particularly in relation to cell walls [18, 24].

Fungi and yeasts have received considerable attention in this area because waste fungal biomass is produced in significant quantities from several industrial fermentations [29, 198]. The majority of bacterial interest concentrates on polysaccharides and siderophores, which are discussed later.

Many fungi have a high chitin content in cell walls, and this polymer of N-acetyl glucosamine is a highly effective metal and radionuclide biosorbent [18, 198, 199]. For uranium, rapid coordination to the amine nitrogen of chitin and simultaneous adsorption in the cell wall chitin structure are followed by slower precipitation of uranyl hydroxide [222]. A free radical on the chitin molecule appears to be involved in the initial coordination of of UO_2^{2+} to nitrogen [223]. The accumulation of hydrolysis products continues until a final equilibrium is achieved [223]. Chitosan and other chitin derivatives can also have significant biosorptive capability. Uranium adsorption by chitin phosphate and chitosan phosphate is rapid and influenced by pH in a similar manner to that of intact biomass. For chitin phosphate, the sequence of efficiency is $UO_2^{2+} \gg Cu^{2+} > Cd^{2+} > Mn^{2+} > Zn^{2+} > Mg^{2+} > Co^{2+} > Ni^{2+} > Ca^{2+}$; for chitosan phosphate, the order is $UO_2^{2+} \gg Cu^{2+} > Zn^{2+} > Mn^{2+} > Co^{2+} > Ni^{2+} > Mg^{2+} > Ca^{2+}$. UO_2^{2+} and Co^{2+} were easily separated from each other using chitin phosphate and can be recovered by sodium carbonate elution. Non-phosphorylated chitin and chitosan are not as effective for biosorption as phosphorylated derivatives [224]. Insoluble chitosan–glucan complexes from *Aspergillus niger* show good biosorptive ability, while glucans possessing amino or sugar acid groups can exhibit enhanced removal of transition metal ions [225]. N-[1,2-dihydroxy-ethyl)tetra-hydrofuryl] chitosan (NDTC) is capable of uranium removal from brines at an order of magnitude greater that intact native biomass of *Rhizopus arrhizus* [18, 225]. After hydroxide treatment of a variety of fungi to expose chitin/chitosan, and other metal-binding ligands, wet-laid papers which incorporate treated mycelium have been found to be efficient metal removal agents in laboratory trials [226].

Fungal phenolic polymers and melanins contain phenolic units, peptides, carbohydrates, aliphatic hydrocarbons and fatty acids. Oxygen-containing groups in these substances, including carboxyl, phenolic and alcoholic hydroxyl, carbonyl and methoxyl groups, may be particularly important in metal binding [3]. Melanin from *Aureobasidium pullulans* has been shown to bind significant amounts of metals such as Cu^{2+} [30] and also organo-metallic compounds, e.g. tributyltin chloride [31].

Other biomolecules of probable interest from fungi include mannans, phosphomannans, polysaccharides, proteins (including metallothioneins and other metal-binding proteins) and externally excreted metal-binding molecules, including siderophores, and citric and oxalic acids [8, 32]. Proteins and polysaccharides may also be important in other microbial groups, particularly bacteria [208].

The metal-binding properties of many bacterial extracellular polymers have been documented for a variety of species, in pure and mixed cultures [206, 208]. A wide range of polymers can be synthesised by bacteria, fitting broadly into major polysaccharide, protein and nucleic acid groups, but with a wide occurrence of hybrid polymers, e.g. glycoproteins and lipopolysaccharides [18, 208]. Potential interactions of metals with these and other biomolecules have been reviewed by Hunt [201], Geesey and Jang [208] and Macaskie and Dean [18]. Metal binding to polymeric material may involve simple ion exchange, with cationic groups involving carboxyl, organic phosphate, organic sulphate and phenolic hydroxyl, and/or complex formation. Additionally, there may be deposition of the metal in a chemically altered form [18, 24].

In activated sludge, the bacterium *Zoogloea ramigera* is important in flocculation because of extensive exopolysaccharide production. This has metal-binding properties, and copper and cadmium have been reported to be taken up to 0.3 and 1 g metal per gram dry wt respectively [227]. Some metal-binding selectivity has been achieved by pH control, with uranium, copper and cadmium being maximally bound at pH values of 3.5, 5.5 and 6.5 respectively [227]. Three strains of *Z. ramigera* have been characterised, each with a differing polymer composition, one being cellulose-like and fibrillar and one containing amino sugars, while one consists of glucose, galactose and pyruvate [228]. However, polymer composition can be influenced by culture conditions, and although in many cases isolated polymer may give enhanced metal uptake it is suggested that any working process should employ whole-cell biomass because of difficulties in quantitative polymer extraction [18]. Extracellular polymer from *Klebsiella aerogenes* bound metal in the order copper, cadmium > cobalt > manganese > nickel [229]. Maximal polymer production occurred in stationary phase cultures, and little binding was evident at pH 4.5 as compared with pH 6.5 [229]. Other bacterial polymers have received attention, e.g. polysaccharides from *Arthrobacter viscosus*, which comprise glucose, galactose and mannuronic acid [27].

The application of a bacterial emulsifying agent, emulsan, and its derivatives, for uranium removal has been described [230]. Emulsan has a polysaccharide backbone comprising three amino sugars, D-galactosamine, D-galactosamine uronic acid and an unidentified hexosamine, with linked fatty acids [18]. Metal-binding ability is linked to the need for divalent cations for full emulsifying activity above pH 6, and this substance has a high capacity for uranium binding. If emulsan is sonicated and dispersed in water/hexadecane, the 'emulsanosol' binds more than 800 mg uranium per gram [230]. The metal-binding significance of extracellular polysaccharides, slime layers, etc. from fungi, algae and cyanobacteria has received less attention, although similar interactions are likely.

Iron is an essential metal for micro-organisms which mainly exist as Fe^{3+} in the environment, and precipitates as insoluble ferric hydroxide at neutral pH values. Many iron-dependent microbes excrete iron-binding siderophores that complex Fe^{3+} [64, 200]. The externally formed ferric chelates subsequently interact with cells so that iron accumulates intracellularly [221]. Siderophores fit into two general classes, the phenolate/catecholate type and the hydroxamate type [18, 64]. The synthesis of both classes is dependent on the iron status of the medium with maximal production occurring under iron-limiting conditions. As well as Fe^{3+}, siderophores and analogous compounds may bind certain other metals, e.g. gallium, nickel, uranium and thorium [3, 18]. A strain of *Debaryomyces hansenii* produces riboflavin, or a related compound, under iron limitation or in the presence of metals such as copper, cobalt and zinc, and this is capable of Fe^{3+} binding [231]. Cyanobacteria and algae may also produce siderophores under iron limitation [3].

8.4.1.3 *Metal transformations.* Although the biotechnological application of microbial metal transformations (see previous sections) has not been fully explored, it has been demonstrated that continuous cultures of Hg^{2+}-resistant bacteria, which catalyse $Hg^{2+} \rightarrow Hg^{\circ}$, are able to volatilise Hg° from sewage at a rate of $2.5 \, mg \, l^{-1} \, h^{-1}$ (98% removal) [232]. Enzymic transformation of arsenic and chromium species is also associated with a decrease in toxicity. Pretreatment of arsenic-loaded sewage with arsenite oxidase-producing bacteria may improve certain arsenic removal methods since arsenate, As^{5+}, is more easily precipitated from waste water by Fe^{3+} than is arsenite, As^{3+} [58]. Biomethylated metal derivatives may also be volatile and lost from a given system, though amounts liberated are usually slow and recovery is unlikely to be economical. Microbial dealkylation of organometallic compounds, which may be an important detoxification mechanism, can result in eventual formation of ionic species which can be removed using a biosorptive process [191].

8.4.1.4 *Metal recovery.* Biotechnological exploitation of microbial metal accumulation may depend on the ease of metal recovery for subsequent reclamation, containment of toxic waste, and biosorbent regeneration for use in multiple biosorption–desorption cycles [29, 198, 199, 233, 234]. The disposal of spent biomass may also raise problems of environment concern [202]. The mechanism of metal recovery may depend on the ease of removal from the biomass, which in turn can depend on the metal involved and the mechanism of accumulation. Metabolism-independent biosorption is frequently reversible by non-destructive methods, whereas energy-dependent intracellular accumulation and compartmentation, binding to induced proteins, etc. is often irreversible, requiring destructive recovery [3]. The latter may be achieved by incineration or dissolution in strong acids or alkalis [186]. Whether metal recovery by these means is envisaged can depend on its commercial value. If cheap, waste biomass is used to recover valuable metals then the economics of destruction may be satisfactory. Most work has concentrated on non-destructive desorption which, for maximum benefit, should be highly efficient, economical and result in minimal damage to the biosorbent. Dilute mineral acids ($\sim 0.1 M$), e.g. nitric acid, hydrochloric acid and sulphuric acid, can be effective for removal of heavy metals although at higher concentrations ($> 1 M$), or with prolonged exposure, there may be damage to the biomass [233]. Sulphate ions in particular can cause irreversible damage to cell wall biosorption sites in fungi which reduces the reuse potential of the biomass [233].

Organic complexing agents may be effective desorbers without affecting biomass integrity. EDTA treatment may be useful in distinguishing surface complex formation from intracellular, unexchangeable metal [235]. For *Chlorella vulgaris*, a selective elution scheme based on pH changes and organic binding agents has been described. In mixtures, metal ions such as Cu^{2+} and Zn^{2+} are desorbed from the algal cells using eluant at pH 2, yet Au^{3+}, Ag^+ and Hg^{2+} remain strongly bound at this pH; Au^{3+} and Hg^{2+} can be selectively eluted using mercaptoethanol [197].

Carbonates and/or bicarbonates are efficient desorption agents with potential for cheap, non-destructive metal recovery. Of several elution systems examined for uranium desorption from *R. arrhizus*, sodium bicarbonate ($NaHCO_3$) exhibited > 90% efficiency of removal and high uranium concentration factors [233]. The solid–liquid ratios can exceed 120:1 (mg:ml) for a 1 M sodium bicarbonate solution with eluate uranium concentrations of at least 1.98×10^4 mg/l [233]. It should be noted that the solid–liquid (S/L) ratio, i.e. the ratio of weight of loaded biomass to the eluant volume required for complete recovery, needs to be maximised in any recovery process in order to yield an eluate of minimal volume yet containing the highest possible metal concentration [199].

8.4.1.5 *Biosorption of radionuclides by fungal biomass.* The high affinity of microbial biomass for radionuclides, including the actinide elements, is of environmental and industrial significance [29, 203]. Binding to biomass is the main route of actinide sedimentation in oceanic systems [236], while metal-loaded bacterial cells can act as nuclei for the formation of a variety of crystalline metal deposits such as phosphates, sulphides and organic–metal complexes in aquatic sediments [7, 12, 24]. Such immobilisation of metals and radionuclides is a significant component of biogeochemical cycling and is also relevant to the fate of metal/radionuclides when discharged into the environment. Ingestion of metals/radionuclide-loaded cells by other organisms can result in transfer along food chains, ultimately to humans [1, 11, 16, 203]. Radionuclide uptake has been examined in a range of fungi and yeasts and, while the uptake mechanism(s) may vary between different elements and species, actinide accumulation appears to be due mainly to metabolism-independent biosorption [237]. The main site of actinide uptake is the cell wall [29, 238], although permeabilisation of cells with carbonates or detergents [193] can increase uptake, indicating that intracellular sites are also capable of binding. The mechanism of biosorption varies between elements. Both adsorption and precipitation of hydrolysis products occurs with uranium [222], while coordination with cell wall nitrogen is the main mechanism of thorium biosorption [223, 239]. Precipitation or crystallisation of radionuclides within or on cell walls may be a feature in some circumstances. In the yeast *Saccharomyces cerevisiae*, uranium is deposited as a layer of needle-like fibrils on cell walls, reaching up to 50% of the dry weight of individual cells [240]. Such precipitation has also been observed for thorium [192, 194, 198]. Under some conditions, thorium may also accumulate intracellularly in *S. cerevisiae* with preferential localisation in the vacuole [194].

Uranium and thorium biosorption can be affected by the external pH. Initial rates of uranium uptake by yeast increase above pH 2.5 [240]. At pH values below 2.5, the predominant species is UO_2^{2+}, but at greater pH values hydrolysis products include $(UO_2)_2(OH)_2^{2+}$, $UO_2(OH)^+$ and $(UO_2)_3(OH)_5^+$. The resulting reduction in solubility favours biosorption [222, 241]. Similar phenomena occur with thorium: at pH values below 2, Th^{4+} is the main species present; at higher pH values, $Th(OH)_2^{2+}$ and other products form which are taken up more efficiently than Th^{4+} [239]. Most previous studies on thorium biosorption have been conducted at pH ranges above 2. This is representative of many natural waters, but certain industrial process streams containing actinide elements are extremely acidic (pH < 1). Furthermore, other cations like Fe^{3+}, which may be present in waste liquors, can interfere with actinide uptake [239]. Consequently, if biomass is to provide a means of removing actinide elements from these waste streams, it must be capable of performing at pH values < 1 and in the presence of potentially competing chemical components. Despite these difficulties, biomass from several fungal species removes thorium from solution in 1 M nitric acid, pH 0–1 [193, 212]. Thorium uptake is saturable with increasing concentration although equilibria do not correspond to any simple adsorption isotherm. Thorium uptake is altered by biomass concentration, uptake per unit biomass being reduced at high biomass concentrations. Thorium uptake is also increased by detergent pretreatment and, in the case of filamentous fungi, varies with culture conditions. This implies that the thorium uptake characteristics of fungal biomass are able to be manipulated by these or similar means for optimum performance [193]. Strains of fungi giving the best performance as thorium biosorbents in batch cultures under acidic conditions have also been examined in several bioreactor configurations [212]. Static- or stirred-bed bioreactors achieve unsatisfactory thorium removal, probably because of poor mixing. However, an air-lift reactor (Figure 8.4) removes approximately 90–95% of the thorium supplied over extended time periods and exhibits a well-defined breakthrough point after biosorbent saturation. This type of reactor provides efficient circulation and effective contact between the thorium solution and the fungal biomass. Of the species tested, *Aspergillus niger* and *Rhizopus arrhizus*, used as mycelial pellets, were the most effective with loading capacities of approximately 0.5 and 0.6 mmol/g respectively (116 and 138 mg per gram dry wt) at an inflow concentration of 3 mM thorium (as nitrate). The efficiency of thorium biosorption by *A. niger* was markedly reduced in the presence of other inorganic solutes, while thorium uptake by *R. arrhizus* was relatively unaffected. Air-life bioreactors containing *R. arrhizus* biomass can remove thorium from acidic solution (1 M nitric acid) over a wide range of initial concentrations (0.1–3 mM) [212]. It can be concluded that biosorption using fungal biomass is a technically feasible method for the removal of thorium from acidic solutions similar to those encountered in industrial process streams. However, the practicality of this approach is very much dependent on economic considerations. The availability of waste biomass or cheap growth substrates may favour such a process, while others, such as the need for handling and transport of a bulky biosorbent or the need for on-site biomass propagation, would act against it. The balance of these factors can only be determined in individual cases by detailed costings.

8.4.1.6 *Industrial and economic considerations.* The industrial potential of biosorption depends on such factors as loading capacities, efficiencies and selectivity, ease of metal recovery and equivalence to traditional physical and chemical treatments in performance, economics and immunity from interference by other effluent components or operating conditions [29]. Clearly, biomass-related processes must be economically competitive with existing technologies. It has been suggested that biosorptive treatments should exhibit > 99% removal with metal loadings > 150 mg per gram biomass for economic success [186, 187], although if decontamination of effluents for environmental discharge is the eventual aim then such high biomass loadings may not be necessary; most economic analyses have been made with the recovery of precious metals in mind. It should be stressed that biosorptive treatments need not necessarily replace existing methodologies but may act as 'polishing' systems to processes that are not completely efficient. Selectivity may be a problem, though this may be minimised depending on the relative concentrations of different metals present in an effluent, by appropriate biosorbent selection and/or choice of appropriate recovery methods [3, 197].

Engineering considerations are also important, with simplicity of use, compactness and cost efficiency being key points. Portability is also important for some commercial applications [202]. Several biosorbent systems described are based on packed- and fixed-bed reactor systems with effluent upflow or downflow (Figure 8.4). Reactors may be used in parallel to avoid saturation and enable biosorbent regeneration of disposal. Other methods include the fluidised- or expanded-bed system [Figure 8.4]. This has effluent upflow with metal-saturated biosorbent being removed from the column base for regeneration [202]. Several other configurations are possible, including air-lift reactors which may be used in series to cope with large flow volumes and provide adequate contact times [199, 202, 212] (Figure 8.4). Separation of immobilised or particulate biosorbent from solution is relatively easy in contrast to non-immobilised biomass. However, in some cases, freely suspended metal-accumulating microbes may be separated by high-gradient magnetic separation (HGMS), a technique developed for extraction of weakly magnetic colloids [202].

A general consensus is that for commercial use immobilised or pelleted preparations are best, with recovery utilising a cheap, replaceable stripping agent which may be recycled [187, 198, 199, 234]. Ideally, the immobilised biomass should be of increased mechanical strength, have a high particle porosity, hydrophilicity, increased resistance to chemicals or adverse operating conditions and be amenable to chosen, engineering designs [202]. However, a high metal-loading capacity and porosity of microbial biosorbent particles are generally incompatible with adequate particle strength and chemical resistance [202], a problem to be overcome with further research. The most publicised commerical process to date is the 'AMT-Bioclaim' system, which is competitive in terms of cost and operational characteristics with traditional treatments, at effluent volumes up to 4×10^4 l/day and requiring a biosorbent volume of $1.5 \, m^3$ [187]. However, significantly higher flow rates $> 10^6$ l/day, which are not unusual in many industries, including those concerned with the provision of nuclear power [203], would require much larger biovolumes which may adversely influence operation and cost [195, 213]. The enzyme-mediated *Citrobacter* process requires much lower biovolumes for high flow rates but has a lower efficiency of metal removal [195, 213]. It should be emphasised that, apart from a few well-documented exceptions, most described biosorptive systems are laboratory-based and have not received full practical or economic assessment under real operating conditions.

A considerable amount of attention has focused on elements of economic significance, e.g. gold and gallium, with rather less industrial attention given to relatively cheap metals, e.g. copper, cadmium, zinc and lead, which are still discharged into the environment on a huge scale. Clearly, there is now great public awareness of the potential dangers of pollution by such elements as well as those radionuclides which arise in waste waters from the nuclear power industry [203]. Fuller exploitation of biomass-based technologies may assist in combating this problem [8, 91]. As described, both living and dead biomass and derived or excreted products can remove heavy metals and radionuclides from solution. Living cells have possibilities in some commercial applications (see previous sections), particularly if growth is possible under metal-contacting conditions or if metabolic products, e.g. hydrogen sulphide, are involved in metal removal [202]. A further example is the use of micro-organisms in purpose-built wetland systems for treatment of acid mine drainage [202]. However, metallothioneins, particulate metal accumulation, extracellular precipitation and complex formation by living cells have received little applied attention [3]. Most biological metal removal systems in commercial production use dead or non-metabolising organisms [29, 186, 187, 199, 202]. Among the reasons for this are difficulties of metal removal if internalised within living cells and toxicity due to the metals, operational conditions or other chemical components of the effluent. Dead biomass is largely immune from these and from the added component of maintenance and nutrient supply. Furthermore, certain biomass types may be obtainable as waste from other industries. It should be evident from this account that there is an abundance of possibilities for microbial metal removal and recovery It is also clear that many aspects of metal–microbe interactions have remained unexploited with little advantage taken of the significant progress made in recent years in molecular biology and genetics; future progress depends on adequate support from government and industry [242].

References

1. G.M. Gadd and A.J. Griffiths, *Microb. Ecol.* **4** (1978) 303.
2. J.M. Wood and H.K. Wang, *Environ. Sci. Technol.* **17** (1983) 582.
3. G.M. Gadd, in *Biotechnology* Vol. 6b *Special Microbial Processes*, ed. H.-J. Rehm, VCH Verlagsgesellschaft, Weinheim (1988) 401.
4. J.S. Thayer, *Organometallic Chemistry, an Overview*, VCH Verlagsgesellschaft, Weinheim (1988).
5. S. Silver, R.A. Laddaga and T.K. Misra, in *Metal–Microbe Interactions* eds. R.K. Poole and G.M. Gadd, IRL Press, Oxford (1989) 49.
6. Y.E. Collins and G. Stotzky, in *Metal Ions and Bacteria*, eds. T.J. Beveridge and R.J. Doyle, Wiley, New York (1989).
7. T.J. Beveridge, *Ann. Rev. Microbiol.* **43** (1989) 147.
8. G.M. Gadd, in *Microbiology of Extreme Environments*, ed. C. Edwards, Open University Press. Milton Keynes (1990) 178.
9. H. Babich and G. Stotzky, *Environ. Res.* **36** (1985) 11.
10. P.J. Craig, (ed.), *Organometallic Compounds in the Environment*, Longman, Harlow (1986).
11. M.N. Hughes and R.K. Poole, *Metals and Micro-organisms*, Chapman & Hall, London (1989).
12. T.J. Beveridge and R.J. Doyle (eds.), *Metal Ions and Bacteria*, Wiley, New York (1989).
13. J.T. Trevors, K.M. Oddie and B.H. Belliveau, *FEMS Microbiol. Rev.* **32** (1985) 39.
14. H. Babich and G. Stotzky, *CRC Crit. Rev. Microbiol.* **8** (1980) 99.
15. H. Babich and G. Stotzky, *Adv. Appl. Microbiol.* **23** (1987) 55.
16. T. Duxbury, in *Advances in Microbial Ecology*, ed. K.C. Marshall, Plenum Press, New York (1985) 185.
17. D. Kaplan, D. Christiaen and S. Arad, *Appl. Environ. Microbiol.* **53** (1987) 2953.
18. L.E. Macaskie and A.C.R. Dean, in *Biosorption of Heavy Metals*, ed. B. Volesky, CRC Press, Boca Raton, FL (1990) 199.
19. V. Harwood-Sears and A.S. Gordon, *Appl. Environ. Microbiol.* **56** (1990) 1327.
20. F.E. Pooley, *Nature* **296** (1982) 642.
21. H. Aiking, H. Govers and J. van't Riet, *Appl. Environ. Microbiol.* **50** (1985) 1262.
22. F.G. Ferris, W.S. Fyfe and T.J. Beveridge, *Chem. Geol.* **63** (1987) 225.
23. F.G. Ferris, W.S. Fyfe and T.J. Beveridge, *Geology* **16** (1988) 149.
24. R.J.C. McLean and T.J. Beveridge, in *Microbial Mineral Recovery*, eds. H.L. Ehrlich and C.L. Brierley, McGraw-Hill, New York (1990) 185.
25. W.C.Ghiorse, *Ann. Rev. Microbiol.* **38** (1984) 515.
26. H. Mann and W.S. Fyfe, *Precamb. Res.* **30** (1985) 337.
27. R.A. Zierenberg and P. Schiffman, *Nature* **348** (1990) 155.
28. A.E. Milodowski, J.M. West, J.M. Pearce, E.K. Hyslop, I.R. Basham and P.J. Hooker, *Nature* **347** (1990) 465.
29. B. Volesky (ed.), *Biosorption of Heavy Metals*, CRC Press, Boca Raton, FL (1990).
30. G.M. Gadd, and L. de Rome, *Appl. Microbiol. Biotechnol.* **29** (1988) 610.
31. G.M. Gadd, P.J. Gray and P.J. Newby, *Appl. Microbiol. Biotechnol.* **34** (1990) 116.
32. G.M. Gadd, in *Microbes in Extreme Environments*, eds. R.A. Herbert and G.A. Codd, Academic Press, London (1986) 83.

33. B.H. Belliveau, M.E. Starodub, C. Cotter and J.T. Trevors, *Biotech. Adv.* **5** (1987) 101.
34. C. White and G.M. Gadd, *FEMS Microbiol. Ecol.* **38** (1986) 277.
35. R.P. Jones and G.M. Gadd, *Enzyme Microb. Technol.* **12** (1990) 402.
36. G.M. Gadd, J.A. Chudek, R. Foster and R.H. Reed, *J. Gen. Microbiol.* **130** (1984) 1969.
37. T.E. Jensen, M. Baxter, J.W. Rauchlin and V. Jani, *Environ. Poll.* **27** (1982) 119.
38. D.P. Higham, P.J. Sadler and M.D. Scawen, *J. Gen. Microbiol.* **132** (1986) 1472.
39. D.P. Higham, P.J. Sadler and M.D. Scawen, *Environ. Health Persp.* **65** (1986) 5.
40. J. Vymazal, *Tox. Assess.* **2** (1987) 387.
41. I. Sakurai, Y. Kawamura, H. Koike, Y. Inoue, Y. Kosako, T. Nakase, Y. Kondou and S. Sakurai, *Appl. Env. Microbiol.* **56** (1990) 2580.
42. L.A. Okorokov, in *Environmental Regulation of Microbial Metabolism*, eds. I.S. Kulaev, E.A. Dawes and D.W. Tempest, Academic Press, London (1985) 339.
43. C. White and G.M. Gadd, *J. Gen. Microbiol.* **133** (1987) 727.
44. R. Olafson, W.K. Abel and R.G. Sim, *Biochem. Biophys. Res. Comm.* **89** (1979) 36.
45. E. Heuillet, F. Guerbette, C. Genou and J.C. Kadar, *Int. J. Biochem.* **20** (1988) 203.
46. N.J. Robinson, *J. Appl. Phycol.* **1** (1989) 5.
47. D.N. Weber, C.F. Shaw and D.H. Petering, *J. Biol. Chem.* **262** (1987) 6962.
48. K. Sakamoto, M. Yagasaki, K. Kirimura and S. Usami, *J. Ferment. Bioeng.* **67** (1989) 266.
49. A.R. Gee and A.W.L. Dudeney, in *Biohydrometallurgy* eds. P.R. Norris and D.P. Kelly, Science and Technology Letters, Kew (1988) 437.
50. R.T. Belly and G.C. Kydd, *Dev. Ind. Microbiol.* **23** (1982) 567.
51. M. Kierans, A.M. Staines, H. Bennett and G.M. Gadd, *Biol. Metals* **4** (1991) 100.
52. J.T. Trevors, *J. Basic Microbiol.* **26** (1986) 499.
53. J.S. Thayer, *Organometallic Compounds and Living Organisms*, Academic Press, New York (1984).
54. J.S. Thayer, *Appl. Organometal. Chem.* **3** (1989) 123.
55. J.J. Cooney, *J. Ind. Microbiol.* **3** (1988) 195.
56. S. Silver and T.K. Misra, *Ann. Rev. Microbiol.* **42** (1988) 717.
57. S. Rapsomanikis and J.H. Weber, in *Organometallic Compounds in the Environment*, ed. P.J. Craig, Longman, Harlow (1986) 279.
58. J.W. Williams and S. Silver, *Enzyme Microb. Technol.* **6** (1984) 530.
59. K. Komori, A. Rivas, K. Toda and H. Ohtake, *Biotechnol. Bioeng.* **35** (1990) 951.
60. E. Fujii, K. Toda and H. Ohtake, *J. Ferment. Bioeng.* **69** (1990) 365.
61. Y. Ishibashi, C. Cervantes and S. Silver, *Appl. Environ. Microbiol.* **56** (1990) 2268.
62. S.J. Blunden and A.H. Chapman, in *Organometallic Compounds in the Environment*, ed. P.J. Craig, Longman, Harlow (1986) 111.
63. J.J. Cooney and S. Wuertz, *J. Ind. Microbiol.* **4** (1989) 375.
64. J.B. Neilands, in *Metal Ions and Bacteria*, eds. T.J. Beveridge and R.J. Doyle, Wiley, New York (1989) 141.
65. G. Winkelmann, D. Van der Helm and J.B. Neilands (eds.), *Iron Transport in Microbes, Plants and Animals*, VCH Verlagsgesellschaft, Weinheim (1987).
66. A. Bagg and J.B. Neilands, *Microbiol. Rev.* **51** (1987) 509.
67. G.M. Gadd, A. Stewart, C. White and J.L. Mowll, *FEMS Microbiol. Lett.* **24** (1984) 231.
68. M. Joho, Y. Imada and T. Murayama, *Microbios* **51** (1987) 183.
69. M. Joho, M. Inouhe, H. Tohoyama and T. Murayama, *FEMS Microbiol. Lett.* **66** (1990) 333.
70. S. Silver, T.K. Misra and R. Laddaga, *Biol. Trace Elem. Res.* **21** (1989) 145.
71. T. Schmidt and H.G. Schlegel, *FEMS Microbiol. Ecol.* **62** (1989) 315.
72. T.J. Foster, *Microbiol. Rev.* **47** (193) 361.
73. T.J. Foster, *CRC Crit. Rev. Microbiol.* **15** (1987) 117.
74. L.S. Tisa and B.P. Rosen, *J. Bioenergetics Biomembr.* **22** (1990) 493.
75. S. Silver, G. Nucifora, L. Chu and T.K. Misra, *Trends Biochem. Sci.* **14** (1989) 76.
76. N.L. Brown, *Trends Biochem. Sci.* **10** (1985) 400.
77. N.L. Brown, P.A. Lund and N. Ni Bhriain, in *Genetics of Bacterial Diversity*, eds. D.A. Hopwood and K.F. Chater, Academic Press, London (1989) 175.
78. B.P. Rosen, *Ann. Rev. Microbiol.* **40** (1986) 263.
79. S. Silver, in *Bacterial Transport*, ed. B.P. Rosen, Dekker, New York (1978) 221.
80. A.O. Summers, *Ann. Rev. Microbiol.* **40** (1986) 607.
81. B.H. Belliveau and J.T. Trevors, *Appl. Organometal. Chem.* **3** (1989) 283.

82. J.B. Robinson and O.H. Tuovinen, *Microbiol. Rev.* **48** (1984) 95.
83. S. Silver and B.P. Rosen (eds.), *Ion Transport in Bacteria*, Academic Press, San Diego (1987).
84. A.O. Summers and S. Silver, *Ann. Rev. Microbiol.* **32** (1978) 637.
85. M. Mergeay, *Trends Biotechnol.* **9** (1991) 17.
86. T.P. Begley, A.E. Walts and C. Walsh, *Biochemistry* **25** (1986) 7186.
87. T.P. Begley, A.E. Walts and C. Walsh, *Biochemistry* **25** (1986) 7192.
88. R. Nakaya, A. Nakamura and Y. Murata, *Biochem. Biophys. Res. Commun.* **3** (1960) 654.
89. T.J. Foster, H. Nakahara, A.A. Weiss and S. Silver, *J. Bacteriol.* **140** (1979) 167.
90. N.L. Brown, S.J. Ford, R.D. Pridmore and D.C. Fritzinger, *Biochemistry* **22** (1983) 4089.
91. N.L. Brown, T.K. Misra, J.N. Winnie, A. Schmidt and S. Silver, *Mol. Gen. Genet.* **202** (1986) 143.
92. S. Silver, in *Changing Metal Cycles and Human Health*, ed. J.O. Nriagu, Springer-Verlag, Berlin (1984) 199.
93. Y. Wang, I. Mahler, H.S. Levinson and H.O. Halvorson, *J. Bacteriol.* **169** (1987) 4848.
94. M.G. Jobling, S.E. Peters and D.A. Ritchie, *FEMS Microbiol. Lett.* **49** (1988) 31.
95. M.J. Bale, J.C. Fry and M.J. Day, *Appl. Environ. Microbiol.* **54** (1988) 972.
96. E. Top, M. Mergeay, D. Springael and W. Verstrate, *Appl. Environ. Microbiol.* **56** (1990) 2471.
97. T. Barkay, D.L. Fouts and B.H. Olson, *Appl. Environ. Microbiol.* **49** (1985) 686.
98. L. Diels and M. Mergeay, *Appl. Environ. Microbiol.* **56** (1990) 1485.
99. W. Witte, L. Green, T.K. Misra and S. Silver, *Antimicrob. Agents Chemother.* **29** (1986) 663.
100. Z. Tynecka, Z. Gos and J. Zajac, *J. Bacteriol.* **147** (1981) 305.
101. Z. Tynecka, Z. Gos and J. Zajac, *J. Bacteriol.* **147** (1981) 313.
102. R.D. Perry and S. Silver, *J. Bacteriol.* **150** (1982) 973.
103. R. Laddaga and S. Silver, *J. Bacteriol.* **162** (1985) 1100.
104. G. Nucifora, L. Chu, S. Silver and T.K. Misra, *J. Bacteriol.* **171** (1989) 4241.
105. K.G. Surowitz, J.A. Titus and R.M. Pfister, *Arch. Microbiol.* **140** (1984) 107.
106. R.A. Laddaga, R. Bessen and S. Silver, *J. Bacteriol.* **162** (1985) 1106.
107. D.P. Higham, P.J. Sadler and M.D. Scawen, *Science* **225** (1984) 1043.
108. D.P. Higham, P.J. Sadler and M.D. Scawen, *J. Gen. Microbiol.* **131** (1985) 2539.
109. M. Mergeay, D. Nies, H.G. Schlegel, J. Gerits, P. Charles and F. Van Gijsegem, *J. Bacteriol.* **162** (1985) 328.
110. D. Nies, M. Mergeay, B. Friedrich and H.G. Schlegel, *J. Bacteriol.* **169** (1987) 4865.
111. H.L.T. Mobley and B.P. Rosen, *Proc. Natl. Acad. Sci. USA* **79** (1982) 6119.
112. B.P. Rosen, U. Weigel, C. Karkaria and P. Gangola, *J. Biol. Chem.* **263** (1988) 3067.
113. A.O. Summers and G.A. Jacoby, *Antimicrob. Agents Chemother.* **13** (1978) 637.
114. D.E. Taylor and A.O. Summers, *J. Bacteriol.* **137** (1979) 1430.
115. L.H. Bopp, A.M. Chakrabarty and H.L. Ehrlich, *J. Bacteriol.* **155** (1983) 1105.
116. H. Ohtake, C. Cervantes and S. Silver, *J. Bacteriol.* **169** (1987) 3853.
117. A. Nies, D.H. Nies and S. Silver, *J. Biol. Chem.* **265** (1990) 5648.
118. M. Ishihara, Y. Kamio and Y. Terawaki, *Biochem. Biophys. Res. Commun.* **82** (1978) 74.
119. T.J. Tetaz and R.K.J. Luke, *J. Bacteriol.* **154** (1983) 1263.
120. C.L. Bender and D.A. Cooksey, *J. Bacteriol.* **165** (1986) 534.
121. C.L. Bender and D.A. Cooksey, *J. Bacteriol.* **169** (1987) 470.
122. D.A. Cooksey, *Appl. Environ. Microbiol.* **56** (1990) 13.
123. M. Mellano and D.A. Cooksey, *J. Bacteriol.* **170** (1988) 2879.
124. F.X. Erardi, M.L. Failla and J.O. Falkinham, *Appl. Environ. Microbiol.* **53** (1987) 1951.
125. D.A. Rouch, B.T.O. Lee and J. Camakaris, in *Metal Ion Homeostasis: Molecular Biology and Chemistry*, eds. D.H. Hamer and D.R. Winge, Alan R. Liss, New York (1989) 439.
126. C. Haefeli, C. Franklin and K. Hardy, *J. Bacteriol.* **158** (1984) 389.
127. J.T. Trevors, *Enzyme Microb. Technol.* **9** (1987) 331.
128. G.M. Gadd, O.S. Laurence, P.A. Briscoe and J.T. Trevors, *Biol. Metals* **2** (1989) 168.
129. J.T. Trevors and M.E. Starodub, *Curr. Microbiol.* **21** (1990) 103.
130. M.E. Starodub and J.T. Trevors, *J. Med. Microbiol.* **29** (1989) 101.
131. S. Silver, in *Biomineralization and Biological Metal Accumulation*, eds. P. Westbroek and E.W. De Jong, Reidel, Dordrecht (1983) 439.
132. D.H. Nies and S. Silver, *J. Bacteriol.* **171** (1989) 896.
133. D.H. Nies, A. Nies, L. Chu and S. Silver, *Proc. Natl. Acad. Sci. USA* **86** (1989) 7351.
134. H. Hennecke, *Mol. Microbiol.* **4** (1990) 1621.

135. D.H. Hamer, *Ann. Rev. Biochem.* **55** (1986) 913.
136. T.R. Butt and D.J. Ecker, *Microbiol. Rev.* **51** (1987) 351.
137. S. Fogel, J.W. Welch and D.H. Maloney, *J. Basic Microbiol.* **28** (1988) 147.
138. C.F. Wright, D.H. Hamer and K. McKenney, *J. Biol. Chem.* **263** (1988) 1570.
139. T.R. Butt, E.J. Sternberg, J.A. Gorman, P. Clark, D. Hamer, M. Rosenberg and S.T. Crooke, *Proc. Natl. Acad. Sci. USA* **81** (1984) 3332.
140. D.R. Winge, K.B. Nielson, W.R. Gray and D.H. Hamer, *J. Biol. Chem.* **260** (1985) 14464.
141. S. Fogel and J.W. Welch, *Proc. Natl. Acad. Sci. USA* **79** (1989) 5342.
142. M. Karin, R. Najarain, A. Haslinger, P. Valenzuela, J. Welch annd S. Fogel, *Proc. Natl. Acad. Sci. USA* **81** (1984) 337.
143. S. Fogel, J.W. Welch and M. Karin, *Curr. Genet.* **7** (1983) 1.
144. R.C.A. Henderson, B.S. Cox and R. Tubb, *Curr. Genet.* **9** (1985) 133.
145. D.H. Hamer, D.J. Thiele and J.E. Lemontt, *Science* **228** (1985) 685.
146. D.J. Thiele and D.H. Hamer, *Mol. Cell. Biol.* **6** (1986) 1158.
147. C-M. Lin and D.J. Kosman, *J. Biol. Chem.* **265** (1990) 9194.
148. D.J. Thiele, *Mol. Cell. Biol.* **8** (1988) 2745.
149. P. Fürst, S. Hu, R. Hackett and D. Hamer, *Cell* **55** (1988) 705.
150. P. Fürst and D. Hamer, *Proc. Natl. Acad. Sci. USA* **86** (1989) 5267.
151. V.C. Culotta, T. Hsu, S. Hu, P. Fürst and D. Hamer, *Proc. Natl. Acad. Sci. USA* **86** (1989) 8377.
152. J. Welch, S. Fogel, C. Buchman and M. Karin, *EMBO J.* **8** (1989) 255.
153. S. Hu, P. Fürst and D. Hamer, *New Biol.* **2** (1990) 544.
154. C. Buchman, P. Skroch, J. Welch, S. Fogel and M. Karin, *Mol. Cell. Biol.* **9** (1989) 4091.
155. J.M. Huibregtse, D.R. Engelke and D.J. Thiele, *Proc. Natl. Acad. Sci. USA* **86** (1989) 65.
156. J.A. Gorman, P.E. Clark, M.C. Lee, C. Debouck and M. Rosenberg, *Gene* **48** (1986) 13.
157. T.R. Butt, E.R. Sternberg, J. Herd and S.T. Crooke, *Gene* **27** (1984) 23.
158. T.R. Butt, M.I. Khan, J. Marsh, D.J. Ecker and S.T. Crooke, *J. Biol. Chem.* **263** (1988) 16364.
159. K. Munger, U.A. German and K. Lerch, *EMBO J.* **4** (1985) 1459.
160. K. Munger, U.A. German and K. Lerch, *J. Biol. Chem.* **262** (1987) 7363.
161. K. Lerch, *Nature* **284** (1980) 368.
162. G.N. George, J. Byrd and D.R. Winge, *J. Biol. Chem.* **263** (1988) 8199.
163. M. Joho, Y. Fujioka and T. Murayama, *J. Gen. Microbiol.* **131** (1985) 3185.
164. M. Inouhe, M. Hiyama, H. Tohoyama, M. Joho and T. Murayama, *Biochim. Biophys. Acta* **993** (1989) 51.
165. D.R. Winge, R.N. Reese, R.K. Mehra, E.B. Tarbet, A.K. Hughes and C.T. Dameron, in *Metal Ion Homeostasis: Molecular Biology and Chemistry*, eds. D.H. Hamer and D.R. Winge, Alan R. Liss, New York (1989) 301.
166. W.E. Rauser, *Ann. Rev. Biochem.* **59** (1990) 61.
167. E. Grill, E.-L. Winnacker and M.H. Zenk, *Experientia* **44** (1988) 539.
168. W. Gekeler, E. Grill, E.-L. Winnacker and M.H. Zenk, *Arch. Microbiol.* **150** (1988) 197.
169. E. Grill, E.-L. Winnacker and M.H. Zenk, *FEBS Lett.* **197** (1986) 115.
170. R.N. Reese, R.K. Mehra, E.B. Tarbet and D.R. Winge, *J. Biol. Chem.* **263** (1988) 4186.
171. A. Murasugi, C. Wada and Y. Hayashi, *J. Biochem.* **90** (1981) 1561.
172. N. Kondo, M. Isobe, K. Imai and T. Goto, *Agric. Biol. Chem.* **49** (1985) 71.
173. Y. Hayashi, C.W. Nakagawa and A. Murasugi, *Environ. Health Perspect.* **65** (1986) 13.
174. E. Grill, E.-L. Winnacker and M.H. Zenk, *Science*, 230 (1985) 674.
175. Y. Hayashi, C.W. Nakagawa, D. Uyakul, K. Imai, M. Isobe and A. Murasugi, *Biochem. Cell Biol.* **66** (1988) 288.
176. R.K. Mehra, E.B. Tarbet, W.R. Gray and D.R. Winge, *Proc. Natl. Acad. Sci. USA* **85** (1988) 8815.
177. R.N. Reese and D.R. Winge, *J. Biol. Chem.* **263** (198) 12832.
178. C.T. Dameron, R.N. Reese, R.K. Mehra, A.R. Kortan, P.J. Carroll, M.L. Steigerwald, L.E. Brus and D.R. Winge, *Nature* **338** (1989) 596.
179. C.T. Dameron and D.R. Winge, *Inorg. Chem.* **29** (1990) 1343.
180. C.T. Dameron, B.R. Smith and D.R. Winge, *J. Biol. Chem.* **264** (1989) 17355.
181. R.K. Mehra, J.R. Garey, T.R. Butt, W.R. Gray and D.R. Winge, *J. Biol. Chem.* **264** (1989)19747.
182. R.K. Mehra, J.R. Garey and D.R. Winge, *J. Biol. Chem.* **265** (1990) 6369.

183. E. Grill, E.-L. Winnacker and M.H. Zenk, *Proc. Natl. Acad. Sci. USA* **84** (1987) 439.
184. J.A. Brierley and C.L. Brierley, in *Biomineralization and Biological Metal Accumulation*, eds. P. Westbroek and E.W. de Jong, Reidel, Dordrecht (1983) 499.
185. S.E. Shumate and G.W. Strandberg, in *Comprehensive Biotechnology*, eds. M. Moo-Young, C.N. Robinson and J.A. Howell, Pergamon Press, New York (1985) 235.
186. C.L. Brierley, D.P. Kelly, K.J. Seal and D.J. Best, in *Biotechnology*, eds. I.J. Higgins and D.J. Best, Blackwell Scientific Publications, Oxford (1985) 163.
187. J.A. Brierley, G.M. Goyak and C.L. Brierley, in *Immobilisation of Ions by Biosorption*, eds. H. Eccles and S. Hunt, Ellis Horwood, Chichester (1986) 201.
188. G.M. Gadd, in *Immobilisation of Ions by Biosorption*, eds. H. Eccles and S. Hunt, Ellis Horwood, Chichester (1986) 135.
189. G.M. Gadd, *Experientia* **46** (1990) 834.
190. G.M. Gadd, in *Microbial Mineral Recovery*, eds. H.L. Ehrlich and C.L. Brierley, McGraw-Hill, New York (1990) 249.
191. G.M. Gadd, *Chem. Ind.* No. 13 (1990) 421.
192. G.M. Gadd and C. White, *Metal–Microbe Interactions* eds. R.K. Poole and G.M. Gadd, IRL Press, Oxford (1989) 19.
193. G.M. Gadd and C. White, *Biotechnol. Bioeng.* **33** (1989) 592.
194. G.M. Gadd and C. White, *Environ. Poll.* **61** (1989) 187.
195. L.E. Macaskie and A.C.R. Dean, in *Biological Waste Treatment*, ed. A. Mizrahi, Alan R. Liss, New York (1989) 159.
196. T.J. Beveridge, in *Metal–Microbe Interactions*, eds. R.K. Poole and G.M. Gadd, IRL Press, Oxford (1989) 65.
197. B. Greene and D.W. Darnall, in *Microbial Mineral Recovery*, eds. H.L. Ehrlich and C.L. Brierley, McGraw-Hill, New York (1990) 277.
198. M. Tsezos, in *Immobilisation of Ions by Biosorption*, eds. H. Eccles and S. Hunt, Ellis Horwood, Chichester (1986) 201.
199. M. Tsezos, in *Microbial Mineral Recovery*, eds. H.L. Ehrlich and C.L. Brierley, McGraw-Hill, New York (1990) 325.
200. L. Birch and R. Bachofen, *Experientia* **46** (1990) 827.
201. S. Hunt, in *Immobilisation of Ions and Biosorption*, eds. H. Eccles and S. Hunt, Ellis Horwood, Chichester (1986) 15.
202. C.L. Brierley, in *Microbial Mineral Recovery*, eds. H.L. Ehrlich and C.L. Brierley, McGraw-Hill, New York (1990) 303.
203. A.J. Francis, *Experientia* **46** (1990) 840.
204. A. McCabe, *Experientia* **46** (1990) 779.
205. J.N. Lester, R.M. Sterritt, T. Rudd and M.J. Brown, in *Microbiological Methods for Environmental Biotechnology*, eds. J.M. Grainger and J.M. Lynch, Academic Press, London (1984) 197.
206. R.M. Sterritt and J.N. Lester, in *Immobilisation of Ions by Biosorption*, eds. H. Eccles and S. Hunt, Ellis Horwood, Chichester (1986) 121.
207. J.A. Scott, S.J. Palmer and J. Ingham, in *Immobilisation of Ions by Biosorption*, eds. H. Eccles and S. Hunt, Ellis Horwood, Chichester (1986) 81.
208. G. Geesey and J. Jang, in *Microbial Mineral Recovery*, eds. H.L. Ehrlich and C.L. Brierley, McGraw-Hill, New York (1990) 223.
209. R.C. Charley and A.T. Bull, *Arch. Microbiol.* **123** (1979) 239.
210. P.A. Goddard and A.T. Bull, *Appl. Microbiol. Biotechnol.* **31** (1989) 308.
211. P.A. Goddard and A.T. Bull, *Appl. Microbiol. Biotechnol.* **31** (1989) 314.
212. C. White and G.M. Gadd, *J. Chem. Technol. Biotechnol.* **49** (1990) 331.
213. L.E. Macaskie, *J. Chem. Technol. Biotechnol* **49** (1990) 357.
214. S.R. Hutchins, M.S. Davidson, J.A. Brierley and C.L. Brierley, *Ann. Rev. Microbiol.* **40** (1986) 311.
215. L. De Rome and G.M. Gadd, *J. Ind. Microbiol.* **7** (1991) 97.
216. S.P. Kuhn and R.M. Pfister, *J. Ind. Microbiol.* **6** (1990) 123.
217. J.S. Watson, C.D. Scott and B.D. Faison, *Appl. Biochem. Biotechnol.* **20/21** (1989) 699.
218. L.E. Macaskie, J.D. Blackmore and R.M. Empson, *FEMS Microbiol. Lett.* **55** (1988) 157.
219. P. Clark, A.J. Butler and L.E. Macaskie, *Biotechnol. Tech.* **4** (1990) 345.
220. L.J. Michel, L.E. Macaskie and A.C.R. Dean, *Biotechnol. Bioeng.* **28** (1986) 1358.

221. K.N. Raymond, G. Muller and B.F. Matzanke, *Top. Curr. Chem.* **123** (1984) 49.
222. M. Tsezos and B. Volesky, *Biotechnol. Bioeng.* **24** (1982) 385.
223. M. Tsezos, *Biotechnol. Bioeng.* **22** (1983) 2025.
224. T. Sakaguchi and A. Nakajima, in *Chitin and Chitosan*, eds. S. Mirano and S. Tokura, Japanese Society of Chitin and Chitosan, Tottori (1982) 177.
225. R.A.A. Muzzarelli, F. Bregani and F. Sigon, in *Immobilisation of Ions by Biosorption*, eds. H. Eccles and S. Hunt, Ellis Horwood, Chichester (1986) 173.
226. D.S. Wales and B.F. Sagar, *J. Chem. Technol. Biotechnol.* **49** (1990) 345.
227. A. Norberg and H. Persson, *Biotechnol. Bioeng.* **26** (1984) 239.
228. D.D. Easson, A.J. Sinskey and O.P. Peoples, *J. Bacteriol.* **169** (1987) 4518.
229. T. Rudd, R.M. Sterritt and J.N. Lester, *J. Chem. Technol. Biotechnol.* **33A** (1983) 374.
230. Z. Zosim, D. Gutnick and E. Rosenberg, *Biotechnol. Bioeng.* **25** (1983) 1725.
231. G.M. Gadd and S.W. Edwards, *Trans. Br. Mycol. Soc.* **87** (1986) 533.
232. C.L. Hansen, G. Zwolinski, D. Martin and J.W. Williams, *Biotechnol. Bioeng.* **26** (1984) 1330.
233. M. Tsezos, *Biotechnol. Bioeng.* **26** (1984) 973.
234. M. Tsezos, R.G.L. McCready and J.P. Bell, *Biotechnol. Bioeng.* **34** (1989) 10.
235. P.S. Lawson, R.M. Sterritt and J.N. Lester *J. Chem. Technol. Biotechnol.* **34B**(1984) 253.
236. N.S. Fisher, J.K. Cochran, S. Krishnaswami and H.D. Livingston, *Nature* **335** (1988) 622.
237. J.M. Tobin, D.G. Cooper and R.J. Neufeld, *Appl. Environ. Microbiol.* **47** (1984) 821.
238. D.P. Weidemann, R.D. Tanner, G.W. Strandberg and S.E. Shumate, *Enzyme Microb. Technol.* **3** (1981) 33.
239. M. Tsezos and B. Volesky, *Biotechnol. Bioeng.* **24** (1982) 955.
240. G.W. Strandberg, S.E. Shumate and J.R. Parrott, *Appl. Environ. Microbiol.* **41** (1981) 237.
241. M. Tsezos and B. Volesky, *Biotechnol. Bioeng.* **25** (1983) 583.
242. G.M. Gadd, *Trends Biotechnol.* **7** (1989) 325.

9 Molecular biology of radiation-resistant bacteria

M.D. SMITH, C.I. MASTERS and B.E.B. MOSELEY

9.1 Introduction

The biosphere is being constantly subjected to a variety of forms and intensities of radiation, many of which are destructive of life because of the damage they cause to DNA. Whilst all bacteria are capable of repairing some of the damage resulting from irradiation, some bacterial genera are much more resistant than others [1–4]. Some of these are resistant to specific types of radiation but sensitive to others, while some bacteria are resistant only under certain circumstances. To illustrate the former, the cyanobacteria are extremely resistant to visible and near-ultraviolet (UV) light but not to far-UV or ionising radiations, which is appropriate since they are photo-synthetic and must be resistant to the harmful effects of the solar radiation that provides their energy [5]. Examples of the latter include bacterial genera such as *Clostridium* and *Bacillus* which are not particularly resistant to far-UV or ionising radiations when growing in the vegetative form but in the resting stage form endospores which are at least an order of magnitude more resistant than vegetative cells [1, 4]. More resistant even than bacterial spores are the vegetative cells of the *Deinococcus* and *Deinobacter* species. These two genera belong to the family Deinobacteriaceae, which is composed of a small but diverse group of bacteria, members of which are characterised by extreme resistance to both far-UV and ionising radiation [2, 6–9]. Few organisms in the world approach their extreme resistance to ionising irradiation and this sets *Deinococcus* and *Deinobacter* apart from other genera [2, 3]. Bacteria belonging to these genera are able to repair large amounts of radiation-induced damage to DNA via biochemical pathways that are not mutagenic [2, 3, 10, 11]. Although the family Deinococcaceae includes both Gram-positive cocci and a Gram-negative rod, most of the species that have been isolated and studied belong to the genus *Deinococcus* and will be referred to simply as 'deinococci' [3, 12–14].

9.2 Types of radiation

There are several types of radiation, which come from various sources and cause different types of damage to biological material. This chapter will focus on bacteria which are resistant to far UV and ionising radiations, but a summary of several types of radiation and the damage they cause may be helpful.

Most of the solar radiation that reaches the earth's surface is in the form of visible light (400–700 nm) which damages cells indirectly because it is absorbed by chromophores [5]. The excited chromophore either transfers the energy to a target molecule or excitation generates radicals such as superoxide which diffuse and degrade a variety of other molecules in the cell. Examples of such chromophores are porphyrin compounds such as porphyrin IV of the electron transport chain in *Escherichia coli* or those involved in photosynthesis in the cyanobacteria. A degree of protection against the harmful effects of light is afforded by bacteria having sufficient amounts of carotenoids and superoxide dismutase which quench free radicals [15–19]. Some bacteria respond to light by producing more pigment and superoxide dismutase than they do in the dark [5, 17, 20]. The resistance of the Deinobacteriaceae to visible light has not been studied.

Near-ultraviolet light (290–400 nm), like visible light, is first absorbed by a chromophore, which then causes damage by transferring energy to the target or by creating radicals [21]. Since near-UV is more energetic than visible light, the damage is greater per photon absorbed. There is less near-UV than visible light in the solar spectrum at the earth's surface [5], but the amount of near-UV flux will increase with the continued depletion of the ozone layer [22].

Far-UV (200–290 nm) is readily absorbed by DNA and RNA and damages both species by dimerising bases and creating other photoproducts which interfere with replication of the DNA and its transcription (for a review, see ref. 2). There is very little far-UV in solar radiation at the surface of the earth. Far-UV is generated by germicidal lamps which are commonly used to sterilise surfaces. A recent report suggests that modulated UV light is more effective at sterilisation than non-modulated [23].

The ionising radiation from the sun and the cosmos is not intense enough to damage microbes significantly. Instead, the greatest concentrations of ionising radiation on earth are from various radioactive deposits and man-made sources. The ionising radiation typically used in laboratories, sterilisation of medical instruments, culture dishes, and food sterilisation are gamma rays from a radioactive ^{60}Co source [2].

When a gamma ray interacts with the irradiated material it transfers energy to electrons, which are raised to a higher energy, or excited, state. If the transferred energy is large enough, the negatively charged electron can leave a molecule, with the result that a positive ion is formed. The ejected electron can move through surrounding material and lose its energy by creating further excited molecules and positive ions. Eventually the ion is captured by a positive ion or is trapped by a structure to form a negative ion. It is this ability to create positive and negative ions that characterises ionising radiation.

Ionising radiations act quite differently from most other forms of chemical or physical agents that are used to inactivate bacteria or other cells in that they do not discriminate between molecules in the irradiated material. The absorption of energy in a particular molecular species will depend, to a first approximation, on the number of electrons, and since biological molecules, including water, are made up of relatively light elements, the number of electrons is proportional to the mass. This means that all molecular species in an irradiated cell are damaged and the scientific puzzle has been to decide which of the bewildering array of damaged molecules are responsible for inactivating cell populations (for reviews, see refs. 24 and 25).

The radiation effect in biological material such as a cell or organism is the sum of two processes—direct and indirect. Direct action describes chemical events resulting from energy being deposited in the target molecule itself, e.g. the ejection of electrons from atoms in the DNA; the indirect effect is a consequence of reactive, diffusible free radicals, mainly hydroxyl radicals formed from the radiolysis of the water in the cells reacting with DNA.

9.3   Radiation resistance of bacterial species

The resistance of bacteria to near-UV, far-UV and ionising radiation varies widely. Figure 9.1 shows survival curves for these radiations and is a composite of data collected from different laboratories. Most vegetative bacteria are sensitive to killing by all three types of radiation (as exemplified by E. coli). Spores of Clostridium perfringens and Bacillus subtilis are resistant to far-UV and X-rays [1, 4]. Various strains of radiation-resistant bacteria have been isolated and not further characterised, such as Pseudomonas radiora [26] and Arthrobacter radiotolerans [27]. The cyanobacteria are extremely resistant to near-UV [5, 28, 29], and some species are resistant to far-UV and gamma radiation (M. Potts, personal communication). The Deinobacteriaceae are extremely resistant to far-UV and gamma radiation [2, 3] but are unusually sensitive to near-UV, even if given a sublethal dose before the challenge dose [30].

9.4   Repair of radiation damage: mutation rates and mutagens

The repair of radiation damage in the deinococci and other bacteria has been covered by reviews [1–4], but a few salient features are important to this discussion. DNA damage does not always result in mutation, and therefore DNA-damaging agents are not always mutagenic. In E. coli, a large amount of radiation induces a repair system that is error-prone, i.e. mutagenic [1, 31]. In contrast, the repair of radiation damage is not mutagenic in the deinococci [10, 11]. Because the deinococci are radiation-resistant and can repair a great

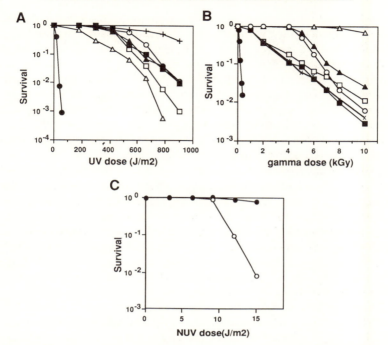

Figure 9.1 Cell survival following irradiation. Fluence is in kilograys (kGy) for ionising radiation or in Joules per square metre (J/m²) for near- and far-UV radiation. -- ○ --, *Deinococcus radiodurans*; -- □ --, *D. radiopugnans*; -- ▲ --, *D. radiophilus*; -- ■ --, *D. proteolyticus*; -- × --, *Deinobacter grandis*; -- + --, *Pseudomonas radiora*; -- △ --, *Arthrobacter radiotolerans*; -- ● --, *E. coli*. (A) Far-UV radiation (254 nm). (B) Gamma radiation from a ⁶⁰Co source. (C) Near-UV radiation (300–400 nm). Data compiled from published data.

deal of DNA damage, they are often mistakenly thought of as having a low spontaneous mutation rate. In fact, the spontaneous mutation rate in the deinococci is about the same as that of *E. coli*, and this rate is not increased by radiation as it is in *E. coli* [11]. Like many bacteria, deinococci can be mutagenised by the powerful mutagen *N*-methyl-*N'*-nitro-*N*-nitrosoguanidine [10, 32] and other nitroso compounds [33, 34]. Strain 302 of *D. radiodurans* is hypermutable by some alkylating agents [35] and has proved useful in the isolation of a large number of DNA repair and auxotrophic mutants [36, 37].

9.5 The biology of the Deinobacteriaceae

Deinococcaceae share several traits which have been carefully described [6–8, 13]. They are radiation- and desiccation-resistant, pigmented, aerobic bacteria that grow well in the laboratory on complex media at 32°C. They are large (2–3 μm diameter), non-motile and do not form spores. The family

Deinococcaceae is one of the 10 major groups of bacteria, based on rRNA cataloguing. It includes a diverse group of phenotypically similar organisms exhibiting sequence divergencies in 16S rRNA equal to the range between *E. coli* and *Streptomyces* [9, 12, 38]. The vast expanse of the Deinococcaceae family is exemplified by the fact that within this group the distinction between 'Gram-positive' and 'Gram-negative' has little meaning: the Gram-positive coccus *D. radiodurans* is more closely related to the Gram-negative rod *Deinobacter grandis* than to the other Gram-positive cocci such as *Deinococcus radiophilus* [14]. Gram positiveness in this case is largely determined by the thickness and complexity of the cell wall [12, 14].

The first known isolate of the Deinobacteriaceae was *Deinococcus radiodurans* strain R1, identified in 1956 as a radiation-resistant contaminant in irradiated canned meat [39]. Because of the extreme resistance of strain R1 to radiation, this strain has been the subject of DNA repair studies for many years [3]. A second *Deinococcus radiodurans* strain (Sark) was isolated 2 years later without radiation selection as a plate contaminant in a hospital [40] and is the focus of studies directed at its unique surface structure and cell division pattern [41–51]. *D. radiodurans* strains R1 and Sark are both naturally transformable [52, 53] and are the most promising candidates for genetic manipulation in the Deinococcaceae. Many other isolates of the Deinococcaceae have been identified, and some have been further characterised [54–66]. Most of the work that has been reported in the Deinobacteriaceae has been on the 'deinococci': *D. radiodurans* and other Gram-positive coccal forms (*Deinococcus radiophilus*, *Deinococcus proteolyticus* and *Deinococcus radiopugnans*). The rod-shaped forms did not receive formal recognition until recently (1987), when Oyaizu *et al.* described *Deinobacter grandis* [14].

Deinococcaceae strains comprise less than 0.1% of the bacteria in soil, but many isolates have been obtained from a variety of sources, often survivors of 'sterilising' doses of radiation of materials [39, 54, 55, 57–62, 64–68]. Environmental study suggests that the Deinococcaceae are most likely soil organisms [67, 68]. In one study, at least 12 Deinococcaceae isolates were obtained from a single soil sample [64, 68]. Phenotypically, these isolates most closely resembled *D. radiopugnans*, but each isolate was a distinct strain, distinguishable from the others by hybridisation with probes for two chromosomal loci (fingerprinting). Three different restriction enzyme activities were found in crude sonicates of each of three isolates [64, 68].

Ionising radiation from natural sources is rarely intense enough to affect the life of even the more sensitive bacteria. To test whether radioresistant organisms would become more prevalent when radiation was part of the environment, a patch of ground in Denmark was exposed to high levels of ionising radiation for a period of months. The relative population of radiation-resistant bacteria increased dramatically, and included deinococcal species [56].

A thermophilic radiation-resistant bacterium, *Arthrobacter radiotolerans*, was isolated from a radioactive hot spring in Japan [27]. *A. radiotolerans* and apparently similar red-pigmented radiation-resistant Gram-positive rods cited in this study and others [56,69] do not seem to be close relatives of *Deinobacter grandis* (R.G.E. Murray, personal communication). Since there is a synergism between radiation and heat damage in *D. radiodurans* [70–73] it would be interesting to explore the relationship between the radioresistant thermophile *Arthrobacter radiotolerans* and the mesophyllic deinococci.

9.5.1   *Radiobiology*

All species of the family Deinococcaceae are extremely resistant to both UV and ionising radiations because they have complex mechanisms for repairing extensive amounts of radiation damage (for reviews, see refs. 2 and 3). At a dose of ionising radiation (6 kGy) which kills 63% of *D. radiodurans* cells and approximates to that required to kill a bacterium, the average genome sustains over 200 double-strand DNA breaks, more than 3000 single-strand breaks, and over 1000 damaged nucleotide bases [3]. At a comparable dose of far-UV ($570 \, J/m^2$), the average genome sustains 12 000 thymine dimers and hundreds of other damaged bases such as thymine glycols [3]. Incredibly, this enormous amount of damage is completely repaired in surviving cells without any induced mutation [10, 33]. It is thought that this repair involves recombination, and a great deal of DNA degradation and resynthesis following damage from massive doses of radiation.

It is curious that the Deinobacteriaceae are most resistant to the kinds of radiation least prevalent on earth (ionising radiation and far-UV) and are unusually sensitive to near-UV, a significant component of sunlight. The resistance of the deinococci to ionising radiation and far-UV radiation may be a reflection of some other adaptation, such as resistance to oxidative damage during desiccation [7,65]. The suggestion that the main lethal effect of near-UV is DNA damage [21] raises the question: 'why are the deinococci, which are able to repair enormous amounts of damage from far-UV and ionising radiation, unable to repair the type of damage done by near-UV?' Laboratory studies on radiation resistance provide controlled measurements of dose and viability, but radiation resistance in the environment may depend on environmental factors such as temperature, oxygen tension, nutrition, growth phase and adaptation. While experiments can measure some of these effects, it is difficult to mimic environmental circumstances in the laboratory even if the typical environment of the bacterium is known.

The deinococci are so resistant to ionising radiation that the survival of other structures besides DNA and RNA may be important for its extreme resistance. For example, it has been suggested that at high radiation doses the lethal effect is delivered, not by even further damage to the DNA but by damage to the lipid membrane [42, 74]. It has been suggested that deinococcal membrane lipids are involved in UV resistance [75, 76] and that the plasma membrane may have properties important to resistance (R.G.E. Murray, personal communication).

9.5.2 *Molecular description of* D. radiodurans

9.5.2.1 *Cell surface.* Species of the Deinococcaceae, Gram-negative and Gram-positive alike, share many features of the complex multilayered surface of the most extensively studied species, *D. radiodurans* (Figure 9.2). Briefly, the plasma membrane and the outer membrane are separated by a peptidoglycan layer and a 'compartmentalised layer' which is loosely constructed and often set into compartment by gooves formed of outer membrane. A number of strains of *D. radiodurans*, but not the other species, are covered by a hexagonal array of protein macromolecules closely associated with the outer membrane [12, 41, 42, 47, 49–51, 77–79]. On some strains the outermost surface is a thick and dense carbohydrate coat [42].

The lipids of the deinococci are complex in composition and distinct from those found in the rest of the bacterial world [16, 80, 81]. The Deinococcaceae seem to lack the usual hydroxy fatty acids and therefore the usual lipopolysaccharide of the outer membrane of Gram-negative bacteria (ref. 14 and R.G.E. Murray, personal communication). *D. radiodurans* possesses none of the conventional phospholipids, and instead its membrane consists of a novel phosphoglycolipid [2′-*O*-(1,2-diacyl-*sn*-glycero-3-phospho)-3′-*O*-(galactosyl)-*N*-D-glyceroyl alkylamine] and several other unidentified glycolipids, glycophospholipids and phospholipids [12, 80–87]. While this alkylamine glycophospholipid appears to be a major component in the plasma membrane in deinococcal strains, there is considerable diversity in the other lipids among the deinococci. Deinococcal fatty acids are a complicated mixture of 15-, 16-, and 17-carbon straight chain and mono-unsaturated acids, and some branched chain fatty acids [80, 81, 84, 85, 88, 89]. Even the monounsaturated acids consist of a mixture of positional isomers differing in the position of the double bond [85].

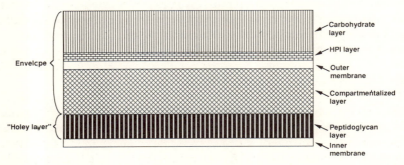

Figure 9.2   Schematic diagram of the cell wall of *D. radiodurans*. See text for further information.

The isoprenoid quinones in the deinococci are unusual in that they are fully unsaturated menaquinones with eight isoprene units, although some strains contain a small amount of seven unit menaquinones [69]. A pink rod-shaped radioresistant organism that shares this trait with the deinococci was cited in the study, lending further credence to the suggestion that a significant number of Deinococcaceae may not be identified as such because they are not cocci.

The thick (14–20 nm) peptidoglycan layer, which probably accounts for the positive reaction of deinococcal cells to the Gram stain, contains alanine, glutamic acid, glycine and, as the diamino acid, L-ornithine instead of the more common lysine or pimelic acid [85, 90, 91]. In *D. radiodurans* and *D. radiopugnans*, holes ('fenestrations') of about 10 nm in diameter appear to pass through this layer [49].

The composition of the compartmentalised layer is unknown. It is partially released from the underlying peptidoglycan layer by treatment with 10% (w/v) sodium chloride, and completely removed by treatment with chloroform or butanol, which makes the cells much more susceptible to the action of lysozyme [79, 92].

The outer membrane contains some of the same lipids as the plasma membrane, but in different proportions. The outer membrane with its attached S-layer can be removed as blebs from *D. radiodurans* strain Sark by suspending the cells in 10% (w/v) sodium chloride or a dilute Tris buffer solution [79].

The outer surface of the cell wall of *D. radiodurans* consists of a protein ($M_r = 104$ kDa) which is arrayed in such a regular paracrystalline lattice that its structure in plan has been determined to 1.5 nm resolution. Three-dimensional image analysis of electron micrographs of isolated fragments demonstrated that the HPI protein (*h*exagonally *p*acked *i*ntermediate) forms one side of a rough hexagon with five other subunits. These proteins are arrayed in an interlocking hexagonal pattern and attach below to the outer membrane and laterally to other units to form the HPI layer [43, 44, 46, 48, 50]. The gene for the major HPI protein has been cloned, characterised and sequenced, and the properties of this gene are discussed below [93, 94]. The HPI protein contains six reducing sugars per polypeptide chain, which may be attached to serine or threonine residues [94]. Fatty acids were also tightly bound to the N-terminal region of the polypeptide in the ratio of about 1.7 : 1 [94].

9.5.3   *Genome of the deinococci*

Species of the Deinococcaceae are characterised by a high GC content, from 65 to 71 mol% [12, 95]. The complexity of the deinococcal chromosome (measured in *D. radiodurans*) is average in size for a bacterium (3000 kb), but the number of identical copies per cell ranges from four in a resting cell to as many as 10 in exponentially growing ones [53, 96–99]. The possibility that the radiation resistance of deinococci is caused by genome multiplicity was raised by the observation that rapidly dividing *E. coli* cells (which have about four chromosomal equivalents per cell) are much more radiation-resistant that *E. coli* cells which are grown slowly in minimal medium and have only one chromosome per cell [100]. It was found, however, that the radiation resistance of *D. radiodurans* does not change as the chromosomal multiplicity is varied from four to 10 [97, 98]. In addition, other bacterial species may have a large number of chromosomal equivalents (e.g. *Azotobacter vinelandii* with 40) and not be radiation-resistant [101]. Thus, chromosomal multiplicity is useful for repair by recombination but is not sufficient on its own to account for radiation resistance. However the deinococci may well have developed a unique recombination mechanism that allows them to use the multichromosome nucleoid for extravagant repair.

The deinococcal chromosome is attached to the plasma membrane, and it has been suggested that these attachments may be important in the process of repair of radiation damage [102–105].

Plasmids have been found in every strain of the Deinococcaceae in which a search has been made (Table 9.1, refs. 106, 107). Although only pUE10 has been cured, no phenotypes have been ascribed to any of the plasmids [106, 108], which are usually low in copy number and larger than 20 kb [106]. The *D. radiophilus* plasmid pUE1 and two *D. radiodurans* plasmids (pUE10 and pUE11) have been cloned in *E. coli* [74, 106, 108]. The pS16 plasmid of *D. radiodurans* strain R1 is isolated in low amounts (0.1 per chromosome) and may be an episome or some other element which occasionally exists as a circle [74, 107]. pS16 bears homology to the *D. radiodurans* chromosome and to the cryptic plasmid pUE11 found in strain Sark [74]. A summary of

Table 9.1 The Deinococcacae.

Isolate	*Deinococcus radiodurans*	*Deinococcus radiopugnans*	*Deinococcus radiophilus*	*Deinococcus proteolyticus*	*Deinobacter grandis*
Morphology	Cocci	Cocci	Cocci	Cocci	Rods
Gram reaction	+	+	+	+	−
G + C content	68	66–68	65	67	69
β-gal assay	−	+	−	−	ND
Transformability	+	−	−	−	ND
Restriction endonucleases	*Mra*I	*Bst*EII* *Pvu*I* *Xho*I*	*Dra*I *Dra*II *Dra*III	ND	ND
Fenestrated peptidoglycan	+	+	−	−	−
Natural plasmids	pS16 pUE10 pUE11	pUE30 pUE31	pUE1 pUE2 pUE3	pUE20 pUE21	pUE15

* Restriction enzymes from isolates of *D. radiopugnans* that are isoschizomers to the enzymes shown here.

ND. Not determined.

Table 9.2 Selected vectors in *Deinococcus radiodurans*.

Plasmid	Description*	Size(kb)	Reference
pUE1	Cryptic, *D. radiophilus*	10.8	106
pUE109†	pUE1:pAT153 chimera	14.8	106
pUE10	Cryptic, *D. radiodurans*	37.0	106
pUE11	Cryptic, *D. radiodurans*	44.9	106
pS11	Duplication insertion	16	107
pEL2	Duplication insertion	14.8	107
pEL5	Duplication insertion *lacZ* fusion	16	109
pUE522†	pAT153:*mtcA mtcB*	14.1	64
pUE200†	pAT153:*uvsC uvsD uvsE*	11.8	68
pUE81†	pAT153::*trp*	9.2	36
pS19	pUE11::pMK20	45	74, 108
pS28	pUE10::pMK20	41.5	74,108
pS30	pUE10::pKK232-8	17.4	110
pEL16	pEL5:pS31 chimera	36	109
pPG100	pS11:*denV*	16.5	111

* pAT153, pKK232-8 and pMK20 are *E. coli* plasmids. Duplication insertion vectors insert by homologous recombination but do not replicate in *D. radiodurans*. pEL5 is a translational fusion between a *D. radiodurans* segment and *lacZ*. pEL16 can replicate in *D. radiodurans* or insert by duplication insertion. pPG100 is a duplication insertion vector.
†These plasmids do not replicate autonomously in *D. radiodurans*.

several plasmids and constructions relevant to *D. radiodurans* is provided in Table 9.2.

The G + C content of the Deinococcaceae is high, ranging from 65 to 71% [6–8]. No methylated bases such as 5-methylcytosine or 6-methyladenine have been identified in the deinococci [112], although some form of modification must protect their DNA from endogenous restriction enzymes such as *Dra*I and *Mra*I [106, 113–115].

Recently the presence of a reiterated sequence ranging in size from 150 to 192 bp in *D. radiodurans* Sark and R1 has been shown (Figure 9.3) [111]. The differences in size depend on the presence or absence of two 21 bp regions within the sequence. There are four areas of dyad symmetry and a direct repeat within the 192 bp length of DNA. The whole sequence is highly conserved and present in at least 40 copies per chromosome in Sark and R1. The possible function of the sequence has not been elucidated but the original sequence was found in one of the DNA damage-inducible *lacZ* fusions which are discussed below. The authors suggest it may be a region involved with regulation of DNA damage-inducible genes. There is also an open reading frame with an upstream Shine–Dalgarno sequence present within the reiterated sequence and this may represent a leader sequence for transport.

9.5.4 *Transformation of the deinococci*

Both the R1 and Sark strains of *D. radiodurans* are naturally transformable and a reproducible and efficient transformation protocol has been established

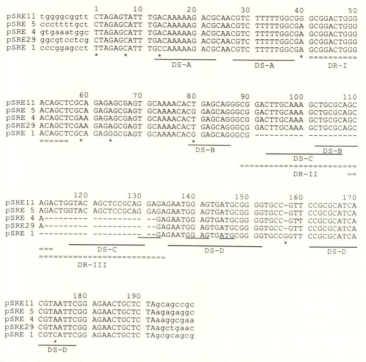

Figure 9.3 Reiterated chromosomal DNA sequence. The reiterated sequence is in uppercase letters. Asterisks denote point differences among the sequences. The dashes in the sequences show the absence of one or both 21-bp regions present in pSRE5 and pSRE11. The dyad symmetries (DS-A, DS-B, DS-C, and DS-D) are indicated by continuous lines below the sequence. The direct repeats (DR-I, DR-II, and DR-III) are indicated by double dashed lines. Proposed Shine–Dalgarno sequences and the ATG start codon are underlined in the sequence. Figure redrawn and adapted from ref. 111.

[53, 53] Studies on the transformation of *D. radiodurans* R1 showed that it becomes competent for transformation when cells are suspended in culture media which contain calcium [53]. Magnesium and strontium are not effective substitutes for calcium, and zinc inhibits transformation. The R1 strain is equally competent throughout logarithmic growth, but stationary phase cells are much less competent [53]. The maintainance of competence throughout logarithmic growth occurs in *Neisseria meningitidis* but is not observed in *Bacillus subtilis, Haemophilus influenzae* or *Streptococcus pneumoniae* [116]. Transformation of *D. radiodurans* is linearly dependent on DNA concentration up to 1 µg/ml. Above 1 µg/ml, transformation was found to increase at a lower rate, such that there was a 5.3-fold increase in transformants as the DNA concentration increased 100-fold. Although single- and double-stranded DNA are taken up with equal efficiency by R1, double-stranded DNA is

10-fold more active in transformation. Uptake is not limited to deinococcal DNA because foreign DNA effectively competes for uptake with deinococcal DNA [53]. In this respect, deinococcal transformation is similar to the transformation of B. subtilis and S. pneumoniae, and unlike the transformation of H. influenzae [116].

D. radiodurans can be grown on a defined medium, and auxotrophic mutants have been isolated [36, 53, 117, 119]. Mutants defective in DNA repair, several drug-resistance mutants, pigment mutants, and temperature-sensitive mutants have also been isolated [32, 52, 53, 119–129]. All these mutations or their wild-type alleles may be used as transformation markers. Transformation frequencies with chromosomal markers vary from 1×10^{-4} to 3×10^{-2}, depending on the marker used [53]. It is therefore possible that R1 has a heteroduplex excision and repair system similar to that described for S. pneumoniae, but this possibility has not been explored [116]. Many important aspects of transformation in D. radiodurans (such as the effect of molecular weight on transformation and the parameters of plasmid transformation) have yet to be explored.

Both Sark and R1 can be transformed by plasmids carrying heterologous drug markers at frequencies ranging from 10^{-4} to 10^{-1} [74, 108].

E. coli plasmids with a segment of cloned deinococcus DNA can transform D. radiodurans by 'duplication insertion' (Figure 9.4), in which the E. coli plasmid integrates into the chromosome by recombination between the cloned

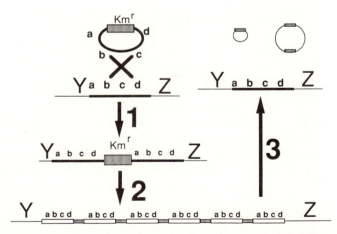

Figure 9.4   Duplication insertion in *Deinococcus radiodurans*. Donor plasmid contains sequences homologous to the deinococcal genome ('abcd') and sequences not homologous (labeled 'Km^r'). (1) Recombination between homologous sequences results in a duplication of the abcd sequence on either side of Km^r. (2) Selection for Km^r (kanamycin resistance) selects for amplification of the abcd-Km^r unit. (3) Recombination between the abcd repeats produces monomeric or multimeric circular forms, and may result in restoration of the wild-type genome. Figure redrawn and adapted from ref. 107.

segment and its homologue in the chromosome [107]. In transformants, the *E. coli* plasmid is flanked by direct repeats of the *Deinococcus* segment. Recombination occurs between the repeats, and there is amplification to give up to 50 copies per chromosome under selection for the drug resistance that is conferred by the *E. coli* plasmid. In the absence of selection, recombination can eliminate the repeats, giving rise to a series of circles which cannot replicate in *D. radiodurans* but can replicate in *E. coli* by virtue of the *E. coli* plasmid the circles contain. This process eventually leads to *D. radiodurans* cells in which the chromosomes have lost the *E. coli* plasmid and the repeats altogether [107]. *D. radiodurans* plasmid pUE11 can also be transformed by duplication insertion if the donor plasmid contains pUE11 sequences and the recipient cell contains pUE11 (M.D. Smith, unpublished observations).

Translational gene fusions were made in *D. radiodurans* by using a *lacZ* vector in duplication insertion experiments [109]. Transformants were screened for fusions which expressed the *lacZ* phenotype at an increased rate after damage by far-UV light. The *lacZ* phenotype in these fusions was not expressed at all when the excised circular forms were transformed into *E. coli*, even though several kilobases of the upstream *D. radiodurans* DNA, and therefore transcriptional and translational start sites, were present in most constructions. When the plasmid DNA from the *E. coli* transformants was reintroduced into *D. radiodurans*, transformants again displayed the damage-inducible *lacZ* phenotype [109].

9.6   Shuttle plasmids between *D. radiodurans* and *E. coli*

Many experiments have failed to transform *D. radiodurans* with any of a variety of cloning vectors from Gram-positive and Gram-negative species. [36, 68, 113, 130, 131]. The pUE1 plasmid from *D. radiophilus* was cloned into *E. coli* plasmid pAT153, and several constructions were devised in an effort to create a shuttle vector, but these were unable to transform *D. radiodurans* in a detectable way [64, 106]. The *cat* and *kan* resistance genes from *B. subtilis* plasmid pBD64 were inserted into *D. radiodurans* plasmid pUE11, but the *D. radiodurans* transformants were not resistant to either chloramphenicol or kanamycin and could only be selected by hybridisation [110]. Shuttle vectors which replicate and express drug resistance in both *D. radiodurans* and *E. coli* were constructed from plasmids pUE10 and pUE11 from *D. radiodurans* strain Sark [74, 107, 110]. The promoter probe plasmids pKK232-8 and pKK175-6 were particularly helpful in the construction of these vectors, for reasons that are described below. These shuttle vectors are useful but are very large, and some effort will be required to develop smaller vectors with more capabilities.

Genetic manipulation via phage or conjugation has not been possible. Attempts, described as 'long and exhaustive', to find a deinococcal phage have been unsuccessful [2, 106]. Conjugation among the deinococci has likewise never been demonstrated [106]. Conjugative plasmids from *Escherichia* (R68.45 and RP4) and *Streptococcus* (pAMβ1, pAD1/pAD2 cointegrate, pAD1::Tn917, pAD1::pAD2::Tn916 cointegrate) were not transferred to derivatives of *D. radiodurans* R1 [36, 131, 132]. No transposition of Tn9, Tn5, or derivatives of Tn10 could be detected after these elements were introduced to *D. radiodurans* by duplication insertion (M.D. Smith, unpublished observations).

9.7   Gene expression in the deinococci

Several results suggest that *D. radiodurans* genes are not expressed well in other bacteria and vice versa. Some examples are the extensive amplification after duplication insertion in *D. radiodurans* [108], the inability to transform *D. radiodurans* with any of a number of plasmids, including the pUE1::pAT153 constructions [36, 64, 106, 131], the inability of the pBD64 genes to express in *D. radiodurans* [110], and the inability of the damage-inducible *lac*Z fusions to express in *E. coli* [109]. In addition, several deinococcal genes have been cloned in *E. coli*, but so far the only deinococcal gene that complements a corresponding *E. coli* mutant is *leuB*, a gene that seems to express in *E. coli* when cloned from any of a variety of sources [113].

Shuttle plasmids able to express drug resistance in *D. radiodurans* were constructed only when *D. radiodurans* sequences were directly upstream from the foreign drug resistance gene [108]. In one experiment, the *E. coli aph* determinant from Tn903 (complete with its *E. coli* promoter) conferred kanamycin resistance to deinococci when placed in one orientation in a site in pUE11, but not when in the opposite orientation at the same site. When drug resistance plasmids are introduced to *D. radiodurans* by duplication insertion in such a way that *D. radiodurans* sequences are positioned directly upstream from the resistance gene, extensive amplification of the duplicated structure is not observed [110]. The suspicion that *Deinococcus* upstream sequences were required for transcription in *Deinococcus* and at the same time not sufficient for transcription in *E. coli* was confirmed when promoterless *cat* and *tet* genes were expressed in *D. radiodurans*, but not in *E. coli*, when *D. radiodurans* sequences were placed upstream from the genes [110].

9.7.1   *The DNA repair genes* mtcA, mtcB, uvsC, uvsC *and* uvsE

D. radiodurans has two independent excision repair systems for repairing far UV damage (ref. 119, for reviews, see refs. 2 and 3). One pathway is also required to repair damage from mitomycin C, and this requires functional *mtcA* and *mtcB* genes. The other pathway requires functional *uvsC*, *uvsD* and *uvsE* genes. Genomic DNA from *D. radiodurans* was cloned into *E. coli* cosmids, and clones which contained these genes were identified by the ability of the cosmid to transform *D. radiodurans* mtc or uvs mutants to wild type [113]. The location of the genes was approximated by subcloning and deletion analysis. The *uvsC*, *uvsD* and *uvsE* genes and the *mtcA* and *mtcB* genes have been located to 8.1 and 10.2 kb DNA fragments respectively. The cloned *uvs* genes were unable to complement a UV-sensitive uvrA⁻ *E. coli* strain [113], which could reflect the problems of expression discussed above.

9.7.2   *The heterologous DNA repair gene* denV *of bacteriophage T4*

The *denV* gene encoding a pyrimidine dimer-DNA glycosylase was cloned into the duplication insertion vector pS11 and transformed into the wild-type and DNA repair mutants of *D. radiodurans* R1 [114]. The *denV* gene had no effect on R1 wild-type or mtc mutants lacking 'UV endonuclease α' but partially complemented uvs mutants lacking the UV endonuclease β [114, 119, 126]. The pattern of complementation suggests that UV endonuclease α does not recognise one or all of the pyrimidine dimers that are recognised by UV endonuclease β and pyrimidine dimer-DNA glycosylase.

9.7.3   *The HPI gene*

The HPI gene has been cloned and sequenced as part of an effort to determine the molecular structure of the surface of *D. radiodurans* [93, 94]. An antiserum was raised to the HPI protein, and the gene for the HPI protein was cloned by screening for *E. coli* recombinants that expressed that antigen [93]. The HPI gene was sequenced and found to code for 1035 amino acids [94]. Amino acid sequencing of about 30% of the HPI protein showed that the N-terminus is blocked and the mature protein contains at least 978 amino acids [94].

The expression of the HPI gene in *E. coli* appears to require an upstream *lac* promoter, although a reasonable fit to the *E. coli* consensus promoter was found in the *Deinococcus* sequence 150 bases upstream from the ATG translation start point [94]. Four bases upstream from the translation start point is a ribosome-binding site GGAGG, which is consistent with the observation that the 3' end of the *D. radiodurans* 16S RNA contains a CCUCC sequence [12, 94].

9.7.4   *The* leuB *gene*

The *leuB* genes from *D. radiodurans* and *D. radiophilus* were cloned on 3.5- and 0.85-kb DNA fragments respectively by selecting for complementation of the *leuB* defect in *E. coli* strain HB101 [113, 131]. These are the only examples to date of complementation of an *E. coli* defect with cloned *Deinococcus* DNA, but the authors noted that the *leuB* gene is complemented by clones from diverse sources such as *Azotobacter vinelandii*, *B. subtilis*, and

Thermus thermophilus. The *leuB* gene has not been used as a marker in *Deinococcus* because no leu⁻ mutants of the deinococci have been isolated.

9.7.5 *The* trp *and* asp *genes*

The *trp* and *asp* genes of *D. radiodurans* have been cloned via complementation of *D. radiodurans* auxotrophic mutants, requiring tryptophan and asparagine respectively [36]. The *trp* gene was subcloned onto a 5.5-kb fragment and the *asp* gene cloned as a 9.3-kb fragment. The *trp* gene did not complement *E. coli* trpA, trpB, trpC, trpD or trpE mutants.

9.8 Prospects in molecular biology

The study of the molecular biology of the radiation-resistant bacteria continues at an accelerating pace. The application of established techniques has proved useful and at other times has been frustrating. The natural transformation of *D. radiodurans* has been central to the understanding of molecular genetics in the deinococci and remains a strong point because highly competent cells are so easy to produce [53]. Non-replicating plasmids can be introduced by transformation and can amplify to 50 copies per chromosome [108]. Duplication insertion affords a great potential here to introduce altered versions of a cloned gene by allelic exchange [116]. Large and cumbersome plasmids are required, at present, to shuttle cloned genes to the deinococci [74, 108, 110].

Continuing advances in cloning and sequencing techniques should improve the prospects for effective genetic systems for cloning, characterising and expressing genes in the deinococci. An array of promoter probes, expression vectors, positive selection vectors and vectors protected by transcription terminators are now available with a variety of restriction sites and selection schemes [133]. The use of promoter probes and the duplication insertion technique can overcome the twin barriers of expression and replication that have plagued the development of molecular genetics in the deinococci.

9.9 Biotechnology of the deinococci

Bacteria are used for many purposes, including the production of speciality chemicals, food additives and enzymes. The properties of the deinococci are described by Murray [6–8], and it appears that the deinococci are attractive choices for several applications. Deinococci are not pathogenic to plants or animals, so they do not require high levels of containment and laboratory personnel do not require extra protection. Deinococci grow to a relatively high cell density in tryptone and yeast extract, and do not require very expensive media components or special conditions such as high temperature, the absence of oxygen or intense illumination. Deinococci grow as individual cells as opposed to mycelial mats and do not differentiate into inert forms such as spores.

9.9.1   Radiation exposure measure

Owing to the desiccation resistance and radiation resistance of the deinococci, it has been proposed that packets of dried deinococci (e.g. *D. radiodurans* strain R1) could be used as a dosimeter for sterilising doses of radiation [60]. The advantage over the present system of using spores is that the actual dose received, rather than a minimum value, is determined because dried deinococci survive 'sterilising' doses of radiation.

9.9.2   Restriction endonucleases

Deinococci are used commercially by several companies to produce restriction endonuclease *Dra*I, which replaced *Aha*III of the same specificity [114]. The unusual feature of *D. radiophilus* is that it produces so much of the enzyme (200 000 units per gram of cells). *Dra*II has also been produced from *D. radiophilus*, but the yield was not published [134, 135]. A Deinococcacal isolate produces a significant amount of *Drd*I (35 000 units/g, ref. 136). Deinococci are also known to produce *Mra*I (an isochizomer of *Sst*II, ref. 115), *Dra*III [134, 135], and isoschizomers of *Bst*EII, *Xho*II and *Pvu*I [64, 68].

The variety of restriction systems sprinkled throughout the bacterial world suggests that genes for restriction systems, which are usually closely linked, may be commonly transmitted between bacteria [137]. The conceptual problem in introducing a new restriction system is that the methylase must protect the cellular DNA before the restriction enzyme is active even though both methylase and endonuclease are transferred together. It seems logical that the deinococci may harbour a wide variety of restriction systems because they may stand a better chance of surviving the double-strand DNA breaks introduced by a new restriction system before the DNA is protected by the methylase.

9.9.3   Z-DNA-binding protein

Alternating d(GC) sequences are the most favoured for forming Z-DNA [138], and thus Z-DNA may be expected to be present in an organism such as *D. radiodurans* (67% G + C, ref. 95). A protein from *D. radiodurans* binds to Z-DNA, and the binding apparently changes the Z-DNA to the B-form [139]. This protein was purified and the N-terminus was determined in the first step of an attempt to clone the gene coding for the protein [140]. The function of the protein has not yet been identified, but it could be labelled and used as a Z-DNA probe.

9.9.4   DNA repair enzymes

A DNA repair enzyme, UV endonuclease β, which incises UV-damaged DNA has been isolated from *D. radiodurans* and partially characterised [141]. The enzyme was purified free of contaminating nuclease and was shown to require manganese for its activity. This activity was not present in uvsC, uvsD, or uvsE mutants of *D. radiodurans* but the gene coding for it may be present on the cosmid which complements these mutations [113, 141].

9.9.5   Membrane-bound exoenzymes

A series of experiments by Mitchel and co-workers characterised two membrane-bound enzymes from *D. radiodurans*. A membrane-bound phosphatase was partially purified from *D. radiodurans* by chromatography in the presence of detergent. This phosphatase (5'-ribonucleotide phosphohydrolase, EC 3.1.3.5) exhibits a strong preference for monophosphorylated deoxynucleotides [142]. A membrane-bound exonuclease (orthophosphoric diester phosphohydrolase, EC 3.1.4.1) is released from radiation-treated *D. radiodurans* cells [143–147]. The unique structure of the *Deinococcus* cell surface allows the isolation of this exonuclease by butanol extraction of whole cells [146]. The outer membrane of deinococci can also be isolated by treatment of whole cells with dilute buffer or 10% (w/v) sodium chloride [47]. The usefulness of these procedures in the production of genetically engineered secreted or membrane-bound products may be a profitable avenue to explore.

9.9.6   DNA polymerase

Several DNA polymerases are produced commercially, and these are used for a variety of purposes. The low mutation rate of DNA repair in *Deinococcus* suggests that it is possible that one or more of the deinococcal repair polymerases may replicate in an error-proof manner.

DNA synthesis during the repair process has been measured in *D. radiodurans* cells damaged in various ways [2, 3, 148]. DNA synthesis and repair has also been studied in *D. radiodurans* cells made permeable by treatment with 20% (w/v) Triton X-100 [149]. Three *D. radiodurans* DNA polymerases were identified as activity peaks from a phosphocellulose column [150, 151]. The major peak of activity was named DNA polymerase I which is associated with both 3'-5' and 5'-3' exonuclease activity that releases 5'-mononucleotides from double-stranded DNA [150]. A start in the cloning of a deinococcal DNA polymerase has been provided by the characterisation of a mutant of *D. radiodurans* which may be temperature-sensitive for a required polymerase [127, 128].

9.9.7 *Manganese*

It has been observed that growth of deinococci is improved in media containing added manganese (R.G.E. Murray, personal communication). Manganese has been reported to affect radiation resistance and induce both cell division and superoxide dismutase levels in deinococci [152, 153]. Many polymerases function poorly in the presence of manganese, but a large amount of manganese is associated with the *Deinococcus* chromosome [154]. The significance of the large amount of deinococcal manganese is unclear at present. The UV endonuclease β from *D. radiodurans* requires manganese [141], but DNA polymerase I of *D. radiodurans* requires magnesium, not manganese [150]. Since most enzymes of DNA metabolism require magnesium, it is usually included in assay buffers, whereas manganese should probably also be included in initial studies of deinococcal enzymes.

9.10   Prospects for biotechnology

The capability of the deinococci to repair accurately enormous amounts of DNA damage must be due to an extraordinary system of DNA metabolism [2, 3]. The enzymes responsible for DNA repair may be found in large quantities, or may have unique properties, or both. Many of these enzymes may prove useful in diagnostic or research procedures. Proteins that are already produced by other organisms besides restriction endonucleases are DNA-binding proteins, endonucleases, exonucleases, phosphatases, glycosidases, topoisomerases, polymerases and ligases. Although the only deinococcal proteins that are commercially produced at the moment are restriction endonucleases, other uses may be developed in the future as more of the properties of these unusual microbes are characterised. There is no predicting what may be found. In the last 5 years, a single thermostable DNA polymerase of the *Thermus aquaticus* replication system has helped to revolutionise DNA amplification, analysis and detection [155].

It is difficult to work with the deinococci, which are not as well characterised as *E. coli* or *B. subtilis*. On the other hand, the rate of research progress is accelerating owing to the application of new techniques and the understanding of mechanisms of gene expression in bacteria. It seems likely, therefore, that in the next few years many interesting and useful tools will emerge from the deinococci as they are further explored.

Acknowledgments

We thank R.G.E. Murray for contributing many helpful comments and unpublished results, Ken Minton and Carol Clark for providing an extensive list of references, and Malcom Potts for information and advice on the cyanobacteria.

References

1. P.C. Hanawalt, P.K. Cooper, A.K. Ganesan and C.A. Smith, *Ann. Rev. Biochem.* **48** (1979) 783.
2. B.E.B. Moseley, *Photochem. Photobiol. Rev.* **7** (1983) 223.
3. B.E.B. Moseley, in *The Revival of Injured Microbes.*, eds. M.H.E. Andrew and A.D. Russell, Academic Press (1984) 149.
4. M.J. Thornley, *J. Appl. Bacteriol.* **26** (1963) 334.
5. L. Van Liere and A.E. Walsby, in *The Biology of Cyanobacteria*, eds. N.G. Carr and B.A. Whitton, University of California Press, Berkeley (1982) 9.
6. R.G.E. Murray, in *Bergeys Manual of Systematic Bacteriology*, Vol. 2, eds. P.A. Sneath, N.S. Mair, M.E. Sharpe and J.H. Holt. Williams & Wilkins, Baltimore (1986) 1035.
7. R.G.E. Murray, in *Perspectives in Microbial Ecology*, eds. F. Megusar and M. Gantor, Slovene Soc. for Microbiol., Ljubljana, *Proc. IV. Int. Symp. on Microbial Ecology* (1986) 153.
8. R.G.E. Murray, in *The Prokaryotes*, 2nd Edn., eds. A. Balows, H.G. Truper, M. Dworkin, W. Harden, and K.H. Schleifer, Springer Verlag, New York (1991).
9. E. Stackebrandt, and C.R. Woese, *Current Microbiol.* **2** (1979) 317.
10. D.M. Sweet and B.E.B. Moseley, *Mutation Res.* **23** (1974) 311.
11. P.R. Tempest and B.E.B. Moseley, *Mutation Res.* **104** (1982) 275.
12. B.W. Brooks, R.G.E. Murray, J.L. Johnson, E. Stackebrandt, C.R. Woese and G.E. Fox, *Int. J. Sys. Bacteriol* **30** (1980) 627.
13. B.W. Brooks and R.G.E. Murray, *Int. J. Sys. Bacteriol.* **31** (1981) 353.
14. H. Oyaizu, E. Stackebrandt, K.H. Schleifer, W. Ludwig, H. Pohla, H. Ito, A. Hirata, Y. Oyaizu and K. Komagata, *Int. J. System. Bacteriol.* **37** (1987) 62.
15. M.A. Carbonneau, A.M. Melin, A. Perromat and M. Clerc, *Arch. Biochem. Biophys.* **275** (1989) 244.
16. N.F. Lewis, M.D. Alur and U.S. Kumta, *Can. J. Microbiol.* **20** (1974) 455.
17. J.M. McCord and I. Fridovich, in *Superoxide and Superoxide Dismutases*, eds. A.M. Michaelson, J.M. McCord and I. Fridovich, Academic Press, London (1977) 1.
18. M.M. Mathews and N.I. Krinsky, *Photochem. Photobiol.* **4** (1965) 813.
19. M.M. Mathews and W.R. Sistrom, *Archiv. Mikrobiol.* **35** (1960) 139.
20. A. Abelovich, D. Kellenberg and M. Shilo, *Photochem. Photobiol.* **19** (1974) 682.
21. A. Eisenstark, *Advanc. Genet.* **26** (1989) 99.
22. M.J. Prather, M.B. McElroy and S.C. Wofsy, *Nature* **312** (1984) 227.
23. H.L. Bank, J. John, M.K. Schmehl and R.J. Dratch, *Appl. Environ. Microbiol.* **56** (1991) 3888.
24. N.E. Gentner and M.C. Paterson, in *Repairable Lesions in Microorganisms*, eds. A. Hurst and A. Nasim, Academic Press, London (1984) 57.
25. F. Hutchinson, *Prog. Nucleic Acid Res. Mol. Biol.* **32** (1985) 115.
26. H. Ito and H. Iizuka, *Agric. Biol. Chem.* **44** (1980) 1315.
27. T. Yoshinaka, K. Yano and H. Yamaguchi, *Agric Biol. Chem.* **37** (1973) 2269.
28. S. Scherer, T.W. Chen and P. Boger, *Plant Physiol.* **88** (1988) 1055.
29. B.A. Whitton, in *Survival and Dormancy in Microorganisms*, ed. Y. Hems, John Wiley & Sons, Chichester (1987) 1.
30. P. Caimi and A. Eisenstark, *Mutation Res.* **162** (1986) 145.
31. J.G. Little and D.W. Mount, *Cell* **29** (1982) 11.
32. N. Rebeyrotte, *Mutation Res.* **108** (1983) 57.
33. D.M. Sweet and B.E.B. Moseley, *Mutation Res.* **34** (1976) 175.
34. P.R. Tempest, *PhD Thesis*, University of Edinburgh (1978).
35. P.R. Tempest and B.E.B. Moseley, *Molec. Gen. Genet.* **179** (1980) 191.
36. G. Al-Bakri, *PhD Thesis*, University of Edinburgh (1985).
37. D.M. Evans, *PhD Thesis*, University of Edinburgh (1984).
38. C.R. Woese, E. Stackebrandt, T.J. Macke and G.E. Fox, *Syst. Appl. Microbiol* **6** (1985) 143.
39. A.W. Anderson, H. Norden, R.F. Cain, G. Parrish and D. Duggan, *Food Technol.* **10** (1956) 575.
40. R.G.E. Murray, and C.F. Robinow, in *VIIth International Congress for Microbiology*, Stockholm, ed. G. Tunevall, Almquist & Wilseus, Uppsala, Sweden (1958) 427.
41. W. Baumeister and O. Kubler, *Proc. Natl. Acad. Sci. USA* **75** (1978) 5525.
42. W. Baumeister, O. Kubler and H.P. Zingsheim, *J. Ultrastruc. Res.* **75** (1981) 60.

43. W. Baumeister, F. Karrenberg, R. Rachel, A. Engel, B.T. Heggeler and W.O. Saxton, *Eur. J. Biochem.* **125** (1982) 535.
44. W. Baumeister, M. Barth, R. Hegerl, R. Guckenberger, M. Hahn and W.O. Saxton, *J. Mol. Biol.* **187** (1986) 241.
45. S.K. Koval and R.G.E. Murray, *Can. J. Biochem. Cell Biol.* **62** (1984) 1181.
46. O. Kubler and W. Baumeister, *Cytobiologie* **17** (1978) 1.
47. P. Lancy Jr. and R.G.E. Murray, *Can. J. Microbiol.* **24** (1977) 162.
48. R. Rachel, U. Jakubowski, H. Tietz, R. Hegerl and W. Baumeister *Ultramicroscopy* **20** (1986) 305.
49. B.G. Thompson and R.G.E. Murray, *Can. J. Microbiol.* **28** (1982) 522.
50. B.G. Thompson, R.G.E. Murray and J.F. Boyce, *Can. J. Microbiol.* **28** (1981) 1081.
51. M.J. Thornley, R.W. Horne and A.M. Glauert, *Archiv. Mikrobiol.* **51** (1965) 267.
52. B.E.B Moseley and J.K. Setlow, *Proc. Natl. Acad. Sci. USA* **61** (1968) 176.
53. S. Tirgari and B.E.B. Moseley, *J. Gen. Microbiol.* **119** (1980) 287.
54. E.A. Christensen and H. Kristensen, *Acta Path. Microbiol. Scand* **89** (1981) 293.
55. N.S. Davis, G.J. Silverman and E.B. Masurovsky, *J. Bacteriol.* **86** (1963) 294.
56. W.H. Eriksen and C. Emborg, *Appl. Env. Microbiol.* **36** (1978) 618.
57. H. Ito, *Agric. Biol. Chem.* **41** (1977) 35.
58. H. Ito, H. Watanabe, M. Takehisa and H. Iizuka, *Agric. Biol. Chem.* **47** (1983) 1239.
59. R.E. Kilburn, W.D. Bellamy and S.A. Terni, *Radiat. Res.* **9** (1958) 207.
60. M Kobatake, S. Tanabe and S. Hasegawa, *C.R. Soc. Paris* **25** (1973) 1506.
61. H. Kristensen and E.A. Christensen, *Acta Path. Microbiol. Scand. B.* **89** (1981) 303.
62. N.F. Lewis, *J. Gen. Microbiol.* **66** (1971) 29.
63. N.F. Lewis and U.S. Kumta, *Biochem. Biophys. Res. Commun.* **47** (1972) 1100.
64. C.I. Masters, *PhD Thesis*, University of Edinburgh (1989).
65. S.W. Sanders, and R.B. Maxcy, *Appl. Environ. Microbiol.* **38** (1979) 436.
67. K.L. Krabbenhoft, A.W. Anderson and P.R. Elliker, *Appl. Microbiol.* **13** (1965) 1030.
66. A.B. Welch and R.B. Maxcy, *J. Food Sci.* **44** (1979) 673.
68. C.I. Masters, R.G.E. Murray, B.E.B. Moseley and K.W. Minton, *J. Gen. Microbiol.* (1991) in press.
69. Y. Yamada, H. Takinami, Y. Tahara and K. Kondo, *J. Gen Appl. Microbiol.* **23** (1977) 105
70. K. Harada, A. Uchida and H. Kadota, *Bull. Japan. Soc. Sci. Fish.* **48** (1982) 415.
71. K. Harada, A. Uchida and H. Kadota, *Bull. Japan. Soc. Sci. Fish.* **50** (1984) 1577.
72. K. Harada, A. Uchida and H. Kadota, *Agric. Biol. Chem.* **48** (1984) 59.
73. A. Uchida, K. Harada and H. Kadota, *Bull. Japan. Soc. Sci. Fish.* **45** (1979) 995.
74. M.D. Smith, L.B. McNeil and K.W. Minton, in *Genetic Transformation and Expression*, eds. L.O. Butler, C. Harwood and B.E.B. Moseley, Intercept, Newcastle upon Tyne (1990) 125.
75. R. Anderson, M. Daya and J. Reeve, *Biochim. Biophys. Acta.* **905** (1987) 227.
76. J. Reeve, L.H. Kligman and R. Anderson, *Appl. Microbiol. Biotechnol.* **33** (1990) 161.
77. U.B. Sleytr, M. Kocur, A.M. Glauert and M.J. Thornley, *Arch. Mikrobiol.* **94** (1973) 77.
78. U.B. Sleytr, T. Silva, M. Kocur and N.F. Lewis, *Arch. Mikrobiol.* **107** (1976) 313.
79. B.G. Thompson and R.G.E. Murray, *Can. J. Microbiol.* **27** (1981) 729.
80. N. Rebeyrotte, P. Rebeyrotte, M.J. Maviel and D. Montaudon, *Annal. Microbiol. (Inst. Pasteur)* **130B** (1979) 407.
81. B.G. Thompson, R. Anderson and R.G.E. Murray, *Can. J. Microbiol.* **26** (1980) 1408.
82. R. Anderson, *Biochim. Biophys. Acta* **753** (1983) 266.
83. R. Anderson, and K. Hansen, *J. Biol. Chem.* **260** (1985) 12219.
84. T.J. Counsell and R.G.E. Murray, *Int. J. Sys. Bacteriol.* **36** (1986) 202.
85. T.M. Embley, A.G. O'Donnell, R Wait and J. Rostron, *System. Appl. Microbiol.* **10** (1987) 20.
86. Y. Huang and R. Anderson, *J. Biol. Chem.* **264** (1989) 18667.
87. D. Montaudon, M-A. Carbonneau, A-M. Melin and N. Rebeyrotte, *Biochimie* **69** (1987) 1243.
88. A.E. Girard, *Can. J. Microbiol.* **17** (1971) 1503.
89. V.A. Knivett, J. Cullen and M.J. Jackson, *Biochem. J.* **96** (1965) 2c.
90. E. Work, *Nature* **201** (1964) 1107.
91. E. Work and H. Griffiths, *J. Bacteriol.* **95** (1968) 641.
92. A.A. Driedger and M.J. Grayston, *Can. J. Microbiol.* **16** (1970) 889.

93. J. Peters and W. Baumeister, *J. Bacteriol.* **167** (1986) 1048.
94. J. Peters, M. Peters, F. Lottspeich, W. Schafer and W. Baumeister, *J. Bacteriol.* **169** (1987) 5216.
95. B.E.B. Moseley and A.H. Schein, *Nature* **203** (1964) 1298.
96. A.A. Driedger, *Can. J. Microbiol.* **16** (1970) 1136.
97. M.T. Hansen, *J. Bacteriol.* **134** (1978) 71.
98. Harsojo, S. Kitayama and A. Matsuyama, *J. Biochem.* **90** (1971) 877.
99. B.E.B. Moseley and D.M. Evans, in *Transformation 1980*, eds. M. Polsinelli and G. Mazza, Cotswold Press, Oxford (1981) 371.
100. F. Krasin and F. Hutchinson, *J. Molec. Biol.* **116** (1977) 81.
101. H.L. Sadoff, B. Shimei and S. Ellis, *J. Bacteriol.* **138** (1979) 871.
102. A.D. Burrell, P. Feldschreiber and C.J. Dean, *Biochim. Biophys. Acta* **247** (1971) 38.
103. M. Dardalhon-Samsonoff and N. Rebeyrotte, *Int. J. Radiat. Biol.* **27** (1975) 157.
104. M. Dardalhon-Samsonoff and D. Averbeck, *Int. J. Radiat. Biol.* **38** (1980) 31.
105. A.A. Driedger, *Can. J. Microbiol.* **16** (1970) 881.
106. M.W. Mackay, G.H. Al-Bakri and B.E.B. Moseley, *Arch. Microbiol.* **141** (1985) 91.
107. M.D. Smith, E. Lennon, L.B. McNeil and K.W. Minton, *J. Bacteriol.* **170** (1988) 2126.
108. M.D. Smith, R. Abrahamson and K.W. Minton, *Plasmid* **22** (1989) 132.
109. E. Lennon and K.W. Minton, *J. Bacteriol.* **172** (1990) 2955.
110. M.D. Smith, C.I. Masters, E.L. Lennon, L.B. McNeil and K.W. Minton, *Gene* **98** (1991) 45.
111. E. Lennon, P.B. Gutman, H. Yao and K.W. Minton, *J. Bacteriol.* **173** (1991) 2137.
112. A. Schein, B.J. Berdahl, M. Low and E. Borek, *Biochim. Biophys. Acta* **272** (1972) 481.
113. G. Al-Bakri, M.W. Mackay, P.A. Whittaker and B.E.B. Moseley, *Gene* **33** (1985) 305.
114. I.J. Purvis and B.E.B. Moseley, *Nucleic Acids Res.* **11** (1983) 5467.
115. A.A. Wani, R.E. Stephens, S.W. D'Ambrosio and R.W. Hart, *Biochim. Biophys. Acta* **697** (1982) 178.
116. G.J. Stewart and C.A. Carlson, *Ann. Rev. Microbiol.* **40** (1986) 211.
117. H.D. Raj, F.L. Duryee, A.M. Deeney, C.H. Wang, A.W. Anderson and P.R. Elliker, *Can. J. Microbiol.* **6** (1960) 289.
118. A. Shapiro, D. DiLello, M.C. Loudis, D.E. Keller and S.H. Hutner, *Appl. Env. Microbiol.* **33** (1977) 1129.
119. D.M. Evans and B.E.B. Moseley, *J. Bacteriol.* **156** (1983) 576.
120. G. Kerszman, *Mutation Res.* **28** (1975) 9.
121. S. Kitayama and A. Matsuyama, *Mutation Res.* **29** (1975) 327.
122. B.E.B. Moseley, *Int. J. Radiat. Biol.* **6** (1963) 489.
123. B.E.B. Moseley, *J. Bacteriol.* **97** (1969) 647.
124. B.E.B. Moseley and H.J.R. Copland, *J. Bacteriol.* **121** (1975) 422.
125. B.E.B. Moseley and H.J.R. Copland, *Molec. Gen. Genet.* **160** (1978) 331.
126. B.E.B. Moseley and D.M. Evans, *J. Gen. Microbiol.* **129** (1983) 2437.
127. B.E.B. Moseley, A. Mattingly and H.J.R. Copland, *J. Gen. Microbiol.* **72** (1972) 329.
128. B.E.B. Moseley, A. Mattingly and M. Shimmin, *J. Gen. Microbiol.* **70** (1972) 399.
129. F. Suhadi, S. Kitayama, Y. Okazawa and A. Matsuyama, *Radiat. Res.* **49** (1972) 197.
130. M.W. Mackay, *PhD Thesis*, University of Edinburgh (1983).
131. I.J. Purvis, *PhD Thesis*, University of Edinburgh (1984).
132. S. Tirgari, *PhD. Thesis*, University of Edinburgh (1977).
133. P.H. Pouwels, B.E. Enger-Valk, W.J. Brammar, in *Cloning Vectors*, eds. Elsevier Science Publishers, Amsterdam (1985) 1.
134. P.B. Gutman, H. Yao and K.W. Minton, *Mutation Research* (1991) in press.
135. C.M. de Wit, B.M.M. Dekker, A.C. Neele and A. de Waard, *FEBS Lett.* **180** (1985) 219.
136. R. Grosskopf, W. Wolf and C. Kessler, *Nucleic Acids Res.* **13** (1985) 1517.
137. C. Polisson and R.D. Morgan, *Nucleic Acids Res.* **17** (1989) 3316.
138. R.J. Roberts, *Nucleic Acids Res.* **18** (1990) 2331.
139. F.M. Pohl and T.M. Jovin, *J. Mol. Biol.* **67** (1972) 375.
140. S. Kitayama, O. Matsumura and S. Masuda, *Biochem. Biophys. Res. Commun.* **130** (1985) 1294.
141. S. Kitayama, O. Matsumura and S. Masuda, *J. Biochem.* **104** (1988) 127.
142. D.M. Evans and B.E.B. Moseley, *Mutation Res.* **145** (1985) 119.
143. R.E.J. Mitchel, *Biochim. Biophys. Acta* **309** (1973) 116.

144. N.E. Gentner and R.E.J. Mitchel, *Radiat. Res.* **61** (1975) 204.
145. R.E.J. Mitchel, *Radiat. Res.* **64** (1975) 321.
146. R.E.J. Mitchel, *Radiat. Res.* **64** (1975) 380.
147. R.E.J. Mitchel, *Biochim. Biophys. Acta* **524** (1978) 362.
148. R.E.J. Mitchel, *Biochim. Biophys. Acta* **621** (1980) 138.
149. B.E.B Moseley and H.J.R. Copland, *J. Gen. Microbiol.* **93** (1976) 251.
150. S. Kitayama and A. Matsuyama, *Biochim. Biophys. Acta* **418** (1976) 321.
151. S. Kitayama, Y. Ishizaka, S. Mujai and A. Matsuyama, *Biochim. Biophys. Acta* **520** (1978) 122.
152. S. Kitayama and A. Matsuyama, *Biochim. Biophys. Acta* **475** (1977) 23.
153. F.I. Chou and S.T. Tan, *J. Bacteriol.* **272** (1990) 2029.
154. J. Wierowski and A.K. Bruce, *Radiat. Res.* **83** (1980) 384.
155. P.J. Leibowitz, L.S. Schwartzberg and A.K. Bruce, *Photochem. Photobiol.* **23** (1976) 45.
156. A. Inris, D.H. Gelfand, J. Sninsky and T.J. White, in *PCR Protocols*, eds. Academic Press, San Diego (1990).
157. D.E. Duggan, A.W. Anderson, P.R. Elliker and R.F. Cain, *Food Res.* **24** (1959) 376.

10 Obligate anaerobes and their biotechnological potential

N.P. MINTON, A. MAULE, P. LUTON and J.D. OULTRAM

10.1 Introduction

To view those organisms which occupy anaerobic niches as extremophiles is, perhaps, surprising, given that the earth occupied by the earliest living organisms is thought to have been almost entirely free of molecular oxygen. Indeed, the organisms which evolved to produce molecular oxygen as a byproduct of photosynthetic activity, and later to consume oxygen during aerobic respiration, have had to adapt to a life in constant contact with the toxic effects of molecular oxygen and its radicals. The basis of the toxicity of oxygen, and its byproducts, to anaerobes has been well studied [1] and refutes the notion that one single causative agent is involved. Rather, aerobic organisms have an array of mechanisms by which the major deleterious effects of exposure are avoided. These may include possession of specific enzymes for the removal of toxic agents (e.g. superoxide dismutase and catalase) as well as an ability to shunt excess reducing power into oxygen removal or protection of key, oxygen-sensitive, cellular components. As with other 'extreme' niches, such as extremes of temperature and salinity, there is some overlap between the true 'extremophile', whose growth optimum lies within extreme conditions, and those facultative organisms whose optimum growth conditions are nearer to the 'normal' range but which can survive or even thrive in extreme environments. In the case of anaerobiosis these include the true or *obligate* anaerobes which cannot grow in the presence of air, through *aerotolerant* anaerobes which can grow, albeit suboptimally, in the presence of air, to the *facultative* anaerobes which are wholly resistant to the toxic effects of oxygen and can flourish either in the presence or absence of air. For the obligate anaerobe even short exposure to air results in a cessation of growth and may be fatal to the organism. For the purpose of this chapter therefore 'extremophilic' anaerobe will be considered to be synonymous with obligate anaerobe, which can be usefully delineated by the criteria of Morris [1] as 'an organism which: (i) generates energy and synthesises its substance without recourse to molecular oxygen: and (ii) demonstrates a singular degree of adverse oxygen-sensitivity which renders it unable to grow under an atmosphere of air'.

With the exception of protozoan species inhabiting the digestive tracts of insects and the herbivore gut, and yeast-like fungi in the rumen (see refs. 2

and 3), by far the majority of extremophilic anaerobes defined thus are bacteria. The main interest which the biotechnologist has in anaerobes stems largely from the unusual fermentations and other biosynthetic reactions which they undertake as a consequence of their inability to utilise oxygen. In the following pages this point will be illustrated by reviewing various aspects of the biotechnological importance of anaerobes, focusing primarily on the clostridia and methanogens.

10.2   The industrial exploitation of anaerobic fermentations

There are three basic mechanisms of energy generation in strictly anaerobic bacteria, namely fermentation, anaerobic respiration and photosynthesis. It is in the first of these, fermentation, that the greatest diversity of substrate and end-products are found. In the absence of oxygen, organic molecules (fermentation intermediates) must act as both primary electron donors and terminal electron acceptors in a series of oxidation–reduction reactions with net generation of ATP. Those products of fermentation which are not subsequently used in biosynthetic reactions are released into the external medium where they can build up to high concentration. It is the great diversity of organic molecules which can serve as fermentation substrates, and the products to which they can be transformed, which offers potential to the biotechnologist.

For instance, during fermentation of hexose and pentose sugars (or following depolymerisation of complex polysaccharides to these simple sugars), saccharolytic clostridia generate ATP by glycolysis to yield pyruvate. It is the subsequent biotransformation of this common intermediate which determines the nature of the fermentation end-products produced. These are species-dependent and can include organic acids such as acetate, butyrate and lactate, and the neutral solvents acetone, ethanol, isopropanol and 1,2-propandiol. In these cases, different fermentation intermediates, formed from a single substrate, act as electron donor and acceptor. Hence, the end-products of the fermentation are divided into those which are more, or less, reduced than the initial substrate. Other molecules which can act as fermentation substrates include organic acids, amino acids, purines and pyrimidines. In many of these reactions (such as the Stickland reaction) two different organic substrates serve as electron donor and acceptor.

10.2.1   The acetone/butanol/ethanol fermentation

The only clostridial pure culture fermentation which has been undertaken on a commercial scale is the acetone/butanol/ethanol fermentation performed by *Clostridium acetobutylicum* and a few closely related strains (e.g. *Clostridium*

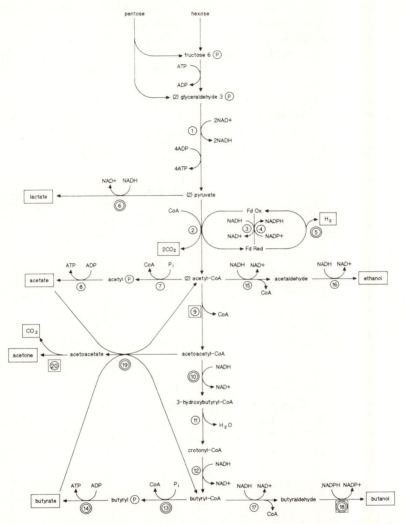

Figure 10.1 The acetone/butanol/ethanol fermentation pathway of *C. acetobutylicum* (adapted from ref. 117). Enzymes are numbered (key below), with those genes which had been cloned in *E. coli* in 1989 [117] circled twice. Genes cloned since have been boxed, and are referenced in Table 10.3. 1, glyceraldehyde-3-phosphate dehydrogenase; 2, pyruvate–ferredoxin oxidoreductase; 3, NADH–ferredoxin oxidoreductase; 4, NADPH–ferredoxin oxidoreductase; 5, hydrogenase; 6, lactate dehydrogenase; 7, phosphate acetyltransferase; 8, acetate kinase; 9, acetyl-CoA acetyl-transferase (thiolase); 10, 3-hydroxybutyryl-CoA dehydrogenase; 11, crotonase; 12, butyryl-CoA dehydrogenase; 13, phosphate butyryltransferase; 14, butyrate kinase; 15, acetaldehyde dehydrogenase; 16, ethanol dehydrogenase; 17, butyraldehyde dehydrogenase; 18, butanol dehydrogenase; 19, acetoacetyl-CoA : acetate/butyrate : CoA transferase; 20, acetoacetate decarboxylase. The second cloned butanol dehydrogenase gene (18) is NADH-dependent. The gene previously reported [117] to encode butyraldehyde dehydrogenase (17) subsequently proved to be lactate dehydrogenase.

saccharoperbutylacetonicum [4]. Industrial interest in the fermentation has fuelled a great deal of research, over many years, into the biochemical pathway by which solvents are formed and the regulatory mechanisms governing the shifts in fermentation pathways which these organisms show. Several authors have reviewed this body of information in detail [5–7], a task which is beyond the scope of this work.

The pathways involved in the fermentation are illustrated in Figure 10.1. Briefly, monosaccharides are catabolised via the Embden–Meyerhof pathway (hexose sugars) or the hexose monophosphate pathway (pentose sugars) to give pyruvate. This is converted to acetyl CoA by pyruvate–ferredoxin oxidoreductase with concomitant release of carbon dioxide and generation of reduced ferredoxin. In the early (acidogenic) phase of the fermentation, either acetyl CoA is converted via acetyl phosphate to acetate with the generation of 1 mole of ATP per acetate produced, or two molecules of acetyl CoA are converted through a series of intermediates to butyryl CoA and on through butyryl phosphate to butyrate (again 1 mole of ATP is formed per mole butyrate produced). During this phase of growth copious hydrogen production is observed, as a method of disposing of excess reducing power, and pH decreases with the accumulation of acid fermentation products. During the second (solventogenic) phase of the fermentation, hydrogen production decreases markedly and the acids are reutilised in the formation of neutral products. The acids are initially converted back to their acyl CoA equivalents by acetoacetyl CoA:acetate/butyrate:CoA transferase, with concomitant conversion of acetoacetyl CoA to acetoacetate, which is then converted to acetone and carbon dioxide by acetoacetate decarboxylase. The acyl CoA moieties can be further reduced to ethanol and butanol by the sequential actions of the appropriate aldehyde and alcohol dehydrogenases. Changes in cell morphology accompany the switch from acid to neutral product formation which are characteristic of cell entering sporulation, and these two properties are frequently lost simultaneously, giving rise to so called degenerate (asolventogenic, asporogenous) cells. The mechanisms governing the transition between these states, and the strength of any link between the two, is the subject of intense research, and no doubt a greater understanding of the regulatory processes which are occurring will be of benefit in the industrial exploitation of the solventogenesis process.

The history and commercial exploitation of this fermentation process have been excellently reviewed by Jones and Woods [5], however some of the salient points of this story are worth repeating. The type strain of *C. acetobutylicum* was isolated by Chaim Weizmann at the University of Manchester between 1912 and 1914 when searching for bacteria which could be used to produce butanol, by fermentation, for use in the production of synthetic rubber. Weizmann, and *C. acetobutylicum*, were called into service by the British government in 1915 when a shortage of calcium acetate, from which acetone was synthesised chemically, threatened the production of cordite for munitions manufacture. Weizmann was able to demonstrate the feasibility of producing acetone on an industrial scale using *C. acetobutylicum*, and over the next 3 years plants dedicated to acetone production were set up in Britain, Canada, the USA and France.

The Weizmann process was used industrially after the war when a market for butanol was created as a solvent for lacquers used in motor car manufacture. The organism was at the centre of the first patent litigation in Britain to involve a biochemical process, when the Commercial Solvents Corporation of the USA acted against a British company which had gone into industrial-scale solvent production using the Weizmann organism. During this period the long-running struggle for economic viability of the process, against acetone synthesis from petroleum, intensified. The isolation of strains with improved fermentation characteristics and the ability to use alternative substrates kept the process viable, indeed expanding, during the interwar years. During the Second World War, the process was again to play a critical role in meeting acetone demand in Britain and the USA. Factories were also established in Japan, Australia, India and South Africa.

The critical factors responsible for the decline in the industrial use of the process after the war were the expansion of the rival petrochemical industry and the competition for molasses for use in cattle feed, which made fermentation feedstock prices prohibitive. In the latter case, it is worth noting that government subsidies to farmers were, at least in part, responsible for the escalation in molasses prices which drove the process out of existence in the USA. In Britain the process was able to continue, using sugarbeet molasses as substrate, until 1957. In South Africa, where molasses are more abundant and petrochemical alternatives more expensive, the process was maintained until 1983.

Although the use of the acetone/butanol fermentation in the western industrialised world has declined, the process still operates competitively in China, where it meets 50% of acetone requirements. Furthermore, a number of factors suggest that a revival of its use in the West is not improbable. The argument that fossil fuels are a finite resource, and that alternative technologies based on renewable resources must be developed, is irrefutable. However, it is likely that socioeconomic forces may represent the deciding factor in the industrial implementation of biocatalytic processes, such as the acetone/butanol fermentation. These factors include heightened public awareness of environmental damage caused by fossil fuel transportation and exploitation and continuing fears of restrictions on access to fossil fuel reserves in the politically unstable oil-producing regions of the world. This optimism for the fortunes of the acetone/butanol process is reflected in the wealth of current research interest which is focused on this fermentation, in fields as diverse as molecular biology, engineering and downstream processing [8–10].

10.2.2   *Methane generation*

The methanogens, as members of the Archaeobacteria, possess many novel enzymes. biochemical pathways and physiological properties. Most investigations into methanogen biochemistry have concentrated on a single unique property, the methano-generating pathway. In this section we will confine ourselves to a description of the biochemistry of methanogenesis. The more practical aspects of methane generation and its exploitation on an industrial scale will be considered in section 10.3.

Although the methanogenic pathway exhibits a high degree of conservation amongst methanogens, slight differences have arisen as a result of evolutionary pressures. Tatiopterin, for example, is present in *Methanogenium tationis* and *Methanogenium thermophilicum* as opposed to methanopterin, which is present in other methanogens. In addition, depending upon their substrate specificity, not all methanogens possess the entire pathway [11]. Although the complete pathway is unique to methanogens, some of its constituents have recently been found in *Archaeglobus fulgidus*, which may be a missing link between methanogens and sulphate reducers [12]. The methanogenic pathway (Figure 10.2) essentially converts carbon dioxide into methane by sequentially reducing the carbon atom from its most oxidised to its most reduced form. Carbon dioxide is passed from one carrier to another down the chain, with the concomitant generation of ATP at the level of methyl CoM reductase. The positioning of ATP generation at this stage in the

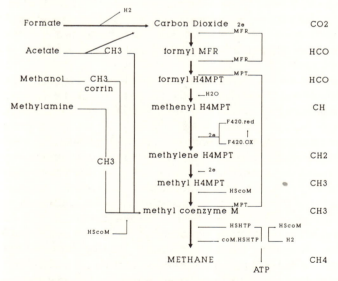

Figure 10.2   Pathway showing the sequence of biochemical transformations in methanogens leading to methane generation. MFR, methanefuran; H$_4$MPT and MPT, methanopterin; HSHTP, mecaptoheptanoyl threonine phosphate; CoM, coenzyme M.

pathway is necessary, as not all methanogens can utilise carbon dioxide and hydrogen as growth substrates, but all do, however, possess methyl CoM reductase.

The generation of ATP itself does not actually occur upon the reduction of methyl CoM to methane, but rather occurs during the reduction of the methyl CoM–mercaptoheptanoyl threonine phosphate (HSHTP), by the methyl CoM reductase complex into its two constituent parts [13, 14]. Methyl CoM reductase is a complex enzyme, consisting of three constituents: A, B and C. Component A is an oxygen-sensitive protein of molecular weight 500 000 daltons [15]; B is an oxygen-sensitive co-factor, presumed to be HSHTP [12]; and component C is CoM reductase. CoM reductase itself is an acidic protein which can account for up to 10% of all cellular protein and is comprised of three subunits in stoichiometrically equal amounts (α, β, γ). The function and genetic composition of proteins α, β and γ are highly conserved across the range of methanogens, as would be expected for such a crucial enzyme [16, 17]. However, two distinct methyl CoM reductases have been discovered in *Methanobacterium thermoautotrophicum*, owing to variation in the gamma gene [18]. Component C, from a range of methanogens, has a tetrahydrocorphin or highly reduced porphinoid ring, factor F430, essential for enzyme function associated with it [19]. The structure and function of co-factors and coenzymes involved in the methanogenic pathway has recently been reviewed by DiMarco *et al.* [20].

Methyl CoM reductase appears to be associated with membranes in all methanogens, providing the opportunity of forming complexes that can produce a transmembrane gradient [21], and thus generation of ATP via a chemiosmotic mechanism. F1F0 ATPases have been reported for methanogen [22], as have membrane-bound corrins, cytochromes and ferredoxins capable of electron transport [14, 23]. That methane formation and ATP generation are coupled at the level of methyl CoM reductase, probably from a central pool of electrons, is well established. The exact mechanisms have yet to be fully elucidated. However, it seems likely that the order Methanomicrobiales has evolved a chemiosmotic process involving hydrogen and sodium ion membrane gradients, while in contrast, the order Methanococcales couples ATP and methane generation directly, avoiding a transmembrane gradient [24].

10.2.3   *Other fermentations*

Several other anaerobic fermentations are close enough to economic viability to warrant interest by the biotechnologist. Among the more attractive of these are the bulk conversion of cellulosic wastes by *Clostridium thermocellum* and the large-scale production of organic acids by the acetogenic clostridia. The thermophilic bulk conversion of cellulose to ethanol by thermophilic clostridia, the best studied of which is *C. thermocellum*, has a relatively long history which has been well documented elsewhere, and various aspects of the enzymology and molecular biology of the process are discussed elsewhere in this chapter.

The homoacetogenic clostridia are of interest because of the scale of demand for acetate (1.3×10^9 kg/year in the USA in 1986), which is set to double if calcium–magnesium acetate (CMA) is adopted as an environmentally safe de-icer for roads in the USA [25]. The physiology and biochemistry of both mesophilic and thermophilic acid producers has been reviewed by Ljungdahl et al. [25], who concluded that CMA could be produced by bulk conversion of corn starch for a cost similar to the current petrochemical production route.

Other less well-characterised fermentations have been described which may be exploitable for the production of specific organic chemicals on a scale sufficient to arouse commercial interest. For instance, Hayashida and Ahn [26] recently reported the isolation of a UV-induced mutant of C. saccharoperbutylacetonicum which failed to produce detectable acids (acetate or butyrate) or solvents (acetone or butanol) during fermentation. Instead, ethanol and isovaleric acid were produced in equimolar ratio (maximum production 23 mM and 20 mM respectively). This fermentation occurred in the absence of added exogenous amino acids (thereby ruling out transamination or oxidative deamination of amino acids as a source of isovalerate), and the authors suggest that this novel fermentation is a result of a build-up of isovalerate as an intermediate of valine biosynthesis from pyruvate. The authors continued their studies [27] to demonstrate that the mutant strain was defective in thiolase activity, and postulate that the excess of reducing power which this lesion generated was shunted into the reductive formation of isovalerate from pyruvate.

Another fermentation pathway which may be exploitable is the novel production of $R(-)$-1,2-propanediol (at greater than 99% enantiomeric excess) and acetol from a variety of sugars (including D-glucose and D-xylose) by Clostridium thermosaccharolyticum [28]. The authors note the potential industrial exploitation of the fermentation as the product is a building block in several organic syntheses, as well as food additive (95 million pounds of racemic 1,2-propanediol, derived from petrochemicals, being used annually in human and pet food). The fermentation is throught to be a variation of the methylglyoxal bypass, wherein dihydroxyacetone phosphate produced during glycolysis is shunted via methylglyoxal into acetol, and subsequently $R(-)$-1,2-propanediol. It is interesting that the three strains which performed the fermentation were all products of mutation and selection regimes for enhanced ethanol production, which may have affected the regulation of alternative fermentation pathways.

10.2.4 Anaerobic mixed culture fermentations

When considering microbial fermentations which might be exploitable by biotechnologists some thought should be given the use of mixed culture fermentations in achieving patterns of fermentations not normally exhibited by a single species. For instance, the anaerobic thermophile *Clostridium thermocellum* has been considered as a candidate for the bulk conversion of cullulosic feedstocks to ethanol, however in co-culture with *Methanobacterium thermoautotrophicum* Weimer and Zeikus [29] found that less ethanol (a reduced end-product) was produced, and that the reducing power was instead used to produce molecular hydrogen, which was used in methanogenesis. This diversion of reducing power allowed the accumulation of acetate as a less reduced end-product of cellulose fermentation by *C. thermocellum*.

Likewise, Yang and Tang [30] demonstrated the conversion of lactate to methane and carbon dioxide, under mesophilic conditions, by a co-culture of *Clostridium formicoaceticum* and *Methanosarcina mazei*. However, in this case, the clostridial strain neither produces nor utilises hydrogen and therefore no interspecies hydrogen transfer can occur. The intermediate used by the methanogen in the co-culture was acetate, and both partners exhibited fermentation characteristics similar to those produced in monoculture under appropriate conditions. The authors note that the co-culture could be linked to the homolactic fermentation of lactose to lactate by a suitable third organism (e.g. *Streptococcus lactis*), to give an overall expected yield of 5.5 mol methane per mol lactose degraded. This is higher than previously found using heterofermentative consortia, and may find application in the biomethanation of lactose (e.g. from cheese whey).

The above examples of anaerobic consortia are representative of fermentations involving the co-culture of defined organisms in the relatively controlled confines of a sterile vessel. However, as will be seen in the next section, the use of complex anaerobic consortia is commonplace worldwide in the disposal of organic waste material.

10.3 Anaerobic disposal of organic wastes

Landfilling and anaerobic digestion are microbially mediated processes used globally to dispose of domestic, agricultural and industrial wastes. Anaerobic digestion of organic material has several advantages over traditional aerobic methods of waste treatment. These include higher solids loading rates, no problems of oxygen limitation, low sludge yield and the end-product (methane) is a valuable energy source [31]. Although the anaerobic processes used to degrade and stabilise wastes vary, they share important features. A heterogeneous population of anaerobic bacteria fungi and protozoa are presented with a diverse range of substrates including carbohydrates, proteins and fats. The degradation of these materials follows four distinct metabolic stages:

(1) *Hydrolyis.* The conversion of complex polymers of insoluble organic compounds into sugars, amino- acids, long chain fatty acids and glycerol.

(2) *Acidogenesis.* Hydrolysis products are converted into organic acids such as acetic acid, propionic acid and butyric acid.
(3) *Acetogenesis.* The products of acidogenesis are converted into acetic acid, hydrogen and carbon dioxide.
(4) *Methanogenesis.* The production of methane from acetic acid or hydrogen and carbon dioxide. Methane may also be formed directly from other substrates such as formic acid and methanol.

The following two sections will describe how anaerobic microbial consortia are used to dispose of and stabilise wastes during landfilling and anaerobic digestion. The pivotal role played by methanogens in this process will receive particular emphasis. Consideration will also be given to the role of methanogens in the breakdown of xenobiotic compounds, which are found in increasingly large quantities in the industrialised world.

10.3.1 Landfilling of wastes

Approximately 20 million tons of solid domestic waste are deposited in landfill annually in the UK. Often, more than 70% of this material is biodegradable, being composed mostly of paper, vegetable matter and putrescibles [32]. Consequently, a landfill site is often considered to be like a large heterogeneous, solid-state fermenter. Initially, just after material is deposited, there is an aerobic phase during which oxygen is depleted and the temperature may begin to rise in the site. At the end of this stage, anaerobic conditions become established and methanogenesis can usually be detected within 200 days. The duration of methanogenesis by a landfill site will be determined by the nature of the refuse and the environmental conditions, and can be anything from 5 to 25 years [33].

Theoretically, landfill sites could yield up to $400\,m^3$ of methane per tonne of refuse, however in practice current yields are of the order of $135\,m^3$/tonne [34]. It seems that only 25% of the available cellulose in refuse is utilised, thus making it a rate-limiting substrate for methanogenesis and ultimately landfill stabilisation [33]. Worldwide there are 242 sites in 20 nations where landfill gas is utilised. Energy savings are believed to be in excess of 2 million tonnes of coal equivalent per annum. In 1989 the USA was the biggest user of landfill gas (1.59 Mtcpa), with the UK in second place (0.162 Mtcpa) [35]. The methane generated in UK landfill sites could satisfy up to 5% of the energy needs [36]. Most of the methane generated is used on site as low calorific value gas for fuelling brick kilns, boilers and gas turbines. However, purification of the gas would allow its storage by liquefaction, conversion to methanol as a chemical feedstock or direct injection into national supply pipelines.

Landfill gas generation can be potentially dangerous and create an explosive hazard if released from the site in an uncontrolled manner. The most notable incident so far was in Loscoe, Derbyshire, UK in 1986, where houses and property were destroyed. The gas had escaped from a landfill site 25 metres from the houses. The site had been filled and capped for 4 years prior to the explosion. Boreholes in the gardens revealed methane concentrations of up to 66%; methane may explode in air at concentrations of 5% or more [37]. It has been estimated that in the UK there are approximately 1800 landfill sites located sufficiently close to housing or other developments to pose a public health hazard should landfill gas migrate off-site [33]. Of these 600 may require the retrofitting of gas control systems to vent off the potentially explosive gas mixture.

In addition to solid domestic waste (SDW) nearly three million tonnes of hazardous industrial wastes are also landfilled each year [38]. Of these hazardous materials about half is landfilled with the SDW in actively methanogenic sites in a process known as co-disposal [39]. In a correctly managed landfill site that is used for co-disposal the chemical and microbiological processes (including methanogenesis) occurring contribute to the degradation or attenuation of hazardous wastes, such that there is no adverse effect on either leachate quality or landfill gas generation. Consequently, it is believed that co-disposal techniques could be used to dispose of a range of compounds including heavy metal wastes, acids, alkalis, phenols, pesticides and polychlorinated biphenyls. However, it must be ensured that the loading rates for these chemicals are within recommended levels. The site must also be actively methanogenic to ensure the maximum stabilisation of the hazardous compounds by the anaerobic community [40]. Thus, it is evident that methanogens play a major role in the degradation of not only domestic organic wastes but also industrial chemicals.

There have been relatively few investigations of the microbiology of landfilling, particularly with reference to the role played by methanogens. This is probably a reflection of not only the difficulties involved in the isolation and enumeration of methanogens from landfill but also the heterogeneous nature of this ecosystem. Sleat et al. [41] studied the numbers of key groups of micro-organisms at different depths within a landfill site. The groups include cellulolytic, hemicellulolytic, sulphate-reducing and methanogenic types. They found that the numbers of all groups, except sulphate reducers, varied dramatically with depth and refuse water content. The hydrolytic bacteria (cellulolytic and hemicellulolytic types) were found in maximum numbers (10^6–10^8/g wet wt refuse) just above the water table. Methanogens, however, both aceticlastic and hydrogen-utilising types, were in maximum numbers (10^6–10^8/g wet wt refuse) at depths below the water table. It is of interest to note that the number of aceticlastic methanogens was considerably lower than that of the hydrogen-utilising types. An analogous situation seems to exist in anaerobic sludge digesters [42]. In another study at the same site (Avely, Essex, UK), it was also shown [43] that methanogenesis was strongly influenced by refuse moisture content. Consequently methanogen activity and biogas generation was found to be much greater at and below the level of the water table than at points above it. They also demonstrated a dramatic reduction in levels of acetate, a major methanogenic substrate at points below the water table.

The first report of successful isolation and characterisation of methanogens from landfill was in 1988 [44]. In this study it proved possible to identify strains of *Methanobacterium formicicum*, *Methanosarcina barkeri* and *Methanobacterium bryantii* using antigenic techniques. However, no indication of the number of methanogens present was given, or even what percentage of the methanogen population the above-named species comprised.

10.3.2   Anaerobic digesters

The principles of the anaerobic digestion of liquid and semiliquid organic materials were first demonstrated on an industrial scale in the late nineteenth century. At this time digesters were built to reduce the biological oxygen demand (BOD) of sewage sludge and some wastewaters [45]. Until the 1950s when better methods became available for studying strictly anaerobic bacteria involved in anaerobic digestion, in particular the methanogens, there were few developments in the process. Up to this time the major changes to the process were based on temperature control and mixing in the reactors to improve digestion rates. A significant finding was that organic acids accumulated in the reactor when the process was functioning suboptimally. The dependence of reactor performance on the efficient conversion of acetic acid by the slow-growing methanogens was thus established [46]. This also helped to explain why reactor operation required long hydraulic retention times, i.e. to prevent washout of the essential methanogens, which can have doubling times in excess of 30 days. Based on evidence such as this, reactors have subsequently been designed which retain or recycle microbial biomass and consequently reduce hydraulic retention times and improve digestion efficiency [31]. Digesters are now available to dispose of a range of materials such as animal wastes from intensive dairy, poultry and pig-rearing units, crop residues and food processing wastes and also to convert agricultural crops into biogas [47].

The types of digester in use and the history of their development have been reviewed [46, 48]. Digesters vary from simple batch-fed static systems to continuous-flow, multistage reactors of advanced design to retain methanogenic biomass. Batch and fed-batch systems are among the most widely used and well-established anaerobic treatment systems. They are easy to construct and maintain, are very effective at treating high solids materials and relatively cheap, which makes them popular in developing countries. In China, India and South Korea, for example, relatively small, simple units are widely used to serve isolated communities. These provide even small settlements with a methane source, and consequently can make them energy self-sufficient. In addition to supplying energy, these units efficiently treat and stabilise wastewater, reduce the number of plant and animal pathogens in the digested material and remove noxious odours. The solid end-product of digestion is a stable, complex organic material rich in nitrogen, phosphorus and potassium which is a good fertiliser and soil conditioner. The importance of anaerobic digestion facilities in the developing world is well recognised, and national biogas development programmes have been set up by the governments in these countries to support this technology. The use of biogas generation systems is particularly widespread in China. In addition to the 5 to 6 million 'family size' units in operation, China also has 100 large-scale biogas power stations with a total output of 8000 kW [48].

Anaerobic digester units installed on farms may greatly reduce pollution incidents related to silage or slurry release into water courses. Not only do these substances cause odour nuisance, they also have far greater polluting power than raw sewage. In 1988 in the UK there were more than 4000 pollution incidents on farms, mostly slurry-related [49]. A Farm and Conservation Grant Scheme operated by the British Government makes available up to 50% of the cost of the farm-based anaerobic digesters. As laws regarding slurry disposal on land become more stringent, the use of farm-based anaerobic digesters is likely to become more widespread, especially as the methane generated would greatly reduce farm energy costs.

Although there is a large body of information on the microbiology of anaerobic digestion of organic matter it is evident that it has not been as extensively researched as other methanogenic environments, such as the rumen [50]. However, from the data available it is apparent that the anaerobic digester is a complex ecological niche, containing large numbers of strictly anaerobic bacteria. Of these, approximately 10% are methanogens [5]. Because the bacteria in the reactors often grow in flocs or clumps and in syntrophic associations (close physical associations where one type of organism provides another type with a growth substrate to their mutual benefit) it is difficult to determine the exact number and type of organisms present. In addition many of the bacteria present may not be actively involved in degradation of the waste compounds in the digester. These have been termed 'passenger' bacteria by Hobson [50] and are generally facultative anaerobes carried over from faecal matter fed into the digester. Although these bacteria have no hydrolytic role they may by scavenging oxygen, metabolise any sugars present, and help maintain a low redox potential suitable for methanogenesis (i.e an E_h of the order of -250 to -300 mV).

The microbial populations which carry out the major transformations during anaerobic digestion have been partially elucidated. The initial stage in anaerobic digestion is the hydrolysis of polymeric substrates, so forming organic acids (including formate, acetate, lactate, propionate and butyrate) ethanol, carbon dioxide and hydrogen. The bacteria involved include members of the following genera: *Clostridium, Bacteroides, Eubacterium, Ruminococcus,*

Butyrivibrio, *Propionibacterium* and *Lactobacillus*. The organic acids lactate, propionate and butyrate are converted to acetic acid and hydrogen in the next stage (acetogenesis) by representatives of the genera *Desulfomonas*, *Desulfovibrio*, *Desulfurolobus*, *Syntrophobacter* and *Syntrophomonas*. These organisms often live in close syntrophic associations with hydrogen-consuming methanogens. The latter maintain a low hydrogen partial pressure, preventing product feedback inhibition of the acetogenic bacteria by hydrogen. This process is called interspecies hydrogen transfer. Substrates for the methanogens may also be supplied by the homoacetogenic bacteria, e.g. *Butyribacterium* and *Acetobacterium*, which synthesise acetate from hydrogen and carbon dioxide.

The methanogens have a limited range of growth substrates. All are able to use hydrogen and carbon dioxide. In addition some can use formate, acetate, methanol and methylamines. In anaerobic digesters about 70% of the methane produced comes from acetate; the remainder is from hydrogen and carbon dioxide. Consequently, most of the methane generated in anaerobic digestion must be by members of the genera *Methanosarcina* and *Methanothrix*, the only methanogens so far shown to utilise acetate [52]. However, these types of methanogen do not appear to be as abundant in digesters as hydrogen or formate utilisers, e.g. *Methanobacterium*, *Methanococcus*, *Methanomicrobium* and *Methanospirillum* [42]. This may be explained by the difficulties in counting members of the genus *Methanosarcina*, which tend to grow in clumps, and consequently their numbers are underestimated by viable counts.

All the data on anaerobic digestion so far presented have referred to growth under mesophilic (20–40°C) conditions. Some work has been conducted under thermophilic conditions, that is temperatures within the range 55–70°C. Although thermophilic anaerobic digestion has been mostly limited to work on sewage sludge, it seems that higher digestion rates and methane yields are possible [52]. However, higher energy inputs are required to maintain thermophilic temperatures, and this may limit its applicability.

From the data available on the microbiology of anaerobic digestion it can be seen that successful operation depends on the interaction of a variety of different bacterial types in a complex consortium. More information on the component members of the anaerobic population will emerge with the advent of the analytical tools described in this chapter. Consequently, this complex anaerobic ecosystem will become better understood, and further optimisation of the process will be facilitated.

10.3.3   Xenobiotic breakdown

Xenobiotic compounds have been appearing in the environment essentially since the beginning of the industrial revolution but in increasing quantity and varying types since the early to mid part of this century. Major sources of this diverse range of compounds are the agrochemical and petrochemical industries. It is estimated that about 300 million tonnes of synthetic organic products are manufactured each year, and a significant percentage of these chemicals will be potentially recalcitrant and disruptive to the environment. Whether released into the environment deliberately such as during agricultural spraying or accidentally because of misuse, these compounds ultimately enter soils, aquatic ecosystems and sediments. Micro-organisms play a major role in the degradation of these compounds in the environment [53]. Thus modern sewage works often treat a mixture of domestic sewage, high in nutrients and biomass and industrial effluent containing chemical cocktails of xeno-biotics. Sewage sludge is therefore a very good source of aerobic and anaerobic bacteria that have been exposed and acclimated to a wide range of chemicals.

The degree to which methanogens are directly responsible for xenobiotic degradation in these anaerobic ecosystems is unknown. However, because methanogens are an integral part of anaerobic digesters, the successful functioning of the digester is dependent on the methanogenic population. The capacity for methanogenic sludge and mixed cultures of methanogens to degrade a wide range of chemicals has been demonstrated. Shelton and Tiedje [54] developed a test system to determine whether a particular chemical could be degraded to carbon dioxide and methane, and this work has since been extended [55] so that a wide range of chemical types are now known to be degraded to various degrees by methanogenic populations. In conjunction with this type of test, systems using fluidised-bed and fixed-film continuous-flow type reactors have been used and similar results obtained with methanogenic consortia from sludges and sediments [56–65].

In most cases, xenobiotic degradation is brought about by mixed consortia in a sequential mechanism and the actual bacterial species responsible are not recorded. However, the fact that methanogens are crucial to the fate of the test compounds is illustrated by Freidman and Gossett [61], who reported that degradation of trichloroethylene stops upon inhibition of methanogenesis with 2-bromoethanesulphonic acid (BES), which acts at the level of methyl CoM reductase, although whether this is due to methanogens directly degrading the compound or acting in an electron flow mechanism with regard to the overall food chain is not known. The existence of a consortium of species probably enhances the degradative process by allowing greater chance of and increased variation in inducible enzymes plus a more tolerant and adaptive system.

Having stated that mixed cultures are probably the most versatile systems for the breakdown of xenobiotics it must be noted that pure cultures of methanogens are capable of degrading some such chemicals; Belay and Daniels [66] first reported that strains of *Methanococcus thermolithotrophicus*, *M. deltae* and *M. thermoautotrophicum* could degrade bromo- and chloro-ethanes, common industrial solvents and pollutants. Reaction mechanisms of hydrolysis, dehydrohalogenation and reductive dehalogenation were proposed to account for the production of ethane, ethylene and acetylene from the parent compounds, with the last two probably being the more important reactions. Mikesell and Boyd [67] also favour a mechanism of reductive dechlorination in their studies in the breakdown of carbon tetrachloride, chloroform and bromoform. The actual mechanism by which methanogens detoxify such chemicals is as yet unknown, although it is reasonable to assume that corrinoid-mediated reactions could be responsible. Methanogens are known to contain relatively large amounts of corrin compounds and, under reducing conditions, corrins dechlorinate certain compounds. For example vitamin B_{12} will catalyse the dehalogenation of the insecticides dieldrin, lindane and DDT. In addition, the organohalogens, carbon tetrachloride, chloroform, bromoform, iodoform and dichloromethane, can be dehalogenated with the production of methane.

A consideration of bioremediation should take into account the fact that most groundwater and soil, from approximately 2 cm below the surface, is anaerobic. Degradation by anaerobic mechanisms is therefore extremely important. The toxicity of polluting compounds, to both consortia and individual species, must also be considered, as well as product inhibition/toxicity (ethylene on methanogens), when dealing with waste clean-up and environment remediation. In this respect, non-living methods, such as corrin-mediated systems, could be of great value. Their production from methanogens is conceivable [68], and a continuous system for the synthesis of corrins using a packed-bed reactor has been described [69]. Interestingly, 5-aminolaevulinic acid (a corrin precursor) may prove useful as a selective herbicide, active against dicotyledons but not monocotyledons [70].

10.4 Anaerobic biotransformations

As has been pointed out earlier, the industrial exploitation of obligate anaerobes in biotechnological processes to date has been limited. As a group, however, they offer considerable potential to the enterprising industrialist. A great deal of the recent interest which has focused on the industrial use of organisms, not only anaerobes, is in the exploitation of biotransformations or bioconversions performed by specific strains or species. While the terms biotransformation and bioconversion can be applied to any process in which the metabolic activity of an organism is used to elicit a change in the composition of a precursor, or mixture of precursors, its usage shall be restricted here so as to exclude the more 'natural' fermentations, i.e. those which the organism is known to perform in its natural environment. Instead, discussion will focus on examples of reactions which exploit the natural metabolism of whole organisms, or isolated enzyme(s) thereof, to facilitate conversions which would not be considered as natural in that sense.

These might include the conversion of precursors which the organism would rarely, if ever, meet in its 'natural' environment, as well as the deliberate manipulation of the natural pathways in order to favour the formation of a desired end-product.

10.4.1 *Fermentation manipulation*

During microbial growth and fermentation many metabolic processes have recourse to finite pools of regenerable intermediates (such as ATP, NAD and NADP) which mediate energy-transfer reactions and supply and store reducing power. During balanced growth, the cell must regulate those catabolic and anabolic processes which would otherwise grossly distort the composition of these pools. Similarly, any externally applied distortion of these pathways must evoke a compensating change in the metabolism of the organism, in order to redress any resultant imbalances. This self-compensation can be used by the biotechnologist to elicit changes in the organism's metabolism to his own benefit. Thus Kim *et al.* [71] observed that inhibition of the hydrogenase of *Clostridium acetobutylicum* ATCC 4259 by carbon monoxide brought about a compensating shift in solventogenesis toward the formation of butanol, which requires a greater input of reducing power than acetone formation. During acidogenesis in *C. acetobutylicum* hydrogen evolution can be considered as a means of disposing of excess electrons, and hence reducing power [72], and therefore inhibiting this route of disposal leads to the formation of more reduced fermentation products. Altered electron flow can be achieved in other ways. Rao and Mutharasan [72] observed shifts in fermentation pattern when viologen dyes, which act as artificial electron carriers (and hence sinks of reducing power), were added to steady-state continuous cultures of *C. acetobutylicum* ATCC 824. Reduction of the viologen dyes was concomitant with a cessation of hydrogen production and a shift in fermentation from the formation of acids to a transient formation of butanol and ethanol. Since the viologen dyes are reduced by hydrogenase, and hence compete with the process by which ferrodoxin is reoxidised during hydrogen evolution, the authors conclude that any mechanism which decreases hydrogenase activity favours increased alcohol production. Viologen dyes have also been shown to mediate the formation of methane from carbon monoxide by *Clostridium thermoaceticum* [73].

Another application of this diversion of electron flow is seen when the precursor of a desired end-product is added to a fermenting culture in order to act as alternative electron sink during co-metabolism. Thus Jewell *et al.* [74] demonstrated that a number of volatile fatty acids (propionate, isobutyrate, valerate) and a hydroxyacid (4-hydroxybutyric acid) could be reduced to their respective alcohols (1,4-butanediol in the last case) by *C. acetobutylicum* NRRL 527 growing on glucose. Again, the range of fatty acids which are reduced by this method probably reflects the broad specificity of the enzymes which reduce butyrate to butanol during the solventogenic phase of the cells' natural fermentation.

10.4.2   Stereospecific, reductive biotransformations

One area which is likely to receive increasing interest, as the metabolism of some of the obligate anaerobes becomes better understood, is the bioconversion of compounds to stereospecific products, a process which is notoriously difficult. if not impossible, to perform in the test tube. The imposition of stereospecificity which occurs during these reactions results from the difference in affinity which an enzyme has for its prochiral substrate when presented in one of two, or more, orientations. The degree of this difference determines the predominance, in the resultant product mix, of one enantiomeric form over the other(s). A number of stereospecific reductions, usually involving cleavage of $C=O$ or $C=C$ bonds, have been demonstrated in obligate anaerobes, and some examples of these have been reviewed by Morris [75].

The biotechnologist, in choosing a particular organism, or isolated enzyme, with which to perform such a biotransformation must give thought to a variety of often competing influences. Obviously, when trying to effect the biotransformation of a substrate outside the natural range of the organism or its isolated enzyme, the system used must be broad enough in its specificity that it will give a sufficient rate of conversion of the precursor to product, but the stereospecificity of the reaction with the novel precursor must be such that an acceptable enantiomeric excess of the desired product is achieved. Lamed and Zeikus [76] demonstrated the potential of a thermophilic aldehyde/ketone oxidoreductase from *Thermoanaerobium brockii* for the reduction of linear and branched-chain primary alcohols, linear and cyclic ketones, and acetaldehyde. The authors observed stereospecific reduction of 2-pentanone to $R(+)$-2-pentanol (giving a final R/S ratio of 9:1) by whole cells, crude cell-free extracts and purified enzyme, when NADP was included to mediate the reduction and propanol to act as a sacrificial reductant for the regeneration of NADPH. The thermostability and broad substrate range of enzyme may make it particularly suitable for the single-step preparative reduction of other aldehydes and ketones, and the authors suggest its use as an immobilised enzyme electrode for the detection of alcohols and carbonyl compounds.

Whether a particular biotransformation is best performed *in vivo* or whether some degree of purification, to crude cell extract or isolated enzyme, is required is also often not clear. Beyond the obvious advantages of using whole organisms, such as the saving in time and cost of enzyme purification, other benefits flow from the use of an *in-vivo* reaction system. The reducing power necessary to perform the reaction will be regenerated by a metabolising cell, and any additional co-factors, etc. which may be essential for the reaction would obviously be available.

Factors mitigating against the use of a whole-cell system include possible uptake problems when using non-natural precursors, together with any toxicity problems which might be associated with the substrate. Also, in the whole-cell system there is a complex array of other enzymes which may either compete for the substrate or bring about secondary, undesired, reactions. A clear example of the former was demonstrated by Belan *et al.* [77], who achieved stereospecific reduction of the ketone sulcatone to $S(+)$-sulcatol (at 88% enantiomeric excess) when washed cells of *Clostridium tyrobutyricum*, which had been grown in crotonate-containing medium, were supplied with the substrate together methyl viologen and hydrogen, which supplied the necessary reducing power for the reaction. However, when the cells were pregrown in medium containing glucose, the same reaction conditions gave a yield of $R(-)$-sulcatol at 80% enantiomeric excess. It was postulated that the organism actually contained two alcohol dehydrogenases, of opposite stereospecificity, and the culture conditions during cell growth influenced the final levels of these competing enzymes in the cell.

In addition to the use of whole-cell co-culture methods it has been shown that cell extracts or purified enzymes from one organism can allow biotransformation to be achieved by a second. For instance, a cell-free extract of *Clostridium kluyveri* has been used to generate reducing power (NADH), from exogenously supplied hydrogen, for subsequent use by an extract of aerobically grown *Enterobacter agglomerans* in the stereospecific reduction of 3-hydroxy-3-methyl-butan-2-one to (S)-2,3-dihydroxy-2-methylbutane [78]. Also, Matsunaga *et al.* [79] demonstrated the use of a three-member system for the conversion of phenylpyruvate to phenylalanine under hydrogen high pressure. In this system, immobilised *Clostridium butyricum* regenerated NADH from hydrogen which allowed continuous production of alanine from pyruvate and ammonium chloride by isolated NADH-dependent alanine dehydrogenase from *Streptomyces phaeochromagenes*. The alanine thus produced donated an amino group to phenylpyruvate, to give phenylalanine, catalysed by the phenylalanine transaminase present in co-immobilised *Micrococcus luteus* whole cells.

10.5 Anaerobic enzymes as industrial products

The use of isolated microbial enzymes for bulk conversion in industrial-scale processes is increasing, and yet the obligate anaerobes have been relatively underexploited in this regard. This probably stems as much from the natural reluctance of industries to handle unfamiliar organisms, and particularly those manipulations which are peculiar to anaerobes, as to any innate inferiority of enzymes from anaerobic sources. While the low growth yields of many anaerobes may mitigate against their direct use, the growing sophistication of microbial genetic techniques means that there are now few insurmountable barriers to the use of 'anaerobic' enzymes cloned into more industrially tractable organisms. The anaerobes are a rich source of biocatalysts, largely because of the very wide range of substrates which anaerobic species are able to ferment, and often the degradative enzymes involved possess properties which make them preferable to their aerobic alternatives. It should be noted that, while the enzymes originate in anaerobic organisms, the majority are insensitive to oxygen.

By far the largest group of enzymes currently used in industry are those which bring about depolymerisation of natural substrates by hydrolytic cleavage. These include a variety of amylases, cellulases, pectinases and proteases. Carbohydrate polymers are generally only extensively degraded by the concerted action of a number of enzymes, some of which depolymerise the substrate to a level at which other enzymes are required to bring about the conversion of these oligomeric moieties to simple metabolisable carbohydrates. Many of the enzymes currently used to accomplish industrial-scale hydrolyses have their anaerobe counterparts, a number of which have already been cloned and sequenced. An extensive analysis of the particular enzymatic properties of each is beyond the scope of this work. Table 10.1 lists examples of alternative anaerobic sources for these activities. While the suitability of an enzyme to perform a particular industrial-scale process will depend upon a number of factors including pH and temperature optima, enzyme stability and end-product inhibition characteristics, etc., a number of the 'anaerobic' enzymes studied have already demonstrated potential in some applications.

Thus, Shen et al. [101] isolated a β-amylase from Clostridium thermosulfurogenes which exhibited extreme thermostability compared with other β-amylases. The enzyme had no requirement for metal ions for activity, but the addition of 5 mM calcium chloride greatly increased the enzyme's stability at high temperature (> 75°C) and during prolonged storage at low temperature (4°C). An apparently identical enzyme has been cloned, from the wild-type strain (C. thermosulfurogenes ATCC 33743 [100]), into Bacillus subtilis and sequenced. The predicted protein sequence of the gene showed homology to Bacillus polymyxa (54%), soybean (32%) and barley (32%) β-amylases. Comparison of the sequences should prove useful in determining the properties which convey extreme thermostability upon the clostridial enzyme. The availability of the cloned gene, and therefore the potential for overexpression of its product, increases the industrial potential for this novel enzyme in the degradation of starches for the production of maltose-containing syrups.

The cellulolytic enzyme systems of various clostridia, but, in particular C. thermocellum, have attracted a great deal of industrial interest, and a corresponding wealth of research in laboratories worldwide. The desire for a highly efficient process for bulk conversion of cellulose is not surprising when recent figures suggest [106] that 4×10^{10} tonnes of cellulose are produced annually. Cellulases have many roles outside the bulk conversion

Table 10.1 Clostridial sources of hydrolysing enzymes.

Enzyme	Reference
Cellulose/Xylan-degrading enzymes	
Endo-14-β-D-glucanase	
C. thermocellum	80
C. stercorarium	81
C. josui	82
Clostridium sp. F1	83
C. cellulolyticum	84, 85
C. acetobutylicum	86
Exo-1,4-β-D-glucanase	
C. thermocellum	87
C. stercorarium	88, 89
Clostridium sp. F1	83
1,4-β-D-Glucosidase	
C. thermocellum	80, 90
C. stercorarium	88
Clostridium sp. F1	83
C. acetobutylicum	91
Endo-β-D-xylanase	
C. thermocellum	80, 92, 93
C. stercorarium	83, 94
C. acetobutylicum	95
β-D-Xylosidase	
C. stercorarium	83, 94
Pectin-degrading enzymes	
Pectin methylesterase	
C. thermosulfurogenes	96
C. aurantibutyricum	97
C. puniceum	97, 98
C. multifermentans	99
Polygalacturonate hydrolase	
C. thermosulfurogenes	96
C. felsineum	97
C. roseum	97
Polygalacturonate lyase	
C. multifermentans	99
C. felsineum	97
C. aurantibutyricum	97
C. roseum	97
C. puniceum	97, 98
Starch/pullulan-degrading enzymes	
β-Amylase	
C. thermosulfurogenes	100, 101
C. thermohydrosulfuricum	102
α-Amylase	
C. thermohydrosulfuricum	102
C. acetobutylicum	103
Pullulanase	
C. thermohydrosulfuricum	104
C. thermosaccharolyticum	105

of cellulose to useful biochemicals, and a purified cellulase system could be employed in areas such as food preparation (e.g. increasing the rehydratability of dried vegetables, increasing the utilisable carbohydrate fraction of animal feeds) and the production of glucose/oligosaccharide solutions for subsequent use as industrial feedstocks [107].

The cellulose-degrading enzyme system of *C. thermocellum* has proved particularly interesting to biotechnologists because of its high activity against crystalline forms of cellulose and its thermostability. The organism itself is considered a candidate for the direct bulk conversion of cellulosic materials to ethanol at elevated temperatures. Enzyme studies, allied to molecular genetic techniques, have revealed a surprising degree of complexity in the cellulase system of *C. thermocellum*. At least 15 distinct endoglucanase genes have been isolated from the organism [80], as have several ancillary genes which also play a role in the degradation process (e.g. *β*-glucosidase and xylanase). The endoglucanases isolated differ in the efficiency with which they will degrade various soluble cellulose derivatives and the lower limit of size of the oligosaccharides produced in the reaction, however none is efficient at attacking more native crystalline substrates. Evidence suggests that *C. thermocellum* achieves its very high rates of degradation of native cellulose by the concerted action of a number of enzymes (constituting at least 70–80% of extracellular endoglucanase activity) organised in the form of an extracellular body, the cellulosome [108]. This body is thought to mediate the strong attachment of the organism to its substrate, as well as to perform degradation of highly crystalline substrates, and exhibits true exoglucanase activity, which, until recently, was not performed by any of the enzymes cloned from the organism. Tuka *et al* [87], however, report the cloning and expression of two cellulase genes from *C. thermocellum* which do appear to show exoglucanase (cellobiohydrolase) activity.

Another thermophilic anaerobe, *Clostridium stercorarium*, has attracted interest because of the observation, by Creuzet *et al.* [109], that its cellulolytic system was much simpler than that of *C. thermocellum*. These workers isolated an enzyme with exoglucanase (cellobiohydrolase) activity, which would effect the saccharification of crystalline cellulose substrates when acting in concert with an endoglucanase and a *β*-glucosidase. Bronnenmier and Staudenbauer [81] have isolated an endoglucanase from this organism which is highly active on Avicel (Avicelase I). This remarkable ability to degrade microcrystalline cellulose and the high thermostability of this peculier enzyme may make it attractive in direct saccharification of cellulosic substrates.

An interesting result of the studies on cloned and purified cellulase enzymes is the finding that many, including several from anaerobes, contain separable domains for catalytic activity and for cellulose binding [110]. The isolation of the region of the gene encoding the cellulose-binding domain has allowed its fusion to other proteins and the demonstration that the fusions produced retain enzymatic activity and can bind to cellulose [111]. This has wide potential application in the fields of enzyme immobilisation and protein purification. Examination of Avicelase [112] has shown that the enzyme appears to contain multiple cellulose-binding domains and this may be responsible for the enzyme's enhanced activity against microcrystalline cellulose.

10.6 Molecular genetic of anaerobic bacteria

The potential usefulness to the biotechnologist of the genetic manipulation of anaerobes cannot be overestimated. A thorough knowledge of the biochemistry and associated molecular biology of the diverse anaerobic processes which these organisms undertake should allow a degree of process control and manipulation which could never be achieved without the application of these methods. The application of such procedures has the following three facets:

(1) *Gene cloning in heterologous shosts.*   Genes encoding enzymes of interest may be cloned, physically characterised and overexpressed in an alternative host, usually *Escherichia coli*. This approach provides invaluable information on gene structure, insight into gene regulation and allows a more thorough examination of the encoding protein's enzymic potential.
(2) *Host–vector system development.* The development of recombinant vehicles into which cloned genes may be inserted, and the elucidation of procedures whereby these vectors may be introduced into the anaerobic micro-organism under investigation, is pivotal to considerations of gene manipulation of industrial bacteria.
(3) *Gene manipulation.* Armed with the necessary genetic elements, gene transfer systems and cloned genes, the derivation of strains with improved fermentation characteristics may be attempted. Improvements made could include: extending the range of substrates metabolised (by gene addition), and manipulating the pathways concerned with product formation both by amplifying and regulating the production of rate-limiting enzymes (gene overexpression/regulation) and by abolishing the synthesis of enzymes detrimental to product formation (gene deletion/regulation).

There are now many examples of (1) and (2) in anaerobic bacteria. Instances of (3) are, to date, scarce. In the following section some of the developments will be reviewed. More details may be found in reviews published elsewhere, on *Bacteroides* [113–115], methanogens [116], *C. acetobutylicum* [117], and clostridia generally [115, 118].

10.6.1 *Host–vector system development*

The two principal requirements of a host–vector system are: (1) the availability of a cloning vehicle encoding a readily selectable genetic trait and (2) a means of introducing it into the host organism under investigation. Cloning vehicles are, more often than not, autonomously replicating plasmids. Selectable markers are usually genes specifying resistance to an antibiotic. However, the importance of integrative elements, such as transposons, phages or replication-deficient plasmids, should not be overlooked when considering the development of a recombinant cloning vehicle. Transfer of the vehicle into the host may be achieved directly, by transformation of naked DNA, or indirectly by conjugative transfer from a bacterial donor or through the agency of phage infection.

10.6.1.1 *Transformation.*   Transformation procedures have been formulated for *C. acetobutylicum* using either protoplasts or whole cells. In the former procedure osmotically stabilised cells stripped of their cell walls are induced to take up plasmid or phage DNA by the addition of polyethylene glycol (PEG). Transformed protoplasts are then induced to revert back to bacillary rods by growth on appropriately selective regeneration media. Various workers [119, 120] have applied this procedure to different strains of *C. acetobutylicum*, however the most successful is that described by Reysset *et al.* [121]. The efficiency of their procedure may be largely attributed to the steps taken to limit autolysin production, and more than 10^6 plasmid transformants per microgram of DNA were obtained. Protoplast or, rather, L-phase variant transformation procedures have also been reported for *Clostridium perfringens* [122, 123].

A number of procedures for the transformation of whole cells of anaerobic bacteria have been developed. These include PEG-induced transformation of *Bacteroides fragilis* [124], an alkaline–Tris method for *C. acetobutylicum* and *C. thermohydrosulfuricum* [125, 126], and an electroporation procedure for *C. acetobutylicum* [127]. The last procedure involves the subjection of cell suspensions to a brief (2.5–10 ms) high-voltage electric field pulses (2–12 kV/cm), and is now the method of choice for devising transformation protocols in bacteria. Its application to *C. acetobutylicum* NCIB 8052 in the presence of plasmid DNA results in transformation frequencies of up to 10^4 transformants per μg DNA [127]. Although the adaption of this procedure to other strains of *C. acetobutylicum*, and in particular ATCC 824, initially proved problematical, successful transformation has now been obtained (G. Bennett, H. Blaschek and E. Popousakis, personal communication). Electroporation methodologies for *C. perfringens* and *Bacteroides* spp. have also been described [128–131].

Interestingly, genetic transformation of two species of methanogen, *Methanococcus voltae* PS and *M. thermoautotrophicum* Marburg, has been described [132, 133] and appears to rely on the development of natural competence. In the case of *M. thermoautotrophicum*, competence appeared to be induced by plating, and transforming, the organisms on gellan gum agar. *M. voltae* cells appear to become naturally competent during exponential phase growth on standard media, although Ca^{2+} was shown to be essential for its development. In both cases, the transformed material was chromosomal DNA, with selection for either prototrophy in the case of *M. voltae* auxotrophic mutants, or 5-fluorouracil resistance with wild-type *M. thermoautotrophicum*.

10.6.1.2 *Conjugative transfer.* The possession of conjugative, antibiotic-encoding genetic elements by certain anaerobic species, both of an integrative (transposons) and of an autonomous (plasmid) nature, is commonplace. As the organisms involved (e.g. *Bacteroides* and pathogenic *Clostridium* spp.) are principally inhabitants of the gut of vertebrates, their frequent exposure to antibiotics explains the predominance of such elements in these organisms. Although the ability of *Bacteroides* spp. to act as conjugal recipients has been exploited to develop gene transfer systems (e.g. refs. 134 and 135), conjugation as a means of vector delivery remains unused in *C. perfringens*. Conjugative vector transfer systems have, however, been devised for the industrially important clostridial species *C. acetobutylicum*.

Conjugative transfer to *C. acetobutylicum* was first demonstrated by Oultram and Young [136], when they showed that the broad-host-range enterococcal plasmid pAMβ1 could be transferred from various Gram-positive donors to *C. acetobutylicum* NCIB 8052. Building on these findings, these same workers subsequently showed that pAMβ1 could effect the mobilisation of cloning vectors from *B. subtilis* to *C. acetobutylicum*, in the form of cointegrate molecules [137]. More recently Williams *et al.* [138] have shown that cloning vectors endowed with the origin of transfer of plasmid RK2 may be conjugatively transferred from *E. coli* to *C. acetobutylicum*, provided the necessary transfer functions are supplied *in trans*. This system has proved useful in testing the ability of different plasmid replicons to function in this clostridial species [139]. Similar IncP-dependent mobilisation systems have been derived for *Bacteroides* [134, 140].

In addition to plasmids, conjugal transposons are beginning to play an important role in the genetic analysis of *C. acetobutylicum*. A number of studies have shown that the well-characterised streptococcal transposons Tn*916* and Tn*1545* can be stably transferred and expressed in different strains of *C. acetobutylicum* [141, 142]. Their future use should facilitate the physical and genetic mapping of the genome of this organism, and offers potential for the targeted cloning of selected genes (see ref. 118). Indeed, Tn*916*-induced mutants of *C. acetobutylicum* defective in solvent production have recently been obtained [143]. The chromosomally located conjugative elements found in *Clostridium* spp. [144, 145] have not, aside from their demonstrable ability to transfer to *B. subtilis* [146], been further exploited. In contrast, similar *Bacteroides* elements have found use as the conjugal mobiliser for cloning vectors [134, 135].

10.6.1.3 *Cloning vectors.* In a number of anaerobic bacteria, numerous cloning vectors have now been derived (Table 10.2). These vectors are, in the

Table 10.2 Anaerobe vectors.

Plasmid	Size (kb)	Source of plasmid	Host-range* replicon	Transfer method†	Features/comment‡	Reference
Vectors for *Bacteroides*						
pFD176	7.3	pBII143 (*B. ovatus*)	*Bacteroides* spp. (*E. coli*)	Transformation	Cc^R, [Ap^R], general-purpose cloning vector	147
pBI191	5.3	pBII143 (*B. ovatus*)	*Bacteroides* spp.	Transformation	Cc^4, unable to replicate in *E. coli*	147
pKBF367-12	10.5	pBF367 (*B. fragilis*)	*Bacteroides* spp. (*E. coli*)	IncP-mobilisation	Cc^R, [Km^R], general-purpose cloning vector	148
pDK3	8.5	pCP1 (*B. fragilis*)	*Bacteroides* spp. (*E. coli*)	IncP-mobilisation	Cc^R, [Km^R], general-purpose cloning vector	140
pVAL-1	11.0	pB8-51 (*B. eggerthii*)	*Bacteroides* spp. (*E. coli*)	IncP-mobilisation	Cc^R, [Ap^R, Tc^R], general-purpose cloning vector	135
pVAL-7	9.5	(Rep$^-$ deletion of pVAL-1)	*E. coli*	IncP-mobilisation	Cc^R, [Ap^R], suicide vector, Rep$^-$ in *Bacteroides*	149
Vectors for *Clostridium perfringens*						
pJU12	11.6	pJU121 (*C. perfringens*)	*C. perfringens* (*E. coli*)	Transformation	Tc^R, [Ap^R, Tc^R], rudimentary cloning vector	150
pJU13	12.2	pJU122 (*C. perfringens*)	*C. perfringens* (*E. coli*)	Transformation	Tc^R, [Ap^R, Tc^R], rudimentary cloning vector	150
pJU16	12.2	pJU122 (*C. perfringens*)	*C. perfringens* (*E. coli*)	Transformation	Tc^R, [Ap^R, Tc^R], rudimentary cloning vector	150
pHR106	7.9	pJU122 (*C. perfringens*)	*C. perfringens* (*E. coli*)	Transformation	Cm^R, [Ap^R, Cm^R], general purpose cloning vector	151
pSB92A2	7.9	pCP1 (*C. perfringens*)	*C. perfringens* (*E. coli*)	Transformation	Cm^R, [Ap^R, Cm^R], general purpose cloning vector	130
pAK201	8.0	pHB101 (*C. perfringens*)	*C. perfringens* (*E. coli*)	Transformation	Cm^R, [Ap^R, Cm^R], general purpose cloning vector	152
Vectors for methanogens						
Mip1	7.4	(None)	*M. voltae* (*E. coli*)	Transformation	Pm^R, His$^+$ [Ap^R, His$^+$], Rep$^-$ integrates into host *hisA*	153
Mip2	7.4	(None)	*M. voltae* (*E. coli*)	Transformation	Pm^R, His$^+$ [Ap^R, His$^+$], Rep$^-$ integrates into host *hisA*	153

Vectors for *Clostridium acetobutylicum*

pKNT11	6.5	pIM13 (*B. subtilis*)	*C. acetobutylicum*/*B. subtilis* (*E. coli*)	Transformation	Em^R, [Ap^R], general-purpose cloning vector	154
pKNT14	4.3	pIM13 (*B. subtilis*)	*C. acetobutylicum*/*B. subtilis*	Transformation	Em^R, Rep$^-$ in *E. coli*	154
pCTC1	7.18	pAMβ1 (*E. faecalis*)	*C. acetobutylicum*/*B. subtilis* (*E. coli*)	IncP-mobilisation	Em^R, [Ap^R], broad-host-range cloning vector	138,139
pCTC511	7.85	pCB101 (*C. butyricum*)	*C. acetobutylicum*/*B. subtilis* (*E. coli*)	IncP-mobilisation	Em^R, [Ap^R], general-purpose cloning vector	138,139
pMTL30/31	4.36	[None]	*E. coli*	IncP-mobilisation	Em^R, [Ap^R], potential suicide vector	138
pMTL20/21E	3.60	[None]	*E. coli*	Transformation	Em^R, [Ap^R], clostridial replicon probe vector	156
pCP3	10.40	pCP1 (*C. paraputrificum*)	*C. acetobutylicum* (*E. coli*)	Transformation	Em^R, [Ap^R], general-purpose cloning vector	117,155
pCB3	7.03	pCB101 (*C. butyricum*)	*C. acetobutylicum*/*B. subtilis* (*E. coli*)	Transformation	Em^R, [Ap^R], general-purpose cloning vector	117,155
pCB4	7.10	pCB102 (*C. butyricum*)	*C. acetobutylicum*. (*E. coli*)	Transformation	Em^R, [Ap^R], general-purpose cloning vector	117,155
pCB5	9.50	pCB103 (*C. butyricum*)	*C. acetobutylicum*, (*E. coli*)	Transformation	Em^R, [Ap^R], general-purpose cloning vector	117,155
pMTL500E	6.43	pAMβ1 (*E. faecalis*)	*C. acetobutylicum*/*B. subtilis* (*E. coli*)	Transformation	Em^R, [Ap^R], broad-host-range clonging vector	127
pMTL502E	7.52	pAMβ1 (*E. faecalis*)	*C. acetobutylicum*/*B. subtilis* (*E. coli*)	Transformation	Em^R, [Ap^R], low-copy-number version of pMTL500E	156
pMTL500d	6.69	pAMβ1 (*E. faecalis*)	*C. acetobutylicum*/*B. subtilis* (*E. coli*)	Transformation	Em^R, [Ap^R], expression vector	117
pMTL513	7.29	pAMβ1 (*E. faecalis*)	*C. acetobutylicum*/*B. subtilis* (*E. coli*)	Transformation	Em^R, [Ap^R], stability cloning vector	117
MTL710e	7.38	pAMβ1 (*E. faecalis*)	*C. acetobutylicum*/*B. subtilis* (*E. coli*)	Transformation	Em^R, [Ap^R], promoter probe vector	117

* The majority of vectors have been designed to shuttle between the anaerobic host and *E. coli*. Where this is the case it is indicated by (*E. coli*). It should be noted that vectors based on the pAMβ1 replicon have an extremely broad host range and will function in most Gram-positive bacteria.

† The transfer method indicated is that original employed for introducing the constructed vectors into the anaerobic host. It should be borne in mind that all vectors should be capable of being transferred by transformation procedures, while certain 'transformation' vectors may be capable of being mobilised.

‡ Selection for the vector in *E. coli* is bracketed. R, resistance; Ap, ampicillin; Tc, tetracycline, Cc, clindomycin; Cm, chloramphenicol; Em, erythromycin; Pm, puromycin. The *C. perfringens*-derived Tc^R and Cm^R genes both function in *E. coli*. The *Bacteroides* Cc^R element confers, in certain instances, T^R resistance on *E. coli* cells grown aerobically (see ref. 114). The pAMβ1 Em^R gene also function in *E. coli*. However, because wild-type strains are inherently resistant to high levels of this antibiotic, selection may only be achieved by the use of mutant hosts which exhibit enhanced sensitivity.

main, shuttle plasmids, designed to 'shuttle' between the anaerobic host and a more genetically amenable organism, i.e. *E. coli*. Vectors have been derived either by combining host plasmid replication origins with selectable genetic markers (usually antibiotic resistance genes) or by employing well-established cloning vectors from other bacterial species which fortuitously replicate in the anaerobic host. In the case of methanogens, although transformation procedures have been available for a number of years, suitable autonomously replicating cloning vectors with which to capitalise on this facility have not yet been developed. Plasmids have been isolated from certain species [116, 157, 158], and indeed the entire nucleotide sequence of one particular plasmid, pME2001, has been determined [159]. However, plasmid chimerae constructed with this plasmid [160] have never been shown to replicate in a methanogen host. Recently, Gernhardt *et al.* circumvented this problem by designing integrative vectors, Mip1/2, for *M. voltae* [153]. Chromosomal integration was mediated by provision of a copy of the *M. voltae hisA* gene. Selection was made possible by the addition of the streptomyces puromycin transacetylase gene (*pac*), under the transcriptional control of the *M. voltae mcr* promoter. Although the number of transformants obtained was low, the system offers great promise for the future manipulation of this methanogen.

The failure to derive functional, autonomously replicating methanogen cloning vectors based on pME2001 emphasises the need to understand more fully replication/stability functions of the plasmids used as the source of replication origin. Extensive studies of the plasmidology of aerobic Gram-positive plasmids has shown that the majority of their plasmids belong to a highly interrelated family of genetic elements (ssDNA plasmids) which all replicate via a rolling circle mechanism. This replicative strategy is easily perturbed. Thus the insertion or deletion of DNA, as occurs during vector construction or cloning, or the use of such vectors in an alternative host, leads to structural and segregational instabilities (reviewed in ref. 161). The choice of replicon used in constructing vectors for use in Gram-positive anaerobes is, therefore, of paramount importance if long-term stability of manipulated organisms is to be achieved. In this sense vectors based on the *C. butyricum* plasmid pCB101 [162] may not represent the best choice for *C. acetobutylicum*, as studies have shown that it too belongs to the ssDNA plasmid family [115, 155]. On the other hand, plasmids based on the pAMβ1, appear admirably suited for the future manipulation of this clostridial and other Gram-positive species. Its replication region has been completely sequenced [156], and derivatives vectors based on its origin are 1000-fold more structurally stable than ssDNA plasmids [163]. This stability is believed to be a consequence of its mode of replication, a unidirectional theta mechanism [164].

Problems of segregational instability must also be addressed when evaluating the potential of constructed vectors. For instance, at present there are no reported examples of a clostridial vector which exhibits 100% segregational stability in the absence of selective pressure. It is apparent that determinants of segregational stability will need to be incorporated in future vectors to alleviate such problems. Potential candidates include site-specific recombinases which stabilise plasmids by preventing the formation of multimers [165]. A resolvase gene of this type has previously been noted on a *C. perfringens* plasmid, pIP404 [166]. More recently we have shown that a region contiguous with the pAMβ1 replication region, previously shown to enhance the stability of vectors in *B. subtilis* [163], also encodes a resolvase [155]. Furthermore, insertion of this gene into the clostridial cloning vector pMTL500E (Table 10.2) significantly improves segregational stability in *C. acetobutylicum*. In considering stabilisation of recombinant strains, the virtues of genome integrative systems which do not rely on autonomous replication should be borne in mind.

10.6.2   *Gene cloning*

The experience of researchers over the last decade has indicated that there do not appear to be any particular barriers to the expression of genes derived from anaerobes in *E. coli*. Exceptions include certain antibiotic resistance genes from *Bacteroides* [113], plasmid-encoded clostridial genes [167] and difficulties encountered in the expression of some of the larger clostridial toxin genes, e.g. those encoding tetanus and botulinum neurotoxins. The inefficient expression of these latter genes is believed to be due to the sheer volume of modulator codons present (a consequence of their codon bias, 70% A + T, and gene size, *c*. 4.0 kb), resulting in translational inefficiencies.

A summary of genes cloned in recent years from the major anaerobic genera is given in Table 10.3. It is by no means exhaustive, and does not include those clostridial, methanogen or *Bacteroides* genes cloned prior to 1989, 1989 and 1986 respectively, as these have been listed elsewhere [114,116,118]. In *C. acetobutylicum* attention has continued to focus on genes encoding key enzymes of primary metabolism, commensurate with biotechnological interest in solvent production. Enzymes whose genes have been cloned to date are indicated in Figure 10.2. Given the key role played by methyl CoM reductase in methanogenesis, cloning experiments have concentrated on the genes responsible for its synthesis. As a result, genes (*mcrA*, *mcrB* and *mcrG*) encoding the three subunits (α, β and γ) of the methyl CoM reductase enzyme complex have been isolated and sequenced from a number of different methanogens, including *Methanococcus barkeri* [225], *Methanococcus vannielii* [226], *Methanococcus thermoautrophicum* [227], *Methanococcus voltae* [228] and *Methanothermus fervidus* [229]. The conservation in structural organisation and polypeptide function evident in these different species suggests evolutionary relatedness [229]. Other genes cloned include those encoding enzymes involved in amino acid or purine biosynthetic pathways, protein translation and nitrogen fixation (see ref. 116).

Table 10.3 Examples of genes cloned from various anaerobes.

Organism	Genes cloned	Sequence	Reference
Bacteroides fragilis	Sucrase	−	168
	IS*4351*	+	169
Bacteroides ruminicola	Xylanase	−	170–172
Bacteroides fibrisolvens	Xylosidase (*xylB*)	−	173
Bacteroides succinogenes	Xylanase	−	174
	β-Glucanases	−	175
Ruminococcus albus	Endoglucanase (*celA*)	+	176
	Endoglucanase (*celB*)	+	176
	Endoglucanase (Eg1)	+	177–179
	Cellulases	−	180
Ruminococcus flavefaciens	Xylanase	−	181
	Endoglucanase	−	181
	Xylosidase	−	181
	Cellulase	−	182
Bacteroides nodosus	Serine protease	+	183
	Fimbrial genes	+	184
Bacteroides thetaiotamicron	Chondro-4-sulphatase	−	185
	Pullulanase	−	186
Bacteroides gingivalis	Fimbrial subunit	+	187
Methanococcus vannielii	Elongation factors	+	188
	Ribosomal protein	+	188
	Spectinomycin operon	+	189
Methanococcus voltae	Glutamine synthetase (*glnA*)	+	190
	Tryptophan biosynthesis (*trpBA*)	+	191
Methanobacterium thermoautotrophicum	RNA polymerase	+	192
Clostridium botulinum	Type A neurotoxin	+	193, 194
	Type C neurotoxin	+	195
	Type D neurotoxin	+	196
	Type B neurotoxin	+	S.M. Whelan, M.E. Elmore
	Type E neurotoxin	+	N. Bodsworth and N.P. Minton,
	Type F neurotoxin	+	unpublished data
	Exoenzyme-C3	+	197
Clostridium difficile	Toxin A	+	198, 199
	Toxin B	+	200
Clostridium perfringens	Acid-soluble spore proteins	+	201
	Histidine decarboxylase	+	202
	Enterotoxin	+	see 203
	Alpha-toxin	+	see 204
	Epsilon-toxin	+	see 204
	Theta-toxin	+	205

Table 10.3   (*Contd.*)

Organism	Genes cloned	Sequence	Reference
Clostridium pasteurianum	Nitrogen-fixation (*nifC*)	+	206
	Nitrogen-fixation (*nifV*)	+	207
Clostridium bifermentans	Acid-soluble proteins	+	208
Clostridium cellulolyticum	Endoglucanase	+	85
Clostridium thermocellum	β-Glucosidase (*bglB*)	+	90
	Endoglucanase H (*celH*)	+	209
Clostridium stercorarium	Avicelase-I (*celZ*)	+	112
Clostridium thermohydrosulfuricum	α-Amylase–pullulanase	+	102, 210
Clostridium thermosulfurogenes	Glucose isomerase	+	211
	β-Amylase	+	100
	β-Amylase	−	21
	Xylanase	+	213
Clostridium josui	Endoglucanase	−	82
Clostridium acetobutylicum	Xylanase (*xynB*)	+	214
	β-Galactosidase	+	215
	Flavodoxin	+	216
	Lactate dehydrogenase	+	217
	CoA transferase	+	218
	Acetoacetate decarboxylase	−	219
	Thiolase	−	220
	NADH–butanol dehydrogenase	−	220
	NADPH–butanol dehydrogenase	−	220
Clostridium sordellii	Sialidase	+	221
Clostridium septicum	Sialidase	+	222
Clostridium thermoaceticum	Formyltetrahydrofolate synthetase	+	223
Clostridium cellulovorans	Cyclic AMP-dependent kinase	+	224

The above list is not intended to be exhaustive but to reflect the types of genes recently cloned from the illustrated anaerobes. Genes cloned from *Bacteroides*, *Clostridium* and methanogens, prior to 1986, 1989 and 1989 respectively have not been included as these have been listed elsewhere [114, 116, 118].

Although these studies reveal fascinating insights into gene structure, further progress on understanding gene regulation awaits the development of methanogen host–vector systems.

The innate power of gene-cloning technology is exemplified by the ongoing molecular dissection of the 'organelle' responsible for cellulose degradation by *C. thermocellum*, the so-called cellulosome [208]. As many as 19 different genes encoding proteins concerned with cellulose/hemicellulose degradation have now been isolated from this organism [209]. Of these, 15 (*cel* genes) encode endoglucanases (EG), two (*xyn* genes) encode xylanases (XYN), and the remaining two (*bgl* genes) encode β-glucosidases. The nucleotide sequence determination of the encoding genes has led to the derivation of the primary protein sequences of seven EG proteins (A, B, C, D, E, H and X) and XYNZ. Using a classification system based on hydrophobic cluster analysis [230], EGB, EGC, EGE and EGH all belong to one family (A), whereas EGA, EGD and XYNZ each form an independent grouping. Although belonging to different families, with the exception of EGC all these proteins contain a reiterated 24 amino acid motif. As EGC was shown not to be part of the cellulosome, and because the proteins in which this motif has been deleted by gene modification still exhibit endoglucanase activity, this domain has been suggested to be involved in binding of these proteins in cellulosome scaffold assembly [231].

Continued molecular analysis of the individual components of the cellulosome should lead to a greater insight into cellulosome-mediated degradation of cellulose. However, it is apparent that the most meaningful experiments will require the future development of a gene transfer system for *C. thermocellum*, allowing the directed addition/subtraction of genetically altered cellulosome components *in vivo*. A similar difficulty faces those studies currently directed at the molecular characterisation of cellulose/hemicellulose degradation in rumen anaerobes, where encoding genes have been cloned (e.g. refs. 170, 177, 180–182), but gene transfer systems have yet to be elucidated.

10.6.3 *Recombinant manipulation of anaerobes*

The use of recombinant DNA methodology deliberately to engineer beneficial alterations in the genetic make-up of anaerobic organisms is still in its infancy. Few cloned genes have been reintroduced back into the organism from which they were isolated. In the clostridia, some rudimentary experiments have been undertaken. Thus, a cloned *leuB* gene from *Clostridium pasteurianum* has been introduced into a Leu⁻ auxotroph of *C. acetobutylicum* NCIB 8052 using both conjugal cointegrate transfer [232] and electroporation [127]. In both cases, acquisition of the recombinant plasmid converted the strain to prototrophy. In other experiments, an expression cartridge was constructed based on the promoter region of the *C. pasteurianum* ferredoxin gene, and has been used to express two different reporter genes in *C. acetobutylicum*, namely a promoterless copy of the pseudomonad *xylE* gene and a promoterless pC194 *cat* gene (see refs. 117 and 233).

The above model experiments demonstrated the feasibility of using developed host–vector systems both for effecting stable changes in the genetic make-up of *C. acetobutylicum* and for promoting the expression of foreign genes using clostridial transcription signals. Experiments to test the feasibility of broadening the substrate range of this organism have also been performed. In these studies, an artificial operon was constructed composed of, in sequential order, the *C. thermocellum celC* promoter, the *celC* structural gene, a promoterless copy of the *celA* gene and the *celA* transcriptional terminator. This operon was then introduced into *C. acetobutylicum*, inserted in the cloning vector pMTL500E, and the growth rates of the resultant clone on lichenan as the sole carbon source compared with the plasmid-free host. The plasmid-bearing strain exhibited a significantly higher growth rate compared with the plasmid-free host (see refs. 117 and 155).

The first usable host–vector systems to be developed in anaerobic bacteria were those of the genus *Bacteroides*. It follows that gene manipulation in this organism is more advanced than in any other anaerobe. In particular, the availability of suicide vectors [149, 234] has allowed cloned genes to be used to mutagenise targeted chromosomal regions insertionally, providing insight into gene function and the physiological role of encoded enzymes. Studies to date have concentrated on enzymes involved in polysaccharide degradation, and in particular chondroitin sulphate. This polysaccharide is a component of epithelial tissue, which is constantly being sloughed into the intestinal lumen, and may therefore represent an important source of carbon and energy to colonic bacteria. *Bacteroides thetaiotamicron* possesses two very similar chondroitin lyase activities, responsible for the cleavage of chondroitin sulphate into unsaturated disaccharides. To determine the relative importance of these two enzymes, a cloned copy of the chondroitin lyase II gene was used to inactivate the chromosomal gene [234]. Although the mutant strain obtained no longer produced this particular lyase, as evidenced from western blots, it was still able to grow on chondroitin sulphate as sole carbon source. In further work evidence was obtained which suggested this gene formed part of an operon with at least one other gene, that encoding chondro-4-sulphatase. Insertions in this gene, which resides 5′ to that encoding chondroitin lyase II, abolishes production of the lyase II enzyme [235]. Strains in which part of the chondroitin lyase II gene was deleted were subsequently shown to be at no measurable selective disadvantage, with regard to survival in the intestinal tracts of germ-free mice, over wild-type organisms [236]. The question of why *B. thetaiotamicron* possesses two chondroitin lyase activities is still, therefore, no nearer being resolved.

In view of the complexity of this system for studying polysaccharide degradation in *Bacteroides*, more recent studies have switched to starch degradation. A cloned pullulanase gene was used to inactivate the equivalent chromosomal copy, reducing pullulanase activity by 55%. However, the mutant exhibited no difference in its growth rates on pullulan compared with the wild-type organism, suggesting the existence of two pullulanase enzymes [237]. Indeed, subsequent studies have clearly demonstrated the presence of a second enzyme with pullulanase activity [238]. Other workers studying polysaccharide degradation by *Bacteroides ruminicola*, have, in the absence of gene transfer systems for this rumen organism, turned to the elegant genetic systems developed in colonic *Bacteroides*. Thus, a cloned *B. ruminicola* xylanase gene was inserted into the shuttle vector pVAL-1 (Table 10.2) and transferred into *B. fragilis* and *Bacteroides uniformis*, where it expressed at 1400-fold higher levels than in *B. ruminicola* [239]. Subsequently, the same gene was integrated into the chromosome of *B. thetaiotamicron*, by its directed insertion into the chondroitin lyase II gene [240]. This particular region of the chromosome was chosen because, as described above, inactivation of the chondroitin lyase II gene confers no selective disadvantage on the host. This study also served to illustrate the benefits of inserting cloned genes into the host chromosome. Under continuous culture conditions, no loss of the chromosomally located xylanase gene was detected. In contrast, after 20 generations, 80% of the cell population had lost the pVAL-RX plasmid in which the same gene was inserted. It was also shown that single-copy genome insertions need not dramatically affect expression levels compared with insertion in a multicopy plasmid. Thus, the level of xylanase produced from the chromosomally located gene was only three-fold lower than from plasmid pVAL-RX [240]. However, chromosomal integration of genetic elements need not necessarily result in complete segregation stability, as evidenced from the studies undertaken with *Methanococcus voltae* [153], in which integrated Mip vector was slowly lost in the absence of selective pressure.

10.6.4 *Gene probes for the detection of anaerobes*

A longer standing problem facing microbiologists studying complex bacterial consortia in extreme anaerobic environments is the enumeration of the bacterial species present. Traditionally, anaerobic bacteria from environs such as the gut, rumen or anaerobic digesters have proven difficult to isolate using conventional culture methods. Many anaerobes have fastidious growth requirements, and the low energy yield of their fermentative metabolism means that anaerobes grow relatively slowly and are frequently overgrown by facultative anaerobes. The toxicity of molecular oxygen dictates that special measures must be taken to exclude it during collection and transport of specimens, as well as during isolation and identification of bacteria. For the above reasons, the trend in recent years has been to develop specific tests which may be applied *in situ*, without recourse to culturing. The most promising approach has been to use genera/species-specific genes as probes. Progress with *Bacteroides* spp., rumen bacteria and pathogenic clostridia has recently been reviewed [241–243]. It is anticipated that the practice of using DNA probes for enumeration will increase in the coming years, and furthermore sensitivity will be improved by the use of polymerase chain reaction technology [244], as has recently been described for certain clostridial species [245–247].

References

1. J.G. Morris, *Adv. Microbial Physiol.* **12** (1975) 169.
2. R.E. Hungate, *The Rumen and its Microbes*, Academic Press, New York (1966).
3. R.E. Hespell *Proc. Nutr. Soc.* **46** (1987) 407.
4. S. Ogata and M. Hongo, *Adv. Appl. Microbiol.* **25** (1979) 241.
5. D.T. Jones and D.R. Woods, in *Microbiol. Rev.* **50** (1986) 484.
6. D.T. Jones and D.R. Woods, in *Clostridia*, eds. N.P. Minton and D.J. Clarke, Plenum Press, New York (1989) 105.
7. B. McNeil and B. Kristiansen, *Adv. Appl. Microbiol.* **31** (1986) 61.
8. P.J. Evans and H.Y. Wang, *Appl. Environ. Microbiol.* **54** (1988) 1662.
9. A.S. Afschar and K. Schaller, *J. Biotechnol.* **18** (1991) 255.
10. M.R. Ladisch, *Enz. Microbial Technol.* **13** (1991) 280.
11. P.C. Raemakers-Franken, A.J. Kortsee, C. Van der Drift and G.D. Vogels, *J. Bacteriol.* **172** (1990) 1157.
12. D. Moller-Zinkham, G. Barner and R.K. Thauer, *Arch. Microbiol.* **152** (1989) 362.
13. J. Ellermann, R. Heddereich, R. Böcher and R.K. Thauer, *Eur. J. Biochem.* **172** (1988) 669.
14. J.N. Reeve, G.S. Beckler, D.S. Cram, P.T. Hamilton, J.W. Brown, J.A. Krzycki, A.F. Kolodziej, L. Alex, W.H. Orme-Johnson and C.T. Walsh, *Proc. Natl. Acad. Sci. USA* **86** (1989) 3031.
15. W.L. Ellefson and R.S. Wolfe, *J. Biol. Chem.* **256** (1981) 4259.
16. P.L. Hartzell and R.S. Wolfe, *System Appl. Microbiol.* **7** (1986) 376.
17. C.F. Wiel, B.A. Sherf and J.N. Reeve, *Can. J. Microbiol.* **35** (1989) 101.
18. S. Rospert, D. Linder, J. Ellermann and R.K. Thauer, *Eur. J. Biochem.* **194** (1990) 871.
19. P.E. Rouviére and R.S. Wolfe, *J. Biol. Chem.* **263** (1988) 7913.
20. A.A. DiMarco, A.B. Thomas and R.S. Wolfe, *Ann. Rev. Biochem.* **59** (1990) 335.
21. F. Mayer, M. Rohde, S. Salzmann, A. Jussofie and G. Gottschalk, *J. Bacteriol.* **170** (1988) 1438.
22. K. Inatomi, *J. Bacteriol.* **167** (1986) 837.
23. U. Deppenmeier, M. Blaut and G. Gottschalk, *Eur. J. Biochem.* **186** (1989) 317.
24. J.R. Lancaster Jr, *J. Bioenergetics and Biomembranes* **21** (1989) 717.
25. L.G. Ljungdahl, J. Hugenholtz and J. Wiegel, in *Clostridia*, eds. N.P. Minton and D.J. Clarke, Plenum Press, New York (1989) 145.
26. S. Hayashida and B.K. Ahn, *Agric. Biol. Chem.* **54** (1990) 343.
27. B.K. Ahn and S. Hayashida, *Agric. Biol. Chem.* **54** (1990) 353.
28. D.C. Cameron and C.L. Cooney, *Biotechnology* **4** (1986) 651.
29. B.P.J. Weimer and J.G. Zeikus, *Appl. Environ. Microbiol.* **33** (1977) 289.
30. S-T. Yang and I-C. Tang, *Appl. Microbiol. Biotechnol.* **35** (1991) 119.
31. W.A. Enger and J.T. Heijnen, in *Biotec, 2 Biosensors and Environmental Biotechnology*, eds. C.P. Hallenberg and H. Sahm, Gustav Fishcer, Stuttgart (1988) 89.
32. E. Senior, in *Microbiology of Landfill Sites*, ed. E. Senior, CRC Press, Boca Raton, FL (1990) 1.
33. Anon., *ENDS Report* **159** (1988) 15.
34. Anon., in *Landfill Microbiology: R&D Workshop*, eds. P. Lawson and Y.R. Alston, Harwell Laboratories, UKAEA, (1990) 12.
35. K.M. Richards, *Biodet. Abs.* **3** (1989) 317.
36. K.M. Richards, in *Anaerobic Digestion: a Creditable Source of Energy*, HMSO, London (1984).
37. Anon., *ENDS Report* **135** (1986) 8.
38. I.A. Watson-Craik, in *Microbiology of Landfill Sites*, ed. E. Senior, CRC, Boca Raton, FL (1990) 159.
39. E.E. Finnecy, in *Hazardous Waste: Detection, Control, Treatment*, ed. R. Abbon, Elsevier, Amsterdam (1988) 1199.
40. P.E. Rushbrook, *Resource, Conserv. Recycle* **4** (1990) 33.
41. R. Sleat, C. Harries, I. Vinney and J.F. Rees, in *Sanitary Landfilling: Process Technology and Environmental Impact*, eds. T.H. Christensen, R. Cossu and R. Stagmann, Academic Press, London (1989) 51.

42. R.A. Mah, D.M. Work, L. Baresi and T.L. Glass, *Ann. Rev. Microbiol.* **31** (1977) 309.
43. K.L. Jones, J.F. Rees and J.M. Grainger, *Eur. J. Appl. Microbiol. Biotechnol.* **18** (1983) 242.
44. E.R. Fielding, D.B. Archer, E.C. de Macario and A.J.L. Macario, *Appl. Environ. Microbiol.* **54** (1988) 835.
45. P.L. McCarty, in *Anaerobic Digestion 1981*, eds. D.E. Hughes *et al.*, Elsevier, Amsterdam (1982) 3.
46. L. Van Den Berg, *Can. J. Microbiol.* **30** (1984) 975.
47. P.J. Large, in *Methylotrophy and Methanogenesis*, Van Nostrand Reinhold, London (1983).
48. E.-J. Nyns in *Biotechnology Vol. 8: Microbial Degradations*, ed. W. Schonborn, VCH Publishers, Weinheim (1986) 207.
49. J. Steel, *Farm. Wkly.* **116** (1989) 63.
50. P.N. Hobson, in *Mixed Culture Fermentations*, eds. M.E. Bushell, and J.H. Slater, Academic Press, London (1981) 53.
51. J.L. Garcia, *FEMS Microbiol. Rev.* **87** (1990) 297.
52. E. Zeurier, in *Advances in Biotechnological Processes, Vol. 12: Biological Waste Treatment*, ed. A. Mizrahi, Alan R. Liss, New York (1989) 73.
53. T. Leisenger, R. Hutter, A.M. Cook and J. Nuesch (eds.), *Microbial Degradation of Xenobiotics and Recalcitrant Compounds*, Academic Press, New York (1981).
54. D.R. Shelton and J.M. Tiedje, *Appl. Env. Microbiol.* **47** (1984) 850.
55. N.S. Battersby and V. Wilson, *Appl. Env. Microbiol.* **55** (1989) 433.
56. G. Barrio-Lage, F.Z. Parsons, R.S. Nassar and P.A. Lorenzo, *Environ. Sci. Technol.* **20** (1986) 96.
57. E.J. Bouwer and P.L. McCarty, *Appl. Env. Microbiol.* **45** (1983) 1286.
58. E.J. Bouwer and P.L. McCarty, *Appl. Env. Microbiol.* **50** (1985) 527.
59. S.A. Boyd and D.R. Shelton, *Appl. Env. Microbiol.* **47** (1984) 272.
60. J.D. Boyer, R.C. Ahlert and D.S. Kosson, *Hazardous Waste & Hazardous Materials* **4** (1987) 241.
61. D.L. Freidman and J.M. Gossett, *Appl. Env. Microbiol.* **55** (1989) 2144.
62. M.L Krumme and S.A. Boyd, *Water Res.* **22** (1988) 171.
63. B.R. Sharak Genthner, W.A. Price and P.H. Pritchard, *Appl. Env. Microbiol.* **55** (1989) 1472.
64. T.M Vogel and P.L. McCarty, *Env. Sci. Technol.* **21** (1987) 1208.
65. T.M Vogel and P.L. McCarty, *Appl. Env. Microbiol.* **49** (1985) 1080.
66. N. Belay and L. Daniels, *Appl. Env. Microbiol.* **53** (1987) 1604.
67. M.D. Mikesell and S.A. Boyd, *Appl. Env. Microbiol.* **56** (1990) 1198.
68. J. Konisky, *Tibtech.* **7** (1989) 88.
69. D. Lin, N. Nishio and S. Nagai, *J. Ferm. Bioeng.* **68** (1989) 88.
70. C.A. Rebeiz, A. Montazer-Zoucher, H.J. Hopen and S.M. Wu, *Enz. Microbial. Technol.* **6** (1984) 390.
71. B.H. Kim, P. Bellows, R. Datta and J.G. Zeikus, *Appl. Environ. Microbiol.* **48** (1984) 764.
72. G. Rao and R. Mutharasan, *Appl. Environ. Microbiol.* **53** (1987) 1232.
73. H. White, H. Lebertz, I. Thanos and H. Simon, *FEMS Microbiol. Lett.* **43** (1987) 173.
74. J.B. Jewell, J.B. Coutinho and A.M. Kropinski, *Current Microbiol.* **13** (1986) 215.
75. J.G. Morris, in *Clostridia*, eds. N.P. Minton and D.J. Clarke, Plenum Press, New York (1989) 193.
76. R.J. Lamed and J.G. Zeikus, *Biochem. J.* **195** (1981) 183.
77. A. Belan, J. Bolte, A. Fauve, J.G. Gourcy and H. Veschambre, *J. Org. Chem.* **52** (1987) 256.
78. H. Simon, J. Bader, H. Guntner, S. Neumann and J. Thanos, *Angew. Chem. Int. Ed. Engl.* **24** (1985) 539.
79. T. Matsunaga, M. Higashijima, H. Nakatsugawa, S. Nishimura, T. Kitamura, M. Tsuji, and T. Kawaguchi, *Appl. Microbiol. Biotechnol.* **27** (1987) 11.
80. G.P Hazlewood, M.P.M. Romaniec, K. Davidson, O. Grépinet, P. Beguin, J. Millet, O. Raynaud and J.-P. Aubert, *FEMS Microbiol. Lett.* **51** (1988) 231.
81. K. Bronnenmier and W.L. Staudenbauer, *Enz. Microbiol Technol.* **12** (1990) 431.
82. K. Ohmiya, T. Fujino, J. Sukhumavasi and S. Shimizu, *Appl. Environ. Microbiol.* **55** (1989) 2399.
83. K. Sakka, S. Furuse and K. Shimida, *Agric. Biol. Chem.* **53** (1989) 905.
84. E. Faure, C. Bagnara, A. Belaich and J.-P. Belaich, *Gene* **65** (1988) 51.
85. E. Faure, A. Belaich, C. Bagnara, C. Gaudin and J.-P. Belaich, *Gene* **84** (1989) 39.

86. H. Zappe, W.A. Jones, D.T. Jones and D.R. Woods, *Appl. Environ. Microbiol.* **54** (1989) 1289.
87. K. Tuka, V.V. Zverlov, B.K. Bumazkin, G.A. Velikodvorskaya and A. Ya. Strongin, *Biochem. Biophy. Res. Commun.* **169** (1990) 1055.
88. W.H. Schwarz, S. Jauris, M. Kouba, K. Bronnenmeier, and W.L. Staudenbauer, *Biotechnol. Lett.* **11** (1989) 461.
89. K. Sakka, Y. Kojima, K. Yoshikawa and K. Shimada, *Agric. Biol. Chem.* **54** (1990) 337.
90. F. Gräbnitz, K.P. Rücknagel, M. Seiß and W.L. Staudenbauer, *Mol. Gen. Genet.* **217** (1989) 70.
91. H. Zappe, D.T. Jones and D.R. Woods, *J. Gen. Microbiol.* **132** (1986) 1367.
92. O. Grépinet, M-C. Chebrou and P. Béguin, *J. Bacteriol.* **170** (1988) 4576.
93. C.R. MacKenzie, R.C.A. Yang, C.B. Patel, D. Bilous and S.A. Narang, *Arch. Microbiol.* **152** (1989) 377.
94. F. Grabnitz, M. Seiss, K.P. Rücknagel and W.L. Staudenbauer, *Eur. J. Biochem.* **200** (1991) 301.
95. H. Zappe, D.T. Jones and D.R. Woods, *Appl. Microbiol. Biotechnol.* **27** (1987) 57.
96. B. Schink and J.G. Zeikus, *FEMS Microbiol. Lett.* **17** (1983) 295.
97. B.M. Lund and T.F. Brocklehurst, *J. Gen. Microbiol.* **104** (1978) 59.
98. B.M. Lund, T.F. Brocklehurst and G.M. Wyatt, *J. Gen. Microbiol.* **122** (1981) 17.
99. M.I. Sheiman, J.D. MacMillan, L. Miller and T. Chase Jr, *Eur. J. Biochem.* **64** (1976) 565.
100. N. Kitamoto, H. Yamagata, T. Kato, N. Tsukagoshi and S. Udaka, *J. Bacteriol.* **170** (1988) 5848.
101. G-J. Shen, B.C. Saha, Y-E. Lee, L. Bhatnagar and J.G. Zeikus, *Biochem. J.* **254** (1988) 835.
102. H. Melasniemi, M. Paloheimo and L. Hemio, *J. Gen. Microbiol.* **136** (1990) 447.
103. B. Ensley, J.J. McHugh and L.L. Barton, *J. Gen. Appl. Microbiol.* **21** (1975) 51.
104. B.C. Saha and J.G. Zeikus, *Trends Biotechnol.* **7** (1989) 234.
105. R. Koch, P. Zablowski and G. Antranakian, *Appl. Microbiol. Biotechnol.* **27** (1987) 192.
106. M.P. Coughlan, *Biochem. Soc. Trans.* **13** (1985) 405.
107. M. Mandels, *Biochem. Soc. Trans.* **13** (1985) 414.
108. R. Lamed and E.A. Bayer, *Adv. Appl. Microbiol.* **33** (1988) 1.
109. N. Creuzet, J.-F. Berenger and C. Frixon, *FEMS Microbiol. Lett.* **20** (1983) 347.
110. E. Ong. J.M. Greenwood, N.R. Gilkes, D.G. Kilburn, R.C. Miller Jr and A.J. Warren, *Trends Biotechnol.* **7** (1989) 239.
111. E. Ong, N.R. Gilkes, A.J. Warren, R.C. Miller Jr and D.G. Kilburn, *Biotechnology* **7** (1989) 604.
112. S. Jauris, K.P. Rücknagel, W.H. Schwarz, P. Kratzsch, K. Bronnenmeier and W.L. Staudenbauer, *Mol. Gen. Genet.* **223** (1990) 258.
113. A.A. Salyers, N.B. Shoemaker and E.P. Guthrie, *Crit. Rev. Microbiol.* **14** (1987) 49.
114. N.P. Minton and D.E. Thompson, in *Anaerobes in Human Disease*, eds. B.I. Duerden and B.S. Drassar, Edward Arnold, London (1991) 38.
115. M. Sebald, *Genetics and Molecular Biology of Anaerobic Bacteria*, Springer Verlag, New York (1991).
116. J.W. Brown, C.J. Daniels and J.N. Reeve, *Crit. Rev. Microbiol.* **16** (1989) 287.
117. N.P. Minton, T.-J. Swinfield, J.K. Brehm, S.M. Whelan and J.D. Oultram, in *Genetics and Molecular Biology of Anaerobic Bacteria*, ed. M. Sebald, Springer Verlag, New York (1991) in press.
118. M. Young, N.P. Minton and W.L. Staudenbaur, *FEMS Microbiol. Rev.* **63** (1989) 301.
119. S.J. Reid, E.R. Allcock, D.T. Jones and D.R. Woods, *Appl. Environ. Microbiol.* **45** (1983) 305.
120. Y.-L. Lin and H.P. Blaschek, *Appl. Environ. Microbiol.* **48** (1984) 737.
121. G. Reysset, J. Hubert, L. Podvin and M. Sebald, *Biotechnol. Lett.* **2** (1988) 199.
122. D.L. Heefner, C.H. Squires, R.J. Evans, B.J. Kopp and M.J. Yarus, *J. Bacteriol.* **159** (1984) 460.
123. D.E. Mahony, J.A. Mader and J.R. Dubel, *Appl. Environ. Microbiol.* **54** (1988) 264.
124. C.J. Smith, *J. Bacteriol.* **164** (1985) 466.
125. S. Yoshino, T. Yoshino, S. Hara, S. Ogata and S. Hayashhida, *Agric. Biol. Chem.* **54** (1990) 437.
126. E. Soutschek-Bauer, L. Hartl and W.L. Staudenbauer, *Biotechnol. Lett.* **7** (1985) 705.
127. J.D. Oultram, M. Loughlin, T.-J. Swinfield, J.K. Brehm, D.E. Thompson and N.P. Minton, *FEMS Microbiol. Lett.* **56** (1988) 83.
128. S.P. Allen and H.P. Blaschek, *Appl. Environ. Microbiol.* **54** (1988) 2322.

129. P.T. Scott and J.I. Rood, *Gene* **82** (1989) 327.
130. M.K. Phillips-Jones, *FEMS Microbiol. Lett.* **66** (1990) 221.
131. A.M. Thomson and H.J. Flint, *FEMS Microbiol. Lett.* **61** (1989) 101.
132. G. Bertani and L. Baresi, *J. Bacteriol.* **169** (1987) 2730.
133. V.E. Worrel, D.P. Nagle Jr, D. McCarthy and A. Eisenbraun, *J. Bacteriol.* **170** (1988) 653.
134. N.B. Shoemaker, C.G. Getty, E.P. Guthrie and A.A. Salyers, *J. Bacteriol.* **166** (1986) 959.
135. P.J. Valentine, N.B. Shoemaker and A.A. Salyers, *J. Bacteriol.* **170** (1988) 1319.
136. J.D. Oultram and M. Young, *FEMS Microbiol. Lett.* **27** (1985) 129.
137. J.D. Oultram, A. Davies and M. Young, *FEMS Microbiol. Lett.* **42** (1987) 113.
138. D.R. Williams, D.I. Young, J.D. Oultram, N.P. Minton and M. Young, in *Anaerobes in Medicine and Industry*, eds S.P. Borrielloand, J.M. Hardied, Wrightson Biomedical Press, Petersfield (1990) 239.
139. D.R. Williams, D.I. Young and M. Young, *J. Gen. Microbiol.* **136** (1990) 819.
149. D.G. Guiney, K. Bouic, P. Hasegawa and B. Matthews, *Plasmid* **20** (1988) 17.
141. J. Betram and P. Dürre, *Arch. Microbiol.* **151** (1989) 551.
142. R.C. Wooley, A. Pennock, R.J. Ashton, A. Davies and M. Young, *Plasmid* **22** (1989) 169.
143. J. Betram, A. Kuhn and P. Dürre, *Arch. Microbiol.* **153** (1990) 373.
144. M. Magot, *FEMS Microbiol. Lett.* **18** (1983) 149.
145. H. Hachler, B. Berger-Bächi and F.H. Kayser, *Antimicro. Agents Chemother.* **31** (1987) 1039.
146. P. Mullany, M. Wilks, I. Lamb, C. Clayton, B. Wren and S. Tabaqchali, *J. Gen. Microbiol.* **136** (1990) 1343.
147. C.J. Smith, *J. Bacteriol.* **164** (1985) 294.
148. P. Pheulpin, Y. Tierny, M. Béchet and J-B. Guillaume, *FEMS Microbiol. Lett.* **55** (1988) 15.
149. K.A Smith and A.A. Salyers, *J. Bacteriol.* **171** (1989) 2116.
150. C.H Squires, D.L. Heefner, R.J. Evans, B.J. Kopp and M.J. Yarus, *J. Bacteriol.* **159** (1984) 465.
151. I. Roberts, W.M. Holmes and P.B. Hylemon, *Appl. Environ. Microbiol.* **54** (1988) 268.
152. A.Y. Kim and H.P. Blaschek, *Appl. Environ. Microbiol.* **55** (1989) 360.
153. P. Gernhardt, O. Possot, M. Foglino, L. Sibold and A. Klein, *Mol. Gen. Genet.* **221** (1990) 273.
154. N. Trauffaut, J. Hubert and G. Reysset, *FEMS Microbiol. Lett.* **58**.
155. N.P Minton, J.K. Brehm, J.D. Oultram, T.-J. Swinfield, S. Schimming, S.M. Whelan, D.E Thompson, M. Young and W.L. Staudenbauer, in *Proceedings of the 6th International Symposium on the Genetics of Industrial Microorganisms*, eds. H. Heslot, J. Davies, J. Florent, L. Bobichon, G. Durand and L. Penasse, Société Francaise de Microbiologie, Strasbourg (1990) 759.
156. T.-J Swinfield, J.D. Oultram, D.E. Thompson, J.K. Brehm and N.P. Minton, *Gene* **87** (1990) 79.
157. M. Thomm, J. Altenbuchner and K.O. Stetter, *J. Bacteriol.* **153** (1983) 1060.
158. L. Meile, A. Kiener and T. Leisinger, *Mol. Gen. Genet.* **191** (1983) 480.
159. M. Bokranz, A. Klein and L. Meile, *Nucleic Acids Res.* **18** (1990) 363.
160. L. Meile and J.N. Reeve, *Biotechnology* **3** (1985) 69.
161. A. Gruss and S.D. Ehrlich, *Microbiol. Rev.* **53** (1989) 231.
162. N.P. Minton and J.G. Morris, *J. Gen. Microbiol.* **127** (1981) 325.
163. L. Janniere, C. Bruand and S.D. Ehrlich, *Gene* **87** (1990) 53.
164. C. Bruand, S.D. Ehrlich and L. Janniére, in *Genetics and Biotechnology of Bacilli*, eds. A.T. Ganesan and J.A. Hoch, Academic Press, New York (1990) 123.
165. S.J. Austin, *Plasmid* **20** (1988) 1.
166. T. Garnier, W. Saurin and S.T. Cole, *Mol. Microbiol.* **1** (1987) 371.
167. T. Garnier and S.T. Cole, *Plasmid* **19** (1988) 134.
168. R.R. Scholle, H.E. Steffen, H.J.K. Goodman and D.R. Woods, *Appl. Environ. Microbiol.* **56** (1990) 1944.
169. J.L. Rasmussen, D.A. Odelson and F.L. Macrina, *J. Bacteriol.* **169** (1987) 3573.
170. T.R. Whitehead and R.B. Hespell, *Appl. Environ. Microbiol.* **55** (1989) 893.
171. T.R. Whitehead and R.B. Hespell, *FEMS Microbiol. Lett.* **66** (1990) 61.
172. T.R. Whitehead, M.A. Cotta and R.B. Hespell, *Appl. Environ. Microbiol.* **57** (1991) 277.
173. G.W. Sewell, E.A. Utt, R.B. Hespell, K.F. Mackenzie and L.O. Ingram, *Appl. Environ. Microbiol.* **55** (1989) 306.
174. A. Sipat, K.A. Taylor, R.Y.C. Lo, C.W. Forsberg and P.J. Krell, *Appl. Environ. Microbiol.* **53** (1987) 477.

175. J.E. Irvin and R.M. Teather, *Appl. Environ. Microbiol.* **54** (1988) 2672.
176. D.M. Poole, G.P. Hazlewood, J.I. Laurie, P.J. Barker and H.J. Gilbert, *Mol. Gen. Genet.* **223** (1990) 217.
177. K. Ohmiya, K. Nagashima, T. Kajino, E. Goto, A. Tsukada and S. Shimizu, *Appl. Environ. Microbiol.* **54** (1988) 1511.
178. K. Ohmiya, T. Kajino, A. Kato and S. Shimizu, *J. Bacteriol.* **171** (1989) 6771.
179. K. Ohmiya, H. Deguchi and S. Shimizu, *J. Bacteriol.* **173** (1991) 636.
180. G.T. Howard and B.A. White, *Appl. Environ. Microbiol.* **54** (1988) 1752.
181. H.J. Flint, C.A. McPherson and J. Bisset, *Appl. Environ. Microbiol.* **55** (1989) 1230.
182. M.E.C. Barros and J.A. Thompson, *J. Bacteriol.* **169** (1969) 1760.
183. E.K. Moses, J.I. Rood, W.K. Yong and G.G. Riffkin, *Gene* **77** (1989) 219.
184. J.S. Mattick, B.J. Anderson, P.T. Cox, B.P. Dalrymple, M.M. Bills, M. Hobbs and J.R. Egerton, *Mol. Microbiol.* **5** (1991) 561.
185. E.P. Guthrie and A.A. Salyers, *J. Bacteriol.* **169** (1987) 1192.
186. K.A. Smith and A.A. Salyers, *J. Bacteriol.* **171** (1989) 2116.
187. D.P. Dickinson, M.A. Kubiniec, F. Yoshimura and R.J. Genco, *J. Bacteriol* **170** (1988) 1658.
188. K. Lechner, G. Heller and A. Böck, *J. Mol. Evol.* **29** (1989) 20.
189. J. Auer, G. Spicker and A. Böck, *J. Mol. Biol.* **209** (1989) 21.
190. O. Possot, L. Sibold and J.P. Aubert, *Res. Microbiol.* **140** (1989) 355.
191. S. Libold and M. Henriquet, *Mol. Gen. Genet.* **214** (1988) 439.
192. J. Schallenberg, M. Moes, M. Truss, M. Reise, M. Thomm, K.O. Stetter and A. Klein, *J. Bacteriol.* **170** (1988) 2247.
193. D.E. Thompson, J.K. Brehm, J.D. Oultram, T-J. Swinfield, C.C. Shone, T. Atkinson, J. Melling and N.P. Minton, *Eur. J. Biochem.* **189** (1990) 73.
194. T. Binz, H. Kurazono, M. Willie, J. Frevert, K. Wernars and H. Niemann, *J. Biol. Chem.* **265** (1990) 9153.
195. D. Hauser, M.W. Eklund, H. Kurazono, T. Binz, H. Niemann, D.M. Gill, P. Boquet and M.R. Popoff, *Nucleic Acids Res.* **18** (1990) 4924.
196. T. Binz, H. Kurazono, M.R. Popoff, M.W. Eklund, G. Sakaguchi, S. Kozaki, K. Krieglstein, A. Henscen, D.M. Gill and H. Niemann, *Nucleic Acids Res.* **18** (1990) 5556.
197. M. Popoff, P. Boquet, D.M. Gill and M.W. Eklund, *Nucleic Acids Res.* **18** (1990) 1291.
198. C. von Eichel-Streiber, D. Suckau, M. Wachter and U. Hadding, *J. Gen. Microbiol.* **135** (1989) 55.
199. C.H. Dove, S-Z. Wang, S.B. Proce, C.J. Phelps, D.M. Lyerly, T.D. Wilkins and J.L. Johnson, *Infect. Immun.* **58** (1990) 480.
200. L.A. Barroso, S-Z. Wang, C.J. Phelps, J.L. Johnson and T.D. Wilkins, *Nucleic Acids Res.* **18** (1990) 4004.
201. A. Holck, H. Blom and P.E. Granum, *Gene* **91** (1990) 107.
202. P.D. Vanpoelje and E.E. Shell, *Biochemistry* **29** (1990) 132.
203. P.E. Granum and G.S.A.B. Stewart, in *Genetics and Molecular Biology of Anaerobic Bacteria*, ed. M. Sebald, Springer Verlag, New York (1991) in press.
204. R.W. Titball, H. Yoeman and S.E.C. Hunter, in *Genetics and Molecular Biology of Anaerobic Bacteria*, ed. M. Sebald, Springer Verlag, New York (1991) in press.
205. R. Tweten, *Infect. Immun.* **56** (1988) 3235.
206. S.Z. Wang, J.S. Chen and J.L. Johnson, *Biophys. Res. Commun.* **169** (1990) 1122.
207. S-Z. Wang, D.R. Dean, J-S. Chen and J.L. Johnson, *J. Bacteriol.* **173** (1991) 3041.
208. R.M. Cabrera-Martinez, J.M. Mason, B. Setlow, W.M. Waites and P. Setlow, *FEMS Microbiol. Lett.* **61** (1989) 139.
209. E. Yagüe, P. Béguin, and J.-P. Aubert, *Gene* **89** (1990) 61.
210. H. Melasniemi and M. Paloheimo, *J. Gen. Microbiol.* **135** (1989) 1755.
211. C. Lee, L. Bhatnagar, D.C. Saha, Y-E. Lee, M. Takagi, T. Imanaka, M. Bagdasarian and J.G. Zeikus, *Appl. Environ. Microbiol.* **56** (1990) 2638.
212. K. Haeckel and H. Bahl, *FEMS Microbiol. Lett.* **60** (1989) 333.
213. J.G. Zeikus, C.-Y. Lee, B.C. Saha and M. Bagdasarian, in *Proceedings of the 6th International Symposium on the Genetics of Industrial Microorganisms*, eds. H. Heslot, J. Davies, J. Florent, L. Bobichon, G. Durand and L. Penasse, Société Francaise de Microbiologie, Strasbourg (1990) 771.
214. H. Zappe, D.A. Jones and D.R. Woods, *Nucleic Acids Res.* **18** (1990) 2179.

215. K.R. Hancock, E. Rockman, C.A. Young, L. Pearce, I.S. Maddox and D.B. Scott, *J. Bacteriol.* **173** (1991) 3084.
216. J.D. Santangelo, D.T. Jones and D.R. Woods, *J. Bacteriol.* **173** (1991) 1088.
217. P.R. Contag, M.G. Williams and P. Rogers, *Appl. Environ. Microbiol.* **56** (1990) 3760.
218. J.W. Cary, D.J. Petersen, E. Papoutsakis and G.N. Bennett, *Appl. Environ. Microbiol.* **56** (1990) 1576.
219. D.J. Petersen and G.B. Bennett, *Appl. Environ. Microbiol.* **56** (1990) 3491.
220. G.N. Bennett and D.J. Petersen, in *Genetic and Molecular Biology of Anaerobic Bacteria*, ed. M. Sebald, Springer Verlag, New York (1991) in press.
221. B. Rothe, P. Roggentin, R. Frank, H. Blöcker and R. Schauer, *J. Gen. Microbiol.* **135** (1989) 3087.
222. B. Rothe, B. Rothe, P. Roggentin and R. Schauer, *Mol. Gen. Genet.* **226** (1991) 190.
223. C.R. Lovell, A. Przybyla and L.G. Ljungdahl, *Biochemistry* **9** (1990) 5694.
224. O. Shoseyov, M. Goldstein, F. Foong, T. Hamamoto and R.H. Doi, *Nucleic Acids Res.* **19** (1991) 1710.
225. M. Bokranz and A. Klein, *Nucleic Acids Res.* **15** (1987) 4350.
226. D.S. Cram, B.A. Sherf, R.T. Libby, R.J. Mattaliano, K.L. Ramachandran and J.N. Reeve, *Proc. Natl. Acad. Sci. USA* **84** (1987) 3992.
227. M. Bokranz, G. Bäumer, R. Allmansberger, D. Ankel-Fuchs and A. Klein, *J. Bacteriol.* **170** (1988) 568.
228. A. Klein, R. Allmansberger, M. Bokranz, S. Knaub, H. Müller and E. Muth, *Mol. Gen. Genet.* **213** (1988) 409.
229. C.F Weil, D.S. Cram, B.A. Sherf and J.N. Reeve, *J. Bacteriol.* **170** (1988) 4718.
230. B. Henrissat, M. Claeyssens, P. Thomme, L. Lemesle and J.-P. Mornon, *Gene* **81** (1989) 83.
231. J. Hall. G.P. Hazlewood, P.J. Barker and H.J. Gilbert, *Gene* **69** 29.
232. J.D. Oultram, H. Peck, J.K. Brehm, D.E. Thompson, T.-J. Swinfield and N.P. Minton, *Mol. Gen. Genet.* **214** (1988) 177.
233. N.F. Minton, J.K. Brehm, J.D. Oultram, D.E. Thompson. T.-J. Swinfield, A. Pennock, S. Schimming, S.M. Whelan, U. Vetter, M. Young and W.L. Staudenbauer, in *Clinical and Molecular Aspects of Anaerobes*, ed. S.P. Borriello, Wrightson Biomedical Publishing Ltd, Petersfield (1990) 187.
234. E.P. Guthrie and A.A. Salyers, *J. Bacteriol.* **166** (1986) 966.
235. E.P. Guthrie and A.A. Salyers, *J. Bacteriol.* **169** (1986) 1192.
236. A.A. Salyers and E.P. Guthrie, *Appl. Environ. Microbiol.* **54** (1988) 1964.
237. K.A. Smith and A.A. Salyers, *J. Bacteriol.* **171** (1989) 2116.
238. K.A. Smith and A.A. Salyers, *J. Bacteriol.* **173** (1991) 2962.
239. T.R. Whitehead, M.A. Cotta and R.B. Hespell, *Appl. Environ. Microbiol.* **57** (1991) 277.
240. T.R. Whitehead and R.B. Hespell, *FEMS Microbiol. Lett.* **66** (1990) 61.
241. D.J. Groves, in *Gene Probes for Bacteria*, eds. A.J.L. Macario and E.C. de Macario, Academic Press, New York (1990) 233.
242. J.D. Brooker, R.A. Lockington, G.T. Attwood and S. Miller, in *Gene Probes for Bacteria*, eds. A.J.L. Macario and E.C. de Macario, Academic Press, New York (1990) 390.
243. K. Wernar and S. Notermans, in *Gene Probes for Bacteria*, eds. A.J.L. Macario and E.C. de Macario, Academic Press, New York (1990) 353.
244. M.A. Innis, D.H. Gelfand, J.J. Sninsky and T.J. White, *PCR Protocols: a Guide to Methods and Applications*, Academic Press, New York (1990).
245. B.W. Wren, C.L. Clayton and S. Tabaqchali, *FEMS Microbiol. Lett.* **70** (1990) 1.
246. B.W. Wren, C.L. Clayton and S. Tabaqchali, *Lancet* **335** (1990) 423.
247. T. Barry, R. Powell and F. Gannon, *Biotechnology* **8** (1990) 233.

Subject index

acidophiles 115–139
  chemiosmotic considerations 116
  commercial application 132–135
  definition of 115
  diversity of 127
  genetic aspects 135–138
  ion transport 123–125
  limits of growth 115
  mining of metals 132
  physiological constraints 116
  protein stability 126
alcohol dehydrogenase 71, 299
alcohol production, thermophilic 71, 287
alkaline environments
  man-made 143, 144
  natural 81, 144–146
  saline see alkaline saline environments
alkaline saline environments
  formation 145–147
  soda lakes and deserts
    see soda lakes 144–147
  springs 10, 144
alkaliphilic see alkaliphiles
alkaliphiles
  amylases from 154
  bioenergetics 151
  cell lipids 152
  cell walls 152
  cellulases from 154
  commercial applications 153–159
  cyclomaltodextrin glucotransferase 155
  definition of 143
  diversity of 148–151
  habitats of 143–146
  pectinases 156
  secretion vectors 159
  xylanases 157
alkalitolerant
  definition of 143
Alvinella pompejana
  bacteria associated with 185, 186
amylases 56, 57, 65, 66, 154, 300

anaerobes
  in biotransformations 296
  disposal of organic wastes 289–292
  facultative, definition of 281
  industrial exploitation of 283–289
  in landfills 290, 291
  molecular genetics of 303–313
  obligate, definition of 281
  xenobiotic breakdown 294, 295
anaerobic digesters 292–294
antibiotics 87
aqualysin I
  homology with subtilisins 51
Archaea
  biochemistry of 8, 13–21
  cloning of structural genes 36
  ecology 9
  enzymes 22–23
  extrachromosal DNA 34
  genome structure and organisation 34
  lipids and lipid biosynthesis 29–34
  lipoquinones 20, 31
  molecular biology 34
  morphological characteristics 14
  phylogenetic relationships 2–7
  physiological characteristics 17
  S-layers of 27
  species diversity 8
archaeobacteria 1–4, 92, 127–129, 150, 189–194, 286, 304, 309 see also Archaea
  non-polar lipids of 30
  phylogeny 2
  prenyl transferases 33
  S-dependent 4
Archaeolobus fulgidus 13, 192
  lactate metabolism 19
  sulphate reduction 13, 19
Arthrobacter spp.
  psychrophilic 219
  radiation resistance 260
ATPase 20, 152
Aureobasidium pullulans
  role of melanin in binding of metals 245

Bacillus sp.
    alkaliphilic/alkali tolerant   152, 154–157,
        see also alkaliphiles
    amylases   56, 57, 154
    commercial applications   55–59, 65
    enzymes produced   55–59, 153–157
    genetic aspects   55–57, 159
    licheniformis
        α-amylase   57, 154
    neutrophilic   152
    stearothermophilus
        α-amylase   56, 57
        cyclodextrin glucotransferase   55, 58
        β-galactosidase   55
        glyceraldehyde-3-phosphate dehydrogenase
            56
        lactate dehydrogenase   56
        pullulanase   58
Barophiles, bacterial   167–169
    amphipod gut dwelling   167, 171
    commercial application   194–197
    culture   167
    definition   167
    environment   *see* pressure
    major groups   169, 171
    obligate   167, 169, 171
    occurrence   168–172
    pressure tolerant (barotolerant)   164, 168,
        169
    psychrophilic   163, 167, 168, 169
    thermophilic   *see* vents, hydrothermal
betaine   97, 100
    glycine   97, 100
bioactive compounds   87–89
biofuels   106–108
bioplastics   92, 93
biotechnology
    acidophiles   132–139
    alkaliphiles   153–160
    barophiles   194–196
    environmental   103–108
    halophiles   80–108
    psychrophiles   219–221
    thermophiles   49–72, 282–314
black smoker vents   2, 11, 188, 189
    see also barophiles and hydrothermal vents

cadmium
    resistance to   233, 234
    genes coding for   233
    in *Staphylococcus aureus*   233
caldariellaquinone   20, 31
caldoactive   7
Calvin cycle enzymes   179, 182, 184
Calypotogena magnifica
    bacteria associated with   180, 181, 185
Candida glabrata
    metal detoxification mechanisms   239, 240

β-carotene
    commercial production   82
carotenoid photoprotection   259
cell membrane   *see* membrane
cellulases   59–62, 288, 300, 310–312
chemiosmosis
    and acidophiles   117–121
    and alkaliphiles   151–152
chemolithotrophs   13, 130, 164, 177–185
Clostridium sp.
    acetobutylicum
        acetone/butanol/ethanol fermentation
            282–285
        psychrophilic   207
    thermocellum
        cellulase genes   60, 300, 302
        β-endoglucanase genes   60, 312
        xylanase genes   60, 312
    thermohydrosulfurogenes
        pullulanase   67
        thermophilic   59–71, 299–302
    thermosaccharolyticum
        1,2-propanediol production   288
    thermosulfurogenes
        β-amylase genes   66
        glucose isomerase   71
compatible solutes   94–98
    as cryoprotectants   101, 103
    as stabilisers   100–103
copper resistance   100, 101
    binding proteins   237
    genetic aspects   237
    in *Neurospora crassa*   238
    in yeasts   237–238
corrosion   82–85
Crenarchaota   6
cyanobacteria
    commercial aspects   80, 81, 85, 86, 158
    saline-alkaline inhabiting   81, 85, 158
cyclodextrins
    glycotransferase   55, 155
    production of   155
cyclomaltodextrin   155

deep sea environments
    see also hydrostatic pressure
        characteristics   163
        definition   163
        growth rates   165
Deinococcaceae
    diversity of   261, 262
Deinococcus radiodurans
    cell surface of   264, 265
    commercial uses   274–276
    DNA polymerase   271
    DNA repair enzymes   275
    DNA repair genes   271–273
    genetic aspects   265–270

genome of   265
plasmids   266, 267
radiobiology   263
Dicytoglomus sp.
α-amylase   6
Amy A, B and C genes   66
DNA
alkaliphilic   159
extrachromosal in archaea   35
polymerase in archaea   36
polymerase in *Thermus aquaticus*   45
polymerases from *Deinococcus radiodurans*
275
pyrimidine dimer formation   263
UV damage   259, 260
dump leaching   132–135
Dunaliella
β-carotene production   82
glycerol production   98
growth in open ponds   85
halotolerance   81
protein production   80–82

ectoines   97, 98
electron transport chain systems in archaea   20
enzymes
of acidophiles   153–157
of moderate halophiles   93–94
of thermophiles   22, 23, 49–71, 300–302
thermostability of   24
ethanol, thermophilic production   71

fatty acids and their esters
chain length and branching   213
effects on   211, 213
equivocal responses   211
gel/lipid crystalline state   211, 212
membrane composition, temperature   211
regulatory mechanisms   213
synthesising enzymes   212–214
ferric iron, microbial generation   128, 129,
134
food industry, microbes in
commercial use   77–82, 89
pathogens   204, 207
spoilage, psychrophilic   204, 206–208
fungi, heavy metal response   *see also* metals
cadystin A and B   239
formation of cadmium sulphide colloid
239
metal-binding proteins   237
see also metallothioneins
metal γ-glutamyl peptides   239, 240

β-galactosidase   53, 55
geothermal environments
see vents, hydrothermal
glucose isomerase   71

glyceraldehyde-3-phosphate dehydrogenase
4, 56
glycerol
as compatible solute   96
commercial potential   98, 99
glycine betaine   97, 100

Halobacterium spp.
commercial potential   93
enzyme activity of   94
pharmaceutical potential   88, 89
halophiles
see also halotolerant micro-organisms
definition of   76
halotolerant micro-organisms
amino acid production   99
biofuel production   106
bioplastics   92, 93
biopolymers and surfactants   89–92
compatible solutes   96, 97
definition   76
enzymes   93, 94
fermenter design   85
food fermentations   78
glycerol production   98
in waste treatment   103, 104
outdoor culture   85
pharmaceutical products   87–89
single cell protein   80–82
heap leaching   132, 133
heat tolerant microbes   *see* thermophiles
heavy metals   *see* metals
hemicellulose degradation   62–65
homeoviscous adaptation   211, 212
hydrogen production   106, 107
hydrothermal fluids, composition of   175, 187
hydrothermal vents
East Pacific Rise   11, 173, 179, 190
Galapagos Rift   11, 172, 181
Guaymas Basin   11, 173, 177, 188, 191
Juan de Fuca Ridge   11, 173
Lau Basin   173
Mariana back-arc basin   173
North Fiji Basin   173
hypersaline environments
microbial diversity   149–151

Kenyan soda lakes   146, 147, 150

Lactobacillus spp.
in fermented foods   78–80
leaching processes, metal   132–135
Leuconostoc sp.
food industry application   78
light damage, UV   *see* UV damage
lipids, bacterial membrane
see also fatty acids, phospholipids
archaeobacterial   29–34

lipids, bacterial membrane *cont'd*
  core diether and tetraether 30, 31
  glycerolipids 30
  non-polar lipids (isoprenoids) 30, 31
  temperature effects 31, 32, 211–213

manganese
  oxidising/reducing barophiles 187
  toxicity 187
marine microbes
  environment *see* marine water environment
  psychrophilic 166, 168, 169, 208, 209
marine water environment
  deep (high pressure) *see* pressure
  food sources in *see* nutrients
  sulphide concentrations 186
membrane, cell
  archaebacterial 29, 31
  architecture and fluidity 169, 211–214
  gel/liquid crystalline state 211, 212
  lipid fluidity maintenance 169, 211–213
mercury
  genetics of resistance 230–233
  plasmid mediated resistance 231, 232
mer genes 135, 136, 231, 232
mer operon 231
metal detoxification mechanisms
  biosorption 227, 240, 247, 248, 249
  biotransformations 247
  extracellular complexation 227
  extracellular precipitation and crystallisation 227
metallothioneins
  physiological role 237, 238
metal recovery
  by biosorption 240–242, 247–249
  economics of 250–252
  by immobilised cell systems 242
metals, heavy
  definitions 225
  toxicity, causes 226
  transformation 227–230, 247
metal tolerance of micro-organisms
  commercial aspects 247–252
  definition 226, 227
  example of species 228, 230, 231, 233, 235, 239, 241
  genetic aspects 230–240
methane production 290, 293, 294
methanogens 4, 6, 11, 33, 189, 190, 193, 286, 287, 289, 294, 309
microbial mining 132–135

Natronobacterium 150
Natronococcus 150
Neurospora crassa
  copper binding protein 238
nonitol 30

nucleic acids *see* DNA, RNA
nutrient deficient environments
  see also starvation
  marine waters
    see marine water environment
  microbes *see* oligotrophs
nutrients, marine sources
  at thermal vents 174, 175, 189
  deep sea 163–167
  low concentrations 163
  organic carbon dissolved (DOC) 163
  particulate (POC) 163, 168, 175
  quality 163, 165

ocean water microbes *see* marine waters
oil recovery, microbial enhanced 90
osmophiles
  commercial aspects 78–108
  compatible solutes in 96–98
  examples of 78, 79, 80, 81, 85
  haloalkiphilic 88, 92, 94, 149–151, 152

panose 58
pH control
  in acidophiles 116–123
  in alkaliphiles 151–152
plasmid vectors
  for acidophiles 136–138
  for alkaliphiles 159
  for obligate anaerobes 303–312
  for thermophiles 35, 52, 53, 59
polymerase chain reaction (PCR) 45–47
pressure, high, environments at 164–191
  energy sources *see also* nutrients
  microbes living at *see* barophiles
  and high temperatures
    see hydrothermal vents
  and low temperatures *see* psychrophiles
proteases
  alkaline 55
  comparitive species studies 52
  halophilic 94, 153
  in detergents 55, 56, 153, 154
  serine 49, 50, 94, 154
  thermophilic 52, 55
proteinases *see* proteases
protein stability
  at acid pH 126, 127
  at high temperatures 23–26
Pseudomonas spp.
  fluorescens 208
  food spoilage 207, 208
  fragi 208
  psychrophilic 205, 206, 211
psychrophiles (bacterial)
  see also psychrotrophs
  cold shock proteins 218
  commercial potential 219–221

definition of   203
ecological significance   208–210
genetic basis   216–218
lability of aminoacyl-tRNA synthetases   217
membrane structure temperature effects
  see membranes   211–214
occurrence   206–210
solute uptake   214–215
psychrophiles (non bacterial)   204, 206, 207,
210
psychrotrophs (cold tolerant)
definition   203–204
low temperature effects   204
predominance   208–209
solute uptake   214–215
pullulanase
type I   23, 49, 57, 65, 67–69
type II   67, 69
pyrite oxidation   132, 133, 134

restriction endonucleases, *Taq*I   48
production of   274
Rhizopus arrhizus   245, 248, 249
ribulose-1-5 bisphosphate carboxylase
in deep sea bacteria   179, 182
Riftia pachyptila
bacteria of   179–181
RNA
rRNA of Archaea
  23S   35
  18S   2
  16S   2, 5, 35–36
  7S   35
  5S   4, 35
tRNA genes   35
rusticyanin   126, 130

Saccharomyces cerevisiae
copper binding   237
heavy metal uptake   237, 238, 249
salt stress, adaptive mechanisms   96–100
sewage treatment   103, 292, 293
single cell protein   80, 81, 85, 86
solute uptake
psychrophilic/psychrotrophic   214–215
solvents
production of   282–288
Spirulina sp.
as a food colouring   158
as a food source   80, 81, 85, 86, 158
maxima   85
platensis   81
starvation   see nutrient deficient environments
steel corrosion   82–85
Sulfolobus acidocaldarius
reductive citric acid cycle   20
Sulfolobus solfataricus
α-amylase   30, 31

caldariellaquinone   31
diglycosyl–diglycerol tetraether   30, 31
glycosylglycerol diether   30, 31
Sulfolobus sp.
SSV1 virus of   35
sulphide
hydrogen, utilisation (deep sea)   178, 179,
195
rich environments   173, 175, 186
toxicity   186
surfactants, microbial   89–92

Taq polymerase   45, 46
temperature
and high pressure
  see hydrothermal vents
effects on membranes   211–214
high, microbes preferring
  see thermophiles
low   see psychrophiles and psychrotrophs
tetritol   30
thermal biotopes
acid   9, 10
alkaline   9, 10
thermal rift vents   see hydrothermal vents
thermophiles
acidophilic   6
commercial applications   55–72
definition   8, 9
enzyme stability   23–27
genetics of   34–38, 45–49, 55, 58, 60–71
high pressure dwelling   189–193
major groups   6, 7, 11, 12
methane production   289–292
waste treatment   292–296
Thermophilus neutrophilus
reductive citric acid cycle   20
Thermoplasma acidophilum
glucose oxidation   18
thermaplasmaquinone   31
thermostability of enzymes   23–26
Thermotoga   71, 72
Thermus aquaticus
DNA polymerase   44, 45, 276
L-lactate dehydrogenase   49
molecular biology and genetics   45–47
plasmids in   52
restriction–modification system   47
Thermus caldophilus
L-lactate dehydrogenase   49
Thermus sp.
AMD-33, thermostable pullulanase   49
genetic transfer and plasmids   52–54
promoter regions   54
proteinases   49–52
T2 cloning of α- and β-galactosidase genes
53
rT41a, serine protease   50–51

Thermus thermophilus
   leuB gene   48
   elongation factor G   49
   ribosomal proteins S12 and S7   49
   tufA gene   48
Thiobacillus ferrooxidans
   genomic diversity   127, 128
   in metal leaching   133
   molecular genetics   135–137
   plasmids   137, 138
Trophosome   *see Riftia pachyptila*

ultramicrocells
   deep sea   166, 167
ultrathermophiles   *see also* thermophiles
   controversy over terminology   7
   isolation of   189, 192, 193
ultraviolet light damage   *see* UV damage
underground leaching   132, 133
uranium
   bacterial leaching   132, 133
   biosorption   243, 245, 249

UV damage
   far UV   259
   near UV   259
   repair   260, 271

vents, hydrothermal   2, 11, 188, 189
Vestimentifera
   bacteria of   179–181
Vibrio sp.
   ANT300   166, 167
   pressure survival   167, 169–171
   psychrophilic   166

Wadi Natrun   146, 147, 150
waste treatment   103, 289–296
water stress   *see* salt stress

xenobiotics
   degradation of   294–296
xylanases   62–65, 302, 310

yeast   *see also* fungi
   heavy metal tolerance   237–239, 249
   psychrophilic   210, 214

Species index

Acetobacterium 294
Acholeplasma laidlawii 212
Achromobacter 206, 207
Acidianis 12, 14, 130
   A.brierleyi 8, 23, 28, 128, 129, 134
   A.infernus 8, 16, 129
Acidiphilium 128, 136, 138
   A.facilis 138
   A.organovorum 136
Acinetobacter 208
   A.calcoaceticus 89
Actinomyces 87
Aerobacter 207
Aeromonas 149, 187, 205, 208
Agrobacterium gelatinovorum 94
Alcaligenes 205, 206, 207
   A.eutrophus 92, 93, 233, 234, 235, 236
Alteromonas 187, 207, 208, 209
Anabaenopsis 150
Anacystis nidulans 4, 5, 212
Anclyobacter 149
Aquaspirillum 218
Archaeoglobus 12, 14, 192
   A.fulgidus 5, 8, 13, 16, 19, 22, 192, 286
   A.profundus 8, 192
Artemia salina 5, 102
Arthrobacter 218, 242
   A.radiotolerans 260, 262, 263
   A.viscosus 246
Arthrospira 158
Aspergillus
   A.flavus 207
   A.niger 245, 249
   A.oryzae 68, 78, 79
   A.parasiticus 207
Asteromonas 96
Aureobasidium pullulans 245
Azotobacter vinelandii 266, 272

Bacillus 52, 63, 64, 65, 80, 87, 89, 90, 94, 99, 148, 152, 153, 154, 155, 156, 157, 159, 205, 208, 232, 258, 305
   B.acidocaldarius 120

B.alcalophilus 148
B.amyloliquefaciens 57
B.caldolyticus 2, 56
B.cereus 208, 241
B.circulans 208
B.coagulans 121, 123, 124
B.firmus 148
B.halodenitrificans 104
B.lentus 148, 149
B.licheniformis 57, 65, 154
B.macerans 155
B.megaterium 149
B.polymyxa 67, 300
B.psychrophilus 218
B.sphaericus 149, 208
B.stearothermophilus 2, 24, 45, 53, 55, 56, 57, 58, 59, 68, 211
B.subtilis 4, 5, 24, 53, 55, 56, 57, 58, 59, 61, 67, 71, 91, 152, 153, 234, 260, 268, 269, 270, 272, 276, 300, 305, 307, 309
B.thermoproteolyticus 24, 56
Bacteroides 293, 303, 304, 305, 306, 309, 313, 314
   B.eggerthii 306
   B.fibrisolvens 310
   B.fragilis 304, 306, 310, 314
   B.gingivalis 310
   B.nodosus 310
   B.ovatus 306
   B.ruminicola 310, 314
   B.succinogenes 310
   B.thetaiotamicron 310, 313, 314
   B.uniformis 314
Beggiatoa 173, 177
Brevibacterium 80, 99
Brochothrix thermosphacta 207
Butyribacterium 294
Butyrivibrio 294

Caldariella solfataricus 8
Caldocellum 63, 64, 65, 66, 71
   C.saccharolyticum 61, 62, 63, 65, 68, 70
Caldococcus littoralis 8

Candida 207
  C.albicans 87
  C.glabrata 239, 240
  C.lipolytica 89
Cellulomonas 61, 62, 64
  C.fimi 63
Chlainomonas rubra 204
Chlamydomonas acidophila 115
Chlorella 81, 85, 86
  C.autotrophica 96
  C.minutissima 220
  C.vulgaris 248
Chlorophyceae 85
Chroococcus 150
Citrobacter 241, 242, 244, 251
Cladosporium herbarum 207
Clostridium 60, 61, 80, 205, 208, 258, 293, 305
  C.acetobutylicum 282, 284, 285, 297, 301, 303, 304, 305, 306, 307, 308, 309, 311, 312, 313
  C.aurantibutyricum 301
  C.bifermentans 311
  C.botulinum 207, 310
  C.butyricum 299, 307, 308
  C.cellulolyticum 301, 311
  C.cellulovorans 311
  C.difficile 310
  C.felsineum 301
  C.formicoaceticum 289
  C.josui 301, 311
  C.kluyveri 299
  C.multifermentans 301
  C.pasterianum 99, 310, 312
  C.perfringens 260, 304, 306, 309, 310
  C.puniceum 301
  C.roseum 301
  C.saccharoperbutylacetonicum 284, 288
  C.septicum 311
  C.sordellii 311
  C.sterrorarium 62, 301, 302, 311
  C.thermoaceticum 297, 311
  C.thermocellum 58, 59, 60, 61, 62, 63, 287, 289, 300, 301, 302, 311, 312, 313
  C.thermohydrosulfuricum 67, 301, 304, 311
  C.thermosaccharolyticum 288, 301
  C.thermosulfurogenes 66, 67, 71, 300, 301, 311
  C.tyrobutyricum 299
Colwellia 171
  C.psychroerythrus 171
Corynebacterium 80, 99, 149, 208
  C.acetoacidophilum 99
Cryptococcus 210
  C.friedmannii 210
Cyclotella meneghiniana 96
Cytophaga 88

Debaromyces hansenii 96, 247
Deinobacter 258
  D.grandis 262, 263, 266
Deinococcus 29, 258, 270, 271, 272, 273, 274, 275, 276
  D.proteolyticus 262, 266
  D.radiodurans 261, 262, 263, 264, 265, 266, 267, 268, 269, 270, 271, 272, 273, 274, 275, 276
  D.radiophilus 262, 266, 267, 270, 272, 274
  D.radiopugnans 262, 265, 266
Deleya 77
  D.marina 94
Desulfomonas 294
Desulfovibrio 227, 294
  D.desulfuricans 4
Desulfurococcus 2, 11, 12, 13, 14, 16, 27, 191, 193
  D.amylolyticus 8
  D.mobilis 4, 5, 8, 35, 191
  D.mucosus 8, 22, 23, 24, 191
Desulfurolobus 294
  Desulfurolobus ambivalens 8, 13, 14, 16, 35, 129
Dictylostelium discoideum 4
Dictyoglomus 66, 68
Dunaliella 81, 82, 86, 96, 98, 99, 108
  D.bardawil 82
  D.parva 99
  D.salina 82

Ectothiorhodospira 97, 150
  E.halochloris 151
  E.halophila 151
  E.marismortui 98
  E.mobilis 150
  E.vacuolata 150
Edgeworthia papyrifera 156
Enterobacter agglomerans 299
Enterococcus 207
  E.faecalis 79, 307
Enteromorpha 85
Erwinia 208
  E.herbicola 220
Escherichia coli 4, 5, 35, 38, 45, 46, 47, 48, 49, 50, 51, 53, 55, 56, 57, 58, 62, 64, 65, 66, 67, 68, 71, 72, 87, 105, 137, 138, 159, 168, 169, 171, 207, 211, 212, 217, 218, 230, 235, 236, 259, 260, 261, 262, 266, 269, 270, 271, 272, 276, 303, 305, 306, 307, 308, 309
Exiguobacterium aurantiacum 149

Flavobacterium 80, 88, 196, 205, 207
  F.heparinarum 4
Flexibacter 169

Gallionella 177
Gleocapsa 206

Haemophilus influenzae 268, 269
Halobacterium 77, 80, 88
   H.cutirubrum 4, 5
   H.halobium 88, 94
   H.salinarum 94
   H.volcanii 4, 5, 88
Halococcus 80
   H.acetoinfaciens 94
   H.morrhuae 4, 5
Haloferax mediterranei 88, 90, 93
Halomonas 77
Hemichloris antarctica 210
Hyperthermus 14
   H.butylicum 4, 8, 13, 16
Hyphomicrobium 177, 185
Hyphomonas 177
   H.hirschiana 177
   H.jannaschiana 177

Klebsiellsa aerogenes 70, 246
   K.pneumoniae 70, 105, 137

Lactobacillus 80, 207, 294
   L.delbruckii 79
   L.plantarum 78
Laminaria 85
Leptospirillum ferrooxidans 128, 130, 134
Leptothrix 177
Leuconostoc 207
   L.mesenteroides 78
Leucothrix 177
Listeria 207
   L.monocytogenes 204, 207
Lyngbya 206

Macrocystis 85
Metallosphaera 12, 14, 130
   M.sedula 8, 16, 127, 128, 129, 134
Methanobacterium 294
   M.bryantii 292
   M.formicicum 4, 5, 292
   M.thermoautotrophicum 4, 8, 33, 287, 289, 310
Methanococcus 11, 12, 190, 294
   M.barkeri 309
   M.deltae 295
   M.fervidus 309
   M.igneus 8
   M.jannaschii 8, 189, 190, 193
   M.thermoautotrophicum 295, 304, 305, 309
   M.thermolithotrophicus 8, 193, 295
   M.vannielii 4, 5, 309, 310
   M.voltae 304, 305, 306, 308, 309, 310, 314
Methanogenium
   M.tationis 286
   M.thermophilicum 286

Methanohalophilus 151
Methanomicrobium 294
Methanopyrus 8, 12, 191
Methanosarcina 4, 294
   M.barkeri 292
   M.mazei 289
Methanospirillum 4, 294
   M.hungatei 4, 5
Methanothermus 12
   M.fervidus 8
   M.sociabilis 8
Methanothrix 294
Micrococcus 80, 149, 206, 221
   M.cryophilis 211, 217
   M.halbius 94
   M.luteus 299
   M.varians 94
Microcoleus 206
Microcyclus 177, 205
Monostroma 85
Moraxella 207, 208
Mortierella 220
Mycobacterium scrofulaceum 235

Natronobacterium 150
Natronococcus 150
Navicula 96
Neisseria meningitidis 268
Neurospora crassa 238, 240
Nostoc 206

Oscillatoria 107, 206

Paracoccus 149
Pediococcus 80, 207
   P.cerevisae 78
   P.halophilus 79
Pedomicrobium 177
Phaeophyceae 85, 90
Phormidium 205, 206, 210
Photobacterium phosphoreum 171
Phytophthora cinnamoni 96
Pinus radiata 65
Planococcus 206
Porphyra 85
Porphyridium 82
Prasiola 210
Propionibacterium 294
Pseudomonas 89, 94, 149, 178, 187, 188, 205, 206, 207, 208, 214
   P.aeruginosa 87, 137, 231, 241
   P.atlantica 54
   P.bathycetes 169
   P.fluorescens 91, 169, 208, 211, 220, 235
   P.fragi 208
   P.putida 208, 233, 236
   P.radiora 260
   P.stutzeri 104, 236

Pseudomonas cont'd
  P.syringae 220, 235
  P.vividiflava 220
Pyrobaculum 12, 14
  P.icelandicus 8, 16
  P.organotrophicum 8, 16
Pyrococcus 12, 13, 14, 15, 23
  P.furiosus 8, 21, 22, 23, 24, 36
  P.woesei 8, 22, 27, 37
Pyrodictium spp. 2, 12, 13, 14, 17, 188
  P.abyssum 8
  P.brockii 8
  P.occultum 4, 8, 27

Rhizobium 106
  R.japonicum 106
Rhizopus arrhizus 245, 248, 250
Rhodophyceae 85, 90
Rhodotorula 207
Ruminococcus 293
  R.albus 310
  R.flavefaciens 310

Saccharomyces 80
  S.cerevisiae 4, 5, 60, 62, 230, 237, 238,
    239, 240, 249
  S.rouxii 79
Salmonella typhimurium 99
Scenedesmus 81
Schizosaccharomyces pombe 239, 240
Schizothrix 206
Serratia marcescens 91, 99
Shewanella 171
  S.benthica 171
Shigella 231
Spirulina 80, 81, 85, 86, 150, 158, 159, 167,
  185
  S.maxima 85, 158
  S.platensis 81, 158
Staphylococcus 80
  S.aureus 87, 137, 232, 233, 235
Staphylothermus 12, 14
  S.marinus 8, 190, 191
  S.maritimus 17
Streptococcus 271
  S.faecalis 168
  S.lactis 289
  S.pneumoniae 268, 269
Streptomyces 87, 149, 241, 262
  S.coelicolor 54
  S.lividans 232
  S.phaeochromagenes 299
  S.viridochromogenes 241
Sulfobacillus thermosulfidooxidans 127, 128
Sulfolobus 2, 12, 13, 15, 17, 18, 28, 30, 35,
  36, 128, 129, 130, 131, 134, 135, 136, 139
  S.acidocaldarius 7, 8, 20, 22, 23, 27, 29,
    36, 37, 128, 129, 138, 139

  S.brierleyi 129, 131
  S.shibatae 8, 129, 136
  S.solfataricus 4, 5, 8, 20, 21, 22, 23, 24,
    27, 30, 31, 34, 36, 37, 129, 139
Sulfurococcus mirabilis 129
Synechococcus elongatus 85
Syntrophobacter 294
Syntrophomonas 294

Thermoactinomyces 52
Thermoanaerobacter 71
  T.ethanolicus 71
Thermoanaerobium brockii 67, 298
Thermococcus 12, 13, 15
  T.celer 4, 8, 17, 22, 24, 38, 191
  T.littoralis 8, 35, 36
  T.stetteri 8
Thermodiscus 12, 15
  T.maritimus 8, 17, 22, 27
Thermofilum 12, 15, 17
  T.librum 8
  T.pendens 8
Thermoplasma 1, 4, 12, 13, 15, 17
  T.acidophilum 2, 5, 7, 8, 18, 20, 22, 23,
    27, 29, 31, 32, 34, 35, 36, 120
  T.volcanium 8
Thermoproteus 11, 12, 13, 15
  T.neutrophilus 8, 17, 20
  T.tenax 2, 4, 5, 8, 17, 22, 23, 28, 29, 35,
    37
  T.uzoniensis 8, 17
Thermotoga 71, 72
  T.maritima 5, 72
Thermus 2, 45, 46, 49, 50, 52, 54
  T.aquaticus 24, 44, 45, 46, 47, 48, 49, 50,
    52, 53, 276
  T.caldophilus 47, 49
  T.flavus 47
  T.thermophilus 47, 48, 49, 53, 54, 55, 273
Thiobacillus 178, 180
  T.acidophilus 121, 127, 136, 138
  T.cuprinus 127, 128, 129
  T.denitrificans 130, 131
  T.ferrooxidans 119, 122, 126, 127, 128,
    129, 130, 131, 133, 134, 135, 136, 137,
    138
  T.neapolitanus 139, 180, 204, 241
  T.novellus 137, 138
  T.perometabolis 138
  T.prosperus 127, 128
  T.thiooxidans 128, 131, 134
Thiobacterium 178
Thiomicrospira 178, 180
  T.crunogena 178
  T.pelophila 178
Thiothrix 177
Torulopsis 80, 214
  T.candida 214

T.psychrophila   214
T.verstilis   79
Trichoderma   64
T.reesei   59
Trichosporom pullulans   218
Tritirachium albus   52

Ulva   85
Undaria   85

Vibrio   166, 167, 171, 187, 205, 209, 214, 216, 217

V.harveyi   171
V.marinus   169, 170
V.psychroerythrus   171

Xanthomonas campestris   220

Yersinia enterocolitica   207

Zoogloea   241, 242
Z.ramigera   246